Xin-She Yang and Slawomir Koziel (Eds.)

Computational Optimization and Applications in Engineering and Industry

Studies in Computational Intelligence, Volume 359

Editor-in-Chief
Prof. Janusz Kacprzyk
Systems Research Institute
Polish Academy of Sciences
ul. Newelska 6
01-447 Warsaw
Poland
E-mail: kacprzyk@ibspan.waw.pl

Further volumes of this series can be found on our homepage: springer.com

Vol. 339. Halina Kwasnicka, Lakhmi C. Jain (Eds.)
Innovations in Intelligent Image Analysis, 2010
ISBN 978-3-642-17933-4

Vol. 340. Heinrich Hussmann, Gerrit Meixner, and Detlef Zuehlke (Eds.)
Model-Driven Development of Advanced User Interfaces, 2011
ISBN 978-3-642-14561-2

Vol. 341. Stéphane Doncieux, Nicolas Bredeche, and Jean-Baptiste Mouret(Eds.)
New Horizons in Evolutionary Robotics, 2011
ISBN 978-3-642-18271-6

Vol. 342. Federico Montesino Pouzols, Diego R. Lopez, and Angel Barriga Barros
Mining and Control of Network Traffic by Computational Intelligence, 2011
ISBN 978-3-642-18083-5

Vol. 343. Kurosh Madani, António Dourado Correia, Agostinho Rosa, and Joaquim Filipe (Eds.)
Computational Intelligence, 2011
ISBN 978-3-642-20205-6

Vol. 344. Atilla Elçi, Mamadou Tadiou Koné, and Mehmet A. Orgun (Eds.)
Semantic Agent Systems, 2011
ISBN 978-3-642-18307-2

Vol. 345. Shi Yu, Léon-Charles Tranchevent, Bart De Moor, and Yves Moreau
Kernel-based Data Fusion for Machine Learning, 2011
ISBN 978-3-642-19405-4

Vol. 346. Weisi Lin, Dacheng Tao, Janusz Kacprzyk, Zhu Li, Ebroul Izquierdo, and Haohong Wang (Eds.)
Multimedia Analysis, Processing and Communications, 2011
ISBN 978-3-642-19550-1

Vol. 347. Sven Helmer, Alexandra Poulovassilis, and Fatos Xhafa
Reasoning in Event-Based Distributed Systems, 2011
ISBN 978-3-642-19723-9

Vol. 348. Beniamino Murgante, Giuseppe Borruso, and Alessandra Lapucci (Eds.)
Geocomputation, Sustainability and Environmental Planning, 2011
ISBN 978-3-642-19732-1

Vol. 349. Vitor R. Carvalho
Modeling Intention in Email, 2011
ISBN 978-3-642-19955-4

Vol. 350. Thanasis Daradoumis, Santi Caballé, Angel A. Juan, and Fatos Xhafa (Eds.)
Technology-Enhanced Systems and Tools for Collaborative Learning Scaffolding, 2011
ISBN 978-3-642-19813-7

Vol. 351. Ngoc Thanh Nguyen, Bogdan Trawiński, and Jason J. Jung (Eds.)
New Challenges for Intelligent Information and Database Systems, 2011
ISBN 978-3-642-19952-3

Vol. 352. Nik Bessis and Fatos Xhafa (Eds.)
Next Generation Data Technologies for Collective Computational Intelligence, 2011
ISBN 978-3-642-20343-5

Vol. 353. Igor Aizenberg
Complex-Valued Neural Networks with Multi-Valued Neurons, 2011
ISBN 978-3-642-20352-7

Vol. 354. Ljupco Kocarev and Shiguo Lian (Eds.)
Chaos-Based Cryptography, 2011
ISBN 978-3-642-20541-5

Vol. 355. Yan Meng and Yaochu Jin (Eds.)
Bio-Inspired Self-Organizing Robotic Systems, 2011
ISBN 978-3-642-20759-4

Vol. 356. Slawomir Koziel and Xin-She Yang (Eds.)
Computational Optimization, Methods and Algorithms, 2011
ISBN 978-3-642-20858-4

Vol. 357. Nadia Nedjah, Leandro Santos Coelho, Viviana Cocco Mariani, and Luiza de Macedo Mourelle (Eds.)
Innovative Computing Methods and Their Applications to Engineering Problems, 2011
ISBN 978-3-642-20957-4

Vol. 358. Norbert Jankowski, Włodzisław Duch, and Krzysztof Grąbczewski (Eds.)
Meta-Learning in Computational Intelligence, 2011
ISBN 978-3-642-20979-6

Vol. 359. Xin-She Yang and Slawomir Koziel (Eds.)
Computational Optimization and Applications in Engineering and Industry, 2011
ISBN 978-3-642-20985-7

Xin-She Yang and Slawomir Koziel (Eds.)

Computational Optimization and Applications in Engineering and Industry

Springer

Editors

Xin-She Yang
Mathematics and Scientific Computing,
National Physical Laboratory,
Teddington,
Middlesex TW11 0LW, UK
E-mail: xin-she.yang@npl.co.uk

Slawomir Koziel
Engineering Optimization & Modeling Center,
School of Science and Engineering,
Reykjavik University, Menntavegur 1,
101 Reykjavik, Iceland
E-mail: koziel@ru.is

ISBN 978-3-642-20985-7 e-ISBN 978-3-642-20986-4

DOI 10.1007/978-3-642-20986-4

Studies in Computational Intelligence ISSN 1860-949X

Library of Congress Control Number: 2011927164

Typeset & Cover Design: Scientific Publishing Services Pvt. Ltd., Chennai, India.

Printed on acid-free paper

9 8 7 6 5 4 3 2 1

springer.com

Preface

Computer simulation as well as numerical modelling and optimization are becoming a commonplace in contemporary engineering and science. The complexity of the systems considered in various fields has been constantly growing over the years and the theoretical models offer more and more accurate description of the physical phenomena, structures and devices. However, most of these models are far too complicated to be handled through analytical solutions; computer simulation is required for a majority of real-world applications not only to evaluate the model but also to exploit it in the design process. Advanced state-of-the-art commercial simulation software packages are available and used in everyday design work in mechanical engineering, civil engineering, aerospace industry, electrical engineering, and many others.

Computational optimization has become an essential and, in many cases, critical component of the design process. In almost all applications in engineering and industry it is necessary to maximize performance and efficiency while minimizing the cost, size, weight, or energy consumption at the same time. This is usually a complex task that involves manipulation of available design parameters in order to find satisfactory values of one or more objectives that are evaluated through often computationally expensive computer simulation. In many cases, complex constraints have to be satisfied in the optimization process.

There are several factors due to which the search for optimal design can be complicated even further. One of them is the presence of uncertainties that is common for most real-world systems. In particular, material properties and geometry of the manufactured device may differ from their nominal values as a result of fabrication tolerances. Therefore, the optimization process may seek for the robust design which ensures the highest probability of satisfying the performance requirements under the presence of uncertainties rather than just for the optimal design. Many optimization problems are nonlinear and NP-hard, that is, the solution time for finding optimal design grows exponentially with the problem size. In some cases the designer may face multiple local optima and the global search procedures are necessary. On the other hand, many practical problems have multiple and competing objectives where the best design is obtained through a decision-making process based on a set of Pareto-optimal solutions.

The dependence of contemporary engineering design on computer simulations introduces additional difficulties to optimization. Growing demand for accuracy and ever-increasing complexity of structures and systems result in the simulation process being more and more time consuming. In many engineering fields, the evaluation of a single design can take as long as several days or even weeks so that straightforward approaches by employing high-fidelity simulator directly in the optimization loop are prohibitive. Interestingly, the increasing computational power of today's computers does not alleviate this problem because the availability of faster computers is compromised by the tendency of simulating more and more complex structures and systems with higher and higher accuracy. On the other hand, simulation-based objective functions are inherently noisy, which makes the optimization process even more difficult. Still, simulation-driven design becomes a must for growing number of areas, which creates a need for robust and efficient optimization methodologies that can yield satisfactory designs even at the presence of analytically intractable objectives and limited computational resources. In particular, any technique that improves the efficiency of simulators or reduces the function evaluation count is crucially important. Surrogate-based and knowledge-based optimization uses certain approximations to the objective so as to reduce the cost of objective evaluations. The approximations are often local, while the quality of approximations is evolving as the iterations proceed.

Extensive research conducted in the area of computational optimization and modeling resulted in many techniques that alleviate the difficulties of traditional design optimization methodologies. Many of these techniques address particular issues, such as multiple local optima, multiple objectives or handling computationally expensive cost functions. Substantial progress has been observed in the development of derivative-free optimization techniques, the use of adjoint sensitivities, as well as methods exploiting surrogate models, both function-approximation- and physically-based.

This book is contributed from worldwide experts who are working in these exciting areas, and each chapter is practically self-contained. This book strives to review and discuss the latest developments concerning optimization and modelling with a focus on applications for solving real-world problems in various disciplines of engineering and science, including aerodynamics, oil industry, gas and water transport, microwave engineering, structural engineering, navigation, civil engineering, and others.

We would like to thank our editors, Drs Thomas Ditzinger and Holger Schaepe, and staff at Springer for their help and professionalism. Last but not least, we thank our families for their help and support.

Xin-She Yang
Slawomir Koziel
2011

List of Contributors

Editors

Xin-She Yang
Mathematics and Scientific Computing, National Physical Laboratory, Teddington, Middlesex TW11 0LW, UK. (xin-she.yang@npl.co.uk)

Slawomir Koziel
Engineering Optimization & Modeling Center, School of Science and Engineering, Reykjavik University, Menntavegur 1, 101 Reykjavik, Iceland. (koziel@ru.is)

Contributors

Amir Hossein Alavi
School of Civil Engineering, Iran University of Science and Technology, Tehran, Iran. (ah_alavi@hotmail.com)

I. Aydogdu
Department of Engineering Sciences, Middle East Technical University, Ankara, Turkey. (aydogdu@metu.edu.tr)

Lindsay Chapman
Materials Processing and Performance, National Physical Laboratory, Teddington, Middlesex TW11 0LW, UK. (lindsay.chapman@npl.co.uk)

Ana Cortés
Departament d'Arquitectura de Computadors i Sistemes Operatius, Escola d'Enginyeria, Universitat Autònoma de Barcelona, 08193 Bellaterra (Barcelona), Spain. (ana.cortes@uab.es)

David Echeverría Ciaurri
Department of Energy Resources Engineering, Stanford University, Stanford, CA 94305, USA. (echeverr@stanford.edu)

Mónica Denham
Facultad de Ingeniería, Universidad Nacional de Río Negro, Sede Andina, San Carlos de Bariloche, Argentina.

Pia Domschke
Department of Mathematics, Technische Universität Darmstadt, Dolivostr. 15, D-64293 Darmstadt, Germany. (domschke@mathematik.tu-darmstadt.de)

Louis J. Durlofsky
Department of Energy Resources Engineering, Stanford University, Stanford, CA 94305-2220, USA. (lou@stanford.edu)

Amir Hossein Gandomi
College of Civil Engineering, Tafresh University, Tafresh, Iran. (a.h.gandomi@gmail.com)

Z.W. Geem
Information Technology Program, iGlobal University, Annandale, Virginia, USA. (zwgeem@gmail.com)

Golnar Gharooni-fard
Computer Department of Islamic Azad University, Mashhad Branch, Young Researchers Club, Iran. (golnar.ghf@gmail.com)

O. Hasancebi
Department of Civil Engineering, Middle East Technical University, Ankara, Turkey. (oguzhan@metu.edu.tr)

Momin Jamil
Harman International, Automotive Division, Becker-Göring Str. 16, D-76307 Karlsbad, Germany. (momin.jamil@harman.com)

Oliver Kolb
Department of Mathematics, Technische Universität Darmstadt, Dolivostr. 15, D-64293 Darmstadt, Germany. (kolb@mathematik.tu-darmstadt.de)

Slawomir Koziel
Engineering Optimization & Modeling Center, School of Science and Engineering, Reykjavik University, Menntavegur 1, 101 Reykjavik, Iceland. (koziel@ru.is)

Jens Lang
Department of Mathematics, Technische Universität Darmstadt, Dolivostr. 15, D-64293 Darmstadt, Germany.
Center of Smart Interfaces, Technische Universität Darmstadt, Petersenstr. 30, D-64287 Darmstadt, Germany.
Graduate School Computational Engineering, Technische Universität Darmstadt, Dolivostr. 15, D-64293 Darmstadt, Germany.
(lang@mathematik.tu-darmstadt.de)

Leifur Leifsson
Engineering Optimization & Modeling Center, School of Science and Engineering, Reykjavik University, Menntavegur 1, 101 Reykjavik, Iceland.
(leifurth@ru.is)

Tomàs Margalef
Departament d'Arquitectura de Computadors i Sistemes Operatius, Escola d'Enginyeria, Universitat Autònoma de Barcelona, 08193 Bellaterra (Barcelona), Spain. (tomas.margalef@uab.es)

Clare Matthews
Mathematics and Scientific Computing, National Physical Laboratory, Teddington, Middlesex TW11 0LW, UK. (clare.matthews@npl.co.uk)

Fahime Moein-darbari
Computer Department of Islamic Azad University, Mashhad Branch, Young Researchers Club, Iran. (fahime.md61@gmail.com)

Tapan Mukerji
Department of Energy Resources Engineering, Stanford University, Stanford, CA 94305-2220, USA. (mukerji@stanford.edu)

Stanislav Ogurtsov
Engineering Optimization & Modeling Center, School of Science and Engineering, Reykjavik University, Menntavegur 1, 101 Reykjavik, Iceland. (stanislav@ru.is)

Simon Roberts
National Physical Laboratory, Teddington, Middlesex TW11 0LW, UK.

M.P. Saka
Department of Engineering Sciences, Middle East Technical University, Ankara, Turkey. (mpsaka@metu.edu.tr)

Kerstin Wendt
Departament d'Arquitectura de Computadors i Sistemes Operatius, Escola d'Enginyeria, Universitat Autònoma de Barcelona, 08193 Bellaterra (Barcelona), Spain. (kerstin.wendt@caos.uab.es)

Louise Wright
Mathematics and Scientific Computing, National Physical Laboratory, Teddington, Middlesex TW11 0LW, UK. (louise.wright@npl.co.uk)

Xin-She Yang
Mathematics and Scientific Computing, National Physical Laboratory, Teddington, Middlesex TW11 0LW, UK. (xin-she.yang@npl.co.uk)

Hans-Jürgen Zepernick
Blekinge Institute of Technology, SE-37179 Karlskrona, Sweden. (hans-jurgen.zepernick@bth.se)

Table of Contents

Chapter 1
Adjoint-Based Control of Model and Discretization Errors for Gas and Water Supply Networks

Pia Domschke, Oliver Kolb, and Jens Lang

Abstract. We are interested in the simulation and optimization of gas and water transport in networks. Those networks consist of pipes and various other components like compressor/pumping stations and valves. The flow through the pipes can be described by different models based on the Euler equations, including hyperbolic systems of partial differential equations. For the other components, algebraic or ordinary differential equations are used. Depending on the data, different models can be used in different regions of the network. We present a strategy that adaptively applies the models and discretizations, using adjoint-based error estimators to maintain the accuracy of the solution. Finally, we give numerical examples for both types of networks.

Pia Domschke
Department of Mathematics, Technische Universität Darmstadt
Dolivostr. 15, D-64293 Darmstadt, Germany
e-mail: domschke@mathematik.tu-darmstadt.de

Oliver Kolb
Department of Mathematics, Technische Universität Darmstadt
Dolivostr. 15, D-64293 Darmstadt, Germany
e-mail: kolb@mathematik.tu-darmstadt.de

Jens Lang
Department of Mathematics, Technische Universität Darmstadt
Dolivostr. 15, D-64293 Darmstadt, Germany
Center of Smart Interfaces, Technische Universität Darmstadt
Petersenstr. 30, D-64287 Darmstadt, Germany
Graduate School Computational Engineering, Technische Universität Darmstadt
Dolivostr. 15, D-64293 Darmstadt, Germany
e-mail: lang@mathematik.tu-darmstadt.de

X.-S. Yang & S. Koziel (Eds.): Computational Optimization & Applications, SCI 359, pp. 1–17.
springerlink.com

1.1 Introduction

Nowadays, water coming out of the tap is taken for granted in industrialized countries. Typically, one does not consider the efforts necessary to ensure its delivery. Huge amounts of water have to be routed through miles of networked pipelines. Such complex systems are difficult to operate and cost-intensive. The same holds for gas supply networks. Both are supposed to work reliably and efficiently for economical as well as ecological reasons and play an important role in the public utility infrastructure. Therefore, the support of gas and water suppliers with software tools is of great common interest. While monitoring systems are already quite advanced, efficient simulation and optimization tools are only available to some extent. Of course, before optimization tasks can be considered, reliable simulation algorithms are essential. In this context, reliability implies robustness as well as trustworthy error estimates.

In the field of simulation and optimization of gas and water supply networks, a lot of research has been done in the last years, see for example [5, 8, 9, 10, 12, 13]. Usually, especially for optimization problems, fixed models and also fixed discretizations are considered. Existing software packages like SIMONE [16] allow stationary as well as transient models for the simulation of gas networks. However, for the simulation process, one model has to be chosen in advance. SIMONE is also able to solve optimal control problems, but only steady state models are used here. In [8, 5], where nonlinear programming techniques are used to solve optimal control problems for gas and water supply networks, full a priori discretizations in time and space are applied to the underlying equations. Similarly, in [12, 13], where mixed-integer linear programming is proposed for gas network optimization, fixed models and fixed discretizations are used. Moreover, the applied discretizations are typically quite coarse to keep the complexity of the resulting problems treatable.

While the application of coarse discretizations or simplified models is often adequate in many parts of the considered networks to resolve the dynamics in the daily operation of gas and water supply networks, no information about the quality of the computed solutions is provided in all mentioned approaches. In [3, 4], a posteriori estimates for the modelling and discretization errors are introduced for finite element approximations. There, adjoint calculus is applied to measure the influence of both errors separately on a given quantity of interest. Recently, we have published an algorithm to adaptively control model and discretization errors in simulations for gas supply networks [6, 7]. This is considered to be the first step towards an efficient optimization framework with reliable error estimates. Due to various similarities, the applied concept of adjoint-based error estimators on a network can as well be used for water supply networks, which are also considered here.

This chapter is organized as follows. We begin with a description of the underlying model equations of gas and water supply networks in Section 1.2. Afterwards, we derive error estimators for the model and discretization error with respect to a given quantity of interest (Section 1.3). In Section 1.4, we present an algorithm where these estimators are used to adaptively control the model and discretization errors. Finally, numerical results are presented for a gas and a water supply network in Section 1.5.

1.2 Modelling

In this section, we give a brief introduction into the modelling of gas and water supply networks. We begin with some general aspects concerning the models of both types of networks before we describe the particular components of each.

1.2.1 General Aspects

The flow of gas or water through pipelines is a directed quantity, which can be adequately described in one space dimension. Since we need to give the pipes an orientation, we model gas and water supply networks as a directed graph $\mathscr{G} = (\mathscr{J}, \mathscr{V})$ with arcs $\mathscr{J}$ and vertices $\mathscr{V}$ (nodes, branching points).

Typically, the set of arcs $\mathscr{J}$ mainly consists of pipes $\mathscr{J}_p \subseteq \mathscr{J}$, where we have a hierarchy of models to describe the underlying gas/water dynamics. For the computations, one of these models is chosen for each pipe in each time step. From top to bottom, each model in the hierarchy results from the previous one by making simplifying assumptions. In the case of gas networks, we have a hierarchy of three models, while we consider only two models to describe the flow of water. In both cases, the most complex model consists of a hyperbolic system of partial differential equations (PDEs). Due to the spatial dimension, we define an interval $[x_j^a, x_j^b]$ with $x_j^a < x_j^b$ for each pipe $j \in \mathscr{J}_p$. In the case of gas transport, the considered networks also consist of compressor stations, valves and control valves, while we have pumping stations, valves and tanks in water supply networks. These components are described by algebraic equations or ordinary differential equations. Although there is no continuous spatial dimension for these components, we also use the spatial coordinates x_j^a and x_j^b to describe states at the beginning and end of an arc $j \in \mathscr{J} \setminus \mathscr{J}_p$. Alternatively, we denote ingoing and outgoing states with a subscript.

In addition to the equations on the arcs of the network, it is necessary to specify adequate initial, coupling and boundary conditions, which is not trivial in the case of hyperbolic equations (see for instance [2]). Let $v \in \mathscr{V}$ be an arbitrary node with ingoing arcs δ_v^- and outgoing arcs δ_v^+. Then, Kirchhoff's first rule states that the sum of currents flowing into that node is equal to the sum of currents flowing out of that node (conservation of mass):

$$\sum_{j \in \delta_v^+} q(x_j^a, t) - \sum_{j \in \delta_v^-} q(x_j^b, t) = q(v, t) \quad \forall t > 0 \tag{1.1}$$

with an auxiliary variable $q(v,t)$, which can be used to model feed-in or demand (see below).

For a unique solution, it does not suffice to only claim (1.1). A further condition which is commonly used in practice is the equality of pressure at the node $v \in \mathscr{V}$, that is,

$$\begin{aligned} p(x_j^a, t) &= p(v,t) \quad \forall j \in \delta_v^+ , \\ p(x_i^b, t) &= p(v,t) \quad \forall i \in \delta_v^- \end{aligned} \tag{1.2}$$

with an auxiliary variable $p(v,t)$ for the pressure at the node. In water supply networks, the pressure p is typically replaced by the pressure head h.

In this work, we use equality of pressure (1.2) together with the conservation of mass (1.1). Due to the hyperbolic nature of the underlying partial differential equations, there is one degree of freedom for either boundary of each arc. This means, at a node $v \in \mathscr{V}$ with m ingoing and n outgoing arcs, we have $m+n+2$ degrees of freedom (including $p(v,t)$ and $q(v,t)$) but only $m+n+1$ equations. Thus, we need one further equation at the node v:

$$e(p(v,t), q(v,t)) = 0 .$$

At branching points in the network, we typically have $q(v,t) = 0$, and at boundary nodes, we use $p(v,t) = p_v(t)$ or $q(v,t) = q_v(t)$ with given profiles $p_v(t)$ or $q_v(t)$. According to (1.1), $q(v,t) > 0$ corresponds to a feed-in of gas/water into the network and $q(v,t) < 0$ to a demand.

1.2.2 *Gas Supply Networks*

In this section, we want to have a closer look on how the flow through gas networks is modelled. As mentioned in Sect. 1.2.1, the network consists of pipes, compressor stations, valves and control valves.

1.2.2.1 Pipes

The models describing gas flow in pipelines are based on the *Euler equations*, a hyperbolic system of nonlinear partial differential equations. The system consists of the conservation of mass, momentum and energy together with the equation of state for real gases. The transient flow of gas may be described appropriately by equations in one space dimension, pressure losses due to friction are modelled via a source term. A common simplification, the restriction to isothermal flows, that is, flows with constant temperature, makes the energy equation become redundant. The resulting equations are called *isothermal Euler equations*. If we assume a constant speed of sound and let the pipes be horizontal, the equations result in the *nonlinear model* [2]

$$p_t + \frac{\rho_0 c^2}{A} q_x = 0, \tag{1.3a}$$

$$q_t + \frac{A}{\rho_0} p_x + \frac{\rho_0 c^2}{A} \left(\frac{q^2}{p} \right)_x = -\frac{\lambda \rho_0 c^2 |q| q}{2dAp}. \tag{1.3b}$$

Here, q denotes the flow rate under standard conditions (1 atm air pressure, temperature of 0 °C), p the pressure, c the speed of sound, λ the friction coefficient, d the diameter, A the cross-sectional area of the pipe and ρ_0 the density under standard conditions.

Neglecting the nonlinear term in the spatial derivative of the momentum equation (1.3b) yields the *semilinear model*. This simplification is motivated by the slow velocity of gas in real networks. We get

$$p_t + \frac{\rho_0 c^2}{A} q_x = 0, \tag{1.4a}$$

$$q_t + \frac{A}{\rho_0} p_x = -\frac{\lambda \rho_0 c^2 |q| q}{2dAp}. \tag{1.4b}$$

A further simplification leads to a *(quasi-)stationary model*: Setting the time derivatives in (1.4) to zero results in an ordinary differential equation, which can be solved analytically:

$$q = \text{const.}, \tag{1.5a}$$

$$p(x) = \sqrt{p(x_0)^2 + \frac{\lambda \rho_0^2 c^2 |q| q}{dA^2} (x_0 - x)}\,. \tag{1.5b}$$

Here, $p(x_0)$ denotes the pressure at an arbitrary point $x_0 \in [x_j^a, x_j^b]$. Setting $x_0 = x_j^a$, that is the inbound of the pipe, and $x = x_j^b$, that is the end of the pipe, yields the so-called *algebraic model* [15].

1.2.2.2 Compressor Stations

A compressor station is a facility that increases the pressure of the gas. Running a compressor generates costs, since the compressor station consumes some of the gas, that means,

$$q_{\text{out}} = q_{\text{in}} - F_c(p_{\text{in}}, p_{\text{out}}, q_{\text{in}}). \tag{1.6}$$

The equation for the fuel consumption of the compressor $c \in \mathcal{J}_c \subseteq \mathcal{J}$ is given by

$$F_c(p_{\text{in}}, p_{\text{out}}, q_{\text{in}}) = d_{F,c}\, q_{\text{in}} \left(\left(\frac{p_{\text{out}}}{p_{\text{in}}} \right)^{\frac{\gamma-1}{\gamma}} - 1 \right), \tag{1.7}$$

with $p_{\text{in}} = p(x_c^a, t)$, $p_{\text{out}} = p(x_c^b, t)$, $q_{\text{in}} = q(x_c^a, t)$ and $q_{\text{out}} = q(x_c^b, t)$ [10]. Here, γ is the isentropic coefficient of the gas. The coefficient $d_{F,c}$ is a compressor specific constant. The increase in pressure, performed by the compressor station c, is denoted by

$$\Delta p_c(t) = p_{\text{out}} - p_{\text{in}} \tag{1.8}$$

and depends on the compressor power

$$P_c(p_{\text{in}}, p_{\text{out}}, q_{\text{in}}) = d_{P,c}\, q_{\text{in}} \left(\left(\frac{p_{\text{out}}}{p_{\text{in}}} \right)^{\frac{\gamma-1}{\gamma}} - 1 \right), \tag{1.9}$$

with a compressor specific constant $d_{P,c}$. Either (1.8) or (1.9) is typically used as control variable.

1.2.2.3 Valves

Valves are used to regulate the flow of the gas by opening or closing. In the case of an open valve, the equations

$$\begin{aligned} q_{\text{in}} &= q_{\text{out}}, \\ p_{\text{in}} &= p_{\text{out}} \end{aligned}$$

hold. If the valve is closed, then $q_{\text{in}} = q_{\text{out}} = 0$.

1.2.2.4 Control Valves

Control valves, sometimes also referred to as regulators [8], are valves that reduce the gas pressure by a controlled amount. The behaviour of a control valve is modelled via

$$p_{\text{in}} - p_{\text{out}} \overset{!}{=} u$$

with control variable $u = u(t)$. The ingoing and outgoing flow rates are identical:

$$q_{\text{in}} = q_{\text{out}}.$$

1.2.3 *Water Supply Networks*

Water supply networks feature similar structures as gas supply networks. Here, the main components are pipes, pumps, valves and tanks.

1.2.3.1 Pipes

To describe the dynamics inside the pipes of a water supply network, we consider two different models, which can for instance be found in [1]. The most complex model, considering the elastic effects, is given by the so-called *water hammer equations*,

$$h_t + \frac{a^2}{gA} q_x = 0, \tag{1.10a}$$

$$q_t + gAh_x = -\lambda \frac{q|q|}{2DA}, \tag{1.10b}$$

a *semilinear* hyperbolic system of partial differential equations, where the piezometric head h and the flow rate q are the space and time-dependent state variables. The gravitational constant is denoted by g, a is the speed of sound in the pipe, A and D are the cross-sectional area and the diameter of the pipe, respectively. The right hand side of (1.10b) models the influence of friction, where λ is the friction coefficient.

A simplified model for the water dynamics inside the pipes can be derived by neglecting the time derivatives in (1.10). The resulting *(quasi-)stationary* or *algebraic* model reads

$$q_x = 0, \tag{1.11a}$$

$$h_x = -\lambda \frac{q|q|}{2gDA^2}. \tag{1.11b}$$

Thus, the flow rate is constant in the pipe,

$$q_{\text{in}} = q_{\text{out}}, \tag{1.12}$$

and the entire pressure head loss is given by

$$h_{\text{in}} - h_{\text{out}} = \lambda \frac{L}{2gDA^2} q|q|, \tag{1.13}$$

where L denotes the length of the pipe, and h_{in} and h_{out} the ingoing and outgoing pressure head, respectively.

1.2.3.2 Pumps

Pumps are installed in water supply networks to generate or maintain a certain pressure. Typically, the relation between the flow rate through a pump and the resulting pressure increase is described by a set of *characteristic curves*. A commonly used form for a single curve (see e.g. [5]) is

$$H(q) = h_{\text{out}} - h_{\text{in}} = \alpha_0 - \alpha_r q^r$$

with

$$q = q_{\text{in}} = q_{\text{out}}$$

and $r \in \mathbb{R}_+$. If multiple curves are given, the parameters α_0 and α_r depend on the current speed ω of the pump.

The running costs of a pump result from the power consumption of its motor. Applying an efficiency curve $\eta_c(q)$ as in [14, pp. 36–37] or [9, p. 46], the costs of each pump $c \in \mathscr{J}_c \subseteq \mathscr{J}$ are proportional to

$$P_c(h_{\text{in}}, h_{\text{out}}, q) = \frac{(h_{\text{out}} - h_{\text{in}})\,q}{\eta_c(q)}\,.$$

Again, if multiple curves are given to characterize the pump, the efficiency η additionally depends on the speed of the pump.

1.2.3.3 Valves

There are various kinds of valves installed in water supply networks to control the flow rate and pressure. Here, we consider gate valves, where the opening can be externally controlled.

With $u \in [0,1]$ being the control variable for the fraction of the opening, we apply

$$u^2(h_{\text{in}} - h_{\text{out}}) = \zeta q|q|$$

with the friction loss coefficient ζ and

$$q = q_{\text{in}} = q_{\text{out}}\,.$$

1.2.3.4 Tanks

Water tanks are used to store water at certain positions in the network. The ingoing flow at the bottom of a tank is given by

$$q = C\,\text{sign}(h_{\text{outer}} - h_{\text{inner}})\sqrt{|h_{\text{outer}} - h_{\text{inner}}|} \tag{1.14}$$

with the discharge coefficient C. Here, h_{outer} denotes the outer pressure head in front of the inlet of the tank and h_{inner} is the inner pressure head at the bottom of the tank, which is the sum of the elevation of the tank and the current stage:

$$h_{\text{inner}} \;=\; \text{elevation} + \text{stage}\,.$$

Note that formally

$$q = q(x_j^a, t)\,, \quad h_{\text{outer}} = h(x_j^a, t)\,, \quad h_{\text{inner}} = h(x_j^b, t)$$

for each tank $j \in \mathscr{J}$.

The change of the stage and therewith the change of h_{inner} is modelled by an ordinary differential equation:

$$\frac{d}{dt}h_{\text{inner}} = \frac{1}{A}q\,, \tag{1.15}$$

where A is the cross-sectional area of the tank.

Additionally to the ingoing flow at the bottom of the tank (given by (1.14)), there can be further inflow or outflow openings, e.g. for refilling the tank or overflow. Concerning the model equations, those terms can be simply added to q in (1.15) and formally refer to $q(x_j^b, t)$.

1.3 Error Estimators

For given initial and boundary conditions as well as control states for the controllable elements, the described model equations on the whole network can be solved applying appropriate discretization schemes. For the discretization of the (hyperbolic) PDEs in the pipes, we apply an implicit box scheme [11], which perfectly matches the properties of the underlying equations. The time steps of this scheme are also used to discretize the ordinary differential equations as occurring in the model of water tanks. So far, one step methods are implemented for this purpose and delivered satisfying results.

Now, we are searching for a compromise between the accuracy of the numerical solution and the computational costs. We want to use the more complex models in the pipes only when necessary and to refine the discretizations only where needed. Using the solution of adjoint equations as done in [3, 4, 2, 6, 7], one may deduce model and discretization error estimators to measure the influence of the model and the discretization on a user-defined target functional M. With u being the exact solution of the (most complex) model equations and u^h being the approximate (numerical) solution for some choice of models, the error in the target functional can be approximated by

$$M(u) - M(u^h) \approx \eta_m + \eta_h \,, \tag{1.16}$$

where η_m estimates the model error and η_h the error resulting from the discretization. Concerning the underlying adjoint equations, these error estimators are currently implemented in a first-discretize manner, which will be briefly described in the following. A more detailed description with some hints on the implementation can be found in [7].

Let t_j $(j = 0, \ldots, N)$ be the times of the discretization. Accordingly, we split up the solution of the discretized model equations

$$(u^h)^T = \left((u_0^h)^T, \ldots, (u_N^h)^T\right).$$

Starting with the given initial state u_0^h, we have to solve a system of the form

$$F_j(\underbrace{u_{j-1}^h}_{u_{\text{old}}}, \underbrace{u_j^h}_{u_{\text{new}}}) = 0 \tag{1.17}$$

in each time step. In the adaptive algorithm described in the next section, we will partition the entire simulation horizon in several blocks. Accordingly, we define

$$(U_k^h)^T = \left((u_{j(k-1)+1}^h)^T, \ldots, (u_{j(k)}^h)^T\right)$$

for $k = 1,\dots,N_B$, where $j(k)$ is given via $t_{j(k)} = T_k$ for $k = 0,\dots,N_B$. For later use, we also define

$$(E_k)^T = \left((F_{j(k-1)+1})^T,\dots,(F_{j(k)})^T\right), \tag{1.18}$$

which summarizes the state-defining equations of the block $[T_{k-1},T_k]$.

Now, we can estimate the model error of the kth block with respect to the functional M via

$$\eta_{m,k} = \frac{\partial}{\partial U_k} M(u^h)\Delta U_k^h, \tag{1.19}$$

where $\Delta U_k^h = U_k - U_k^h$. Here, U_k formally denotes a reference solution in the kth block which solves a different system of equations $\tilde{E}_k$ based on more complex or simpler models. Thus, the difference ΔU_k^h results from the differences in the models and can be estimated by

$$\Delta U_k^h \approx -\left(\frac{\partial}{\partial U_k}\tilde{E}_k(u^h)\right)^{-1}\Delta\tilde{E}_k \tag{1.20}$$

with

$$\Delta\tilde{E}_k = \tilde{E}_k(U_k^h) - \underbrace{\tilde{E}_k(U_k)}_{=0} = \tilde{E}_k(U_k^h). \tag{1.21}$$

Inserting (1.20) and (1.21) in (1.19) finally gives

$$\eta_{m,k} = -\frac{\partial}{\partial U_k} M(u^h)\left(\frac{\partial}{\partial U_k}\tilde{E}_k(u^h)\right)^{-1}\tilde{E}_k(U_k^h) = -\xi_k^T\tilde{E}_k(U_k^h)$$

with ξ_k being the solution of the *adjoint equation*

$$\left(\frac{\partial}{\partial U_k}\tilde{E}_k(u^h)\right)^T\xi_k = \left(\frac{\partial}{\partial U_k}M(u^h)\right)^T. \tag{1.22}$$

Instead of $\frac{\partial}{\partial U_k}\tilde{E}_k(u^h)$ one may also apply $\frac{\partial}{\partial U_k}E_k(u^h)$ in (1.22) to get an error estimation. This way, it suffices to solve one system of adjoint equations (per block) for the estimators with respect to higher and lower models and also with respect to the discretization. Moreover, note that (1.22) can be solved very efficiently due to the special structure of $\tilde{E}_k$ and E_k (see [7]).

Regarding the derivation of the model error estimator $\eta_{m,k}$ above, one may observe that an error estimator for discretization errors can be deduced in exactly the same way. Here, the reference solution U_k resulting from solving a modified system of equations $\tilde{E}_k$ must refer to another discretization. In our implementation, the residual $\tilde{E}_k(U_k^h)$ is estimated by comparing the single terms in the applied discretization scheme with reconstruction formulas of higher order. This way, separate error estimators for the temporal and the spatial error (for each element in each block) can be evaluated:

$$\eta_{h,k} = \eta_{x,k} + \eta_{t,k}, \tag{1.23}$$

where $\eta_{x,k}$ denotes the estimator for the spatial error and $\eta_{t,k}$ for the temporal error in the kth block [6, 7].

1.4 Adaptive Error Control

In the last section, we have developed error estimators for model and discretization errors. With these estimators, we can now control the computational errors inside the network. Since in practice the dynamic behaviour in the network varies, we want to control the relative error resulting from the choice of the models and the discretization in blocks of several time steps. Thus, we divide the time interval $[0,T]$ into blocks of equal size $[T_{k-1},T_k]$, $k=1,\ldots,N_B$. Regarding one subinterval $[T_{k-1},T_k]$, we can compute the forward as well as the backward/adjoint solution and evaluate the error estimators locally, which yields

$$M_k(u)-M_k(u^h)\approx \eta_{m,k}+\eta_{t,k}+\eta_{x,k}\,.$$

Given a tolerance $\mathtt{TOL}$ for the relative error, we can approximate the exact error by the estimators, giving

$$\frac{|M_k(u)-M_k(u^h)|}{|M_k(u)|}\approx\frac{|\eta_{m,k}+\eta_{t,k}+\eta_{x,k}|}{|M_k(u^h)|}\overset{!}{\leq}\mathtt{TOL}\,. \tag{1.24}$$

We first examine the discretization error to ensure the discretization to be adequate. Then we consider the model error.

Check Discretization Error. First, the discretization is checked. Given the tolerance $\mathtt{TOL}$ as above, we ensure the discretization error to be small enough by decreasing $\mathtt{TOL}$ by a user-defined factor $0<\kappa<1$ giving $\mathtt{TOL}_h:=\kappa\cdot\mathtt{TOL}$. We demand the discretization error estimator to satisfy

$$\left|\eta_{t,k}+\eta_{x,k}\right|\leq\mathtt{TOL}_h\cdot\left|M_k(u^h)\right|\,.$$

If the error estimator exceeds the given upper bound, the temporal and spatial discretization errors are treated individually, that is,

$$\left|\eta_{t,k}\right|\leq\frac{1}{2}\mathtt{TOL}_h\cdot\left|M_k(u^h)\right|\quad\text{and}\quad\left|\eta_{x,k}\right|\leq\frac{1}{2}\mathtt{TOL}_h\cdot\left|M_k(u^h)\right|\,.$$

Check Temporal Discretization Error. If the temporal error estimator exceeds the given tolerance, the time step size is marked for refinement. After checking the spatial discretization error, the time interval $[T_{k-1},T_k]$ has to be computed again. If, in contrast, the error estimator $\left|\eta_{t,k}\right|$ is much smaller than the upper bound, the time step size is marked for coarsening. If the current time interval has to be recomputed due to spatial or model errors, the temporal coarsening is not applied.

Check Spatial Discretization Error. Now, the spatial discretization error is estimated locally for each pipe,

$$\eta_{x,k} = \sum_{j \in \mathcal{J}_p} \eta_{x,k,j} \,.$$

Thus, we want to satisfy

$$\left| \sum_{j \in \mathcal{J}_p} \eta_{x,k,j} \right| \leq \frac{1}{2} \mathrm{TOL}_h \cdot \left| M_k(u^h) \right| \,.$$

For this inequality to hold, it suffices to claim

$$\sum_{j \in \mathcal{J}_p} \left| \eta_{x,k,j} \right| \leq \frac{1}{2} \mathrm{TOL}_h \cdot \left| M_k(u^h) \right| \,.$$

In order to get an upper bound for each pipe itself, we uniformly distribute the target functional, i.e., we divide it by the number of pipes $|\mathcal{J}_p|$, giving

$$\left| \eta_{x,k,j} \right| \leq \frac{1}{2} \mathrm{TOL}_h \cdot \frac{\left| M_k(u^h) \right|}{\left| \mathcal{J}_p \right|} \qquad \forall j \in \mathcal{J}_p \,.$$

If $|\eta_{x,k,j}|$ exceeds the given tolerance, the pipe is marked for refinement. If, instead, the error estimator is much smaller than the right hand side, the pipe is marked for coarsening. The time interval $[T_{k-1}, T_k]$ is computed again with a finer discretization where needed.

Check Total Error. If the discretization error is small enough, the total error estimator $\eta_{m,k} + \eta_{t,k} + \eta_{x,k}$ is evaluated. If

$$\left| \eta_{m,k} + \eta_{t,k} + \eta_{x,k} \right| > \mathrm{TOL} \cdot \left| M_k(u^h) \right| \,,$$

that is, the total error does not fulfill the desired tolerance while the discretization error did, the model error is checked.

Check Model Error. If the discretization error is small enough, but the total error is not, the model errors of all pipes are checked. Again, we uniformly distribute the target functional over all pipes. If the error estimator exceeds the given tolerance, that is,

$$\left| \eta_{m,k,j} \right| > \mathrm{TOL}_m \cdot \frac{\left| M_k(u^h) \right|}{\left| \mathcal{J}_p \right|} \,,$$

with $\mathrm{TOL}_m := (1 - \kappa) \cdot \mathrm{TOL}$, the pipe is supposed to use the model above subject to the hierarchy. The time interval $[T_{k-1}, T_k]$ is computed again with the adjusted models.

Coarsen Temporal and/or Spatial Discretization and Switch Down Models. If the total error fulfills the desired tolerance, the time interval $[T_{k-1}, T_k]$ is accepted and k is increased. If the time step size or any pipes were marked for coarsening, the coarsening is applied. Then, the estimators with respect to the lower models are computed. If the error estimator is much less than the given tolerance, that is,

$$\left|\eta_{m,k,j}\right| \leq s \cdot \mathrm{TOL}_m \cdot \frac{\left|M_k(u^h)\right|}{\left|\mathscr{I}_P\right|},$$

with a "shift down factor" $s \ll 1$ (e.g. 10^{-1} or 10^{-2}), the pipe can use the lower model for the next calculations and we go on to the next interval.

1.5 Numerical Examples

In this section, we give numerical results for a medium sized real life gas network and a water supply network. All presented computations were done on an AMD Athlon^{TM} 64 X2 Dual Core 6000+.

1.5.1 Gas Supply Network

We begin with a gas supply network, which is shown in Fig. 1.1. The considered network consists of twelve pipes (P01 – P12, with lengths between 30km and 100km), two sources (S01 – S02), four consumers (C01 – C04), three compressor stations (Comp01 - Comp03) and one control valve (CV01).

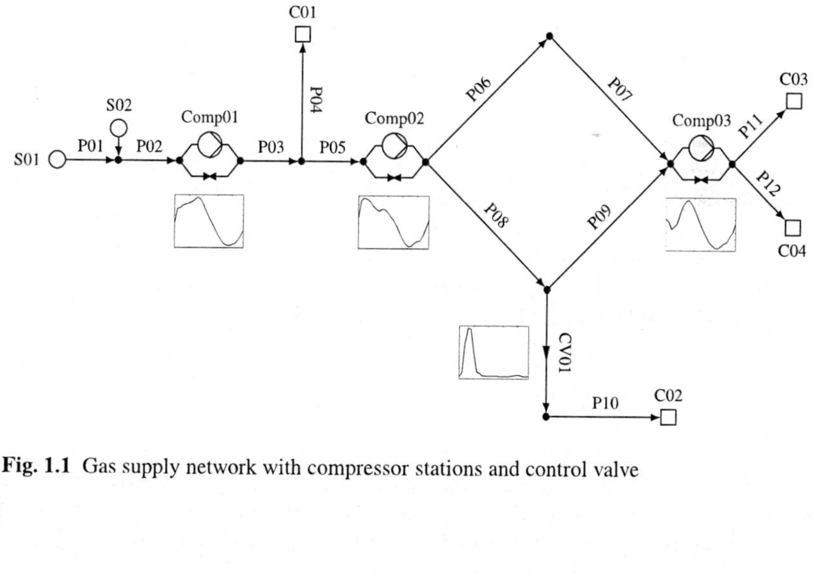

Fig. 1.1 Gas supply network with compressor stations and control valve

The simulation starts with stationary initial data. The boundary conditions and the control for the compressor stations and the control valve are time-dependent. Plots of the control functions are given in Fig. 1.1. The target functional is given by the total fuel gas consumption of the compressors, i.e.

$$M(u) = \sum_{c \in \mathscr{J}_c} \int_0^T F_c(t)\,\mathrm{d}t.$$

The simulation time is 86,400 seconds (24 hours) with an initial time step size $\Delta t = 3,600$ seconds. The subintervals are 7,200 seconds (2 hours) each. The initial spatial step size is $\Delta x = 10,000$m. The factor κ is set to 10^{-1} and the shift down factor $s = 10^{-1}$. The tolerance TOL is set to values between 10^{-1} and 10^{-4}.

Table 1.1 shows the maximal relative error in the target functional

$$\text{rel.err.} = \max_k \frac{|M_k(u) - M_k(u^h)|}{|M_k(u)|}, \tag{1.25}$$

the total target functional, the maximal and the minimal time and spatial step size used subject to the tolerance TOL and the running time. As an approximation of the exact solution we computed a solution with the nonlinear model and a finer discretization than used in the adaptive algorithm, which is shown in the last row.

Table 1.1 Results using different values for TOL

TOL	rel.err.	$M(u^h)$	max/min Δt	max/min Δx	time [s]
1e-01	1.690905e-01	5.0480603810e+01	3600/900	33,333.3/10,000	2.7e-01
1e-02	1.756343e-02	4.8408860265e+01	900/450	33,333.3/10,000	1.0e+00
1e-03	1.288994e-03	4.8487439184e+01	225/28.125	16666.7/1,250	3.5e+01
1e-04	4.010694e-05	4.8486374440e+01	14.0625/1.7578	16666.7/312.5	1.0e+03
reference solution		4.8485402013e+01	1	312.5	4.2e+03

Generally, we observe that the maximal relative error decreases with the tolerance TOL. We can also see that the error estimators do not provide a sharp upper bound for the error.

Besides the discretization, it is also interesting how the model switching part works depending on TOL. Table 1.2 shows how often which model is used during the simulation. The trend is the same as for the discretization. For smaller tolerances, the share of the more complex models is higher.

1.5.2 Water Supply Network

As a second example, we consider the water supply network shown in Fig. 1.2. The network consists of sixteen pipes (P01 – P16, with lengths between 500m and 20km), two suppliers (S01 – S02), six consumers (C01 – C06), three pumps (Pump01 - Pump03), four tanks (T01 – T04) and two valves (V01 – V02).

Table 1.2 Models used during simulation for different values of TOL

TOL	ALG	LIN	NL
1e-01	100%	0%	0%
1e-02	46.7%	53.3%	0%
1e-03	13.7%	85%	1.3%
1e-04	2.5%	73.7%	23.7%

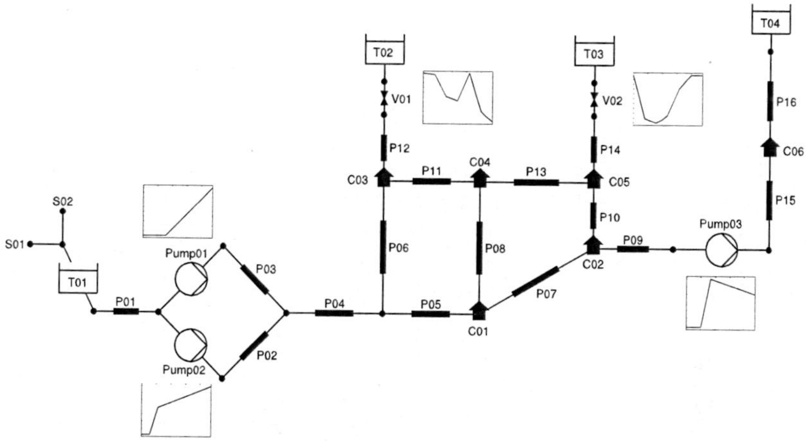

Fig. 1.2 Water supply network with pumps, valves and tanks

The simulation starts with approximately stationary data. The boundary conditions as well as the control for the pumps and the valves are time-dependent. As above, plots of the control functions are given in Fig. 1.2. The target functional is given by the total energy consumption of the pumps, which is proportional to

$$M(u) = \sum_{c \in \mathscr{I}_c} \int_0^T P_c(t)\,\mathrm{d}t.$$

The simulation time is 86,400 seconds (24 hours) with an initial time step size $\Delta t = 3,600$ seconds. The subintervals are 7,200 seconds (2 hours) each. The initial spatial step size has been chosen as coarse as possible such that the applied (spatial) error estimators can be evaluated, that is exactly four grid points per pipe. Similar to above, the factor κ is set to 10^{-1} and the shift down factor $s = 10^{-1}$. The tolerance TOL is set to values between 10^{-1} and 10^{-4}.

Table 1.3 shows the maximal relative error in the target functional according to (1.25), the total target functional, the maximal and the minimal time and spatial step size used subject to the tolerance TOL and the running time. As an approximation of the exact solution we computed a solution with the time-dependent model

Table 1.3 Results using different values for TOL

TOL	rel.err.	$M(u^h)$	max/min Δt	max/min Δx	time [s]
1e-01	1.228465e-02	8.3184539798e+01	3600/900	6666.7/166.7	5.4e-01
1e-02	1.491658e-03	8.3079443781e+01	900/225	6666.7/166.7	1.9e+00
1e-03	9.460297e-04	8.3014423132e+01	112.5/28.125	6666.7/166.7	9.7e+00
1e-04	7.546011e-05	8.3037462607e+01	14.0625/3.5156	6666.7/166.7	1.6e+02
reference solution		8.3038197026e+01	2	100	1.5e+03

and a finer discretization than used in the adaptive algorithm, which is shown in the last row.

As above, we observe that the maximal relative error in each block decreases with TOL. Moreover, the error tolerance is satisfied here. Obviously, the time discretization plays the crucial role in this example. The initial spatial discretization is not refined. Additionally, the difference between the two models only becomes important for the smallest tolerance, which can be seen from Table 1.4.

Table 1.4 Models used during simulation for different values of TOL

TOL	ALG	LIN
1e-01	100%	0%
1e-02	100%	0%
1e-03	100%	0%
1e-04	42.2%	57.8%

1.6 Conclusion and Outlook

In this chapter, we have presented an algorithm to adaptively control model and discretization errors for the simulation of gas and water flow through networked pipelines. The gas and water dynamics in the pipes are described by a hierarchy of models, ranging from partial differential to algebraic equations. Further network components are modelled by algebraic and ordinary differential equations. Using adjoint equations, we introduced error estimators to measure the influence of the discretization in time and space and the applied models with respect to a given target functional. With these estimators, we developed an algorithm to adaptively control the different errors within a given tolerance.

We gave examples for both types of networks to show the applicability of the algorithm. In both cases, it could be seen that the actual errors decreased with the prescribed tolerance. By construction, the error estimators do not provide an upper bound but are a first order approximation of the true error. For the considered water supply network, all error bounds were maintained. For the gas network, the actual errors were slightly larger than the given tolerance.

The results achieved so far (also in [6, 7]) make us confident that the presented techniques to solve simulation tasks can build a reliable basis to address optimal control problems for gas and water supply networks. In particular, the sensitivity information computed for the evaluation of the error estimators can be used to compute gradient information for derivative-based optimization. There, we will have to consider multiple quantities of interest. These are the objective function of the given task and all constraints, which are supposed to be evaluated within given tolerances as well.

Another part of our future work is the extension of the presented approach to further applications. The principle of adjoint-based control of model and discretization errors does not stick to gas and water supply networks. For instance, other transport processes on networks can be considered. Here, one application we have in mind is traffic flow on networks.

References

1. Abreu, J., Cabrera, E., Izquierdo, J., García-Serra, J.: Flow modeling in pressurized systems revisited. Journal of Hydraulic Engineering 125(11), 1154–1169 (1999)
2. Bales, P., Kolb, O., Lang, J.: Hierarchical modelling and model adaptivity for gas flow on networks. In: Allen, G., Nabrzyski, J., Seidel, E., van Albada, G.D., Dongarra, J., Sloot, P.M.A. (eds.) ICCS 2009. LNCS, vol. 5544, pp. 337–346. Springer, Heidelberg (2009)
3. Becker, R., Rannacher, R.: An optimal control approach to a posteriori error estimation in finite element methods. Acta numerica 10, 1–102 (2001)
4. Braack, M., Ern, A.: A posteriori control of modeling errors and discretization errors. SIAM Journal of Multiscale Modeling and Simulation 1(2), 221–238 (2003)
5. Burgschweiger, J., Gnädig, B., Steinbach, M.C.: Optimization models for operative planning in drinking water networks. Optimization and Engineering 10(1), 43–73 (2009)
6. Domschke, P., Kolb, O., Lang, J.: An adaptive model switching and discretization algorithm for gas flow on networks. In: Procedia Computer Science ICCS 2010, vol. 1(1), pp. 1325–1334 (2010)
7. Domschke, P., Kolb, O., Lang, J.: Adjoint-based control of model and discretisation errors for gas flow in networks. Int. J. Mathematical Modelling and Numerical Optimisation 2(2), 175–193 (2011)
8. Ehrhardt, K., Steinbach, M.C.: Nonlinear optimization in gas networks. In: Bock, H.G., Kostina, E.A., Phu, H.X., Ranacher, R. (eds.) Modeling, Simulation and Optimization of Complex Processes, pp. 139–148. Springer, Heidelberg (2005)
9. Hähnlein, C.: Numerische Modellierung zur Betriebsoptimierung von Wasserverteilnetzen. PhD thesis, TU Darmstadt (2008)
10. Herty, M.: Modeling, simulation and optimization of gas networks with compressors. Networks and Heterogeneous Media 2(1), 81–97 (2007)
11. Kolb, O., Lang, J., Bales, P.: An implicit box scheme for subsonic compressible flow with dissipative source term. Numerical Algorithms 53(2), 293–307 (2010)
12. Martin, A., Möller, M., Moritz, S.: Mixed integer models for the stationary case of gas network optimization. Mathematical Programming 105, 563–582 (2006)
13. Moritz, S.: A Mixed Integer Approach for the Transient Case of Gas Network Optimization. PhD thesis, TU Darmstadt (2006)
14. Rossman, L.A.: EPANET 2 users manual. U.S. Environmental Protection Agency, Cincinnati, OH (2000)
15. Sekirnjak, E.: Transiente Technische Optimierung (TTO-Prototyp). Technical report, PSI AG (November 2000)
16. SIMONE Research Group, `http://www.simone.eu`

Chapter 2
Derivative-Free Optimization for Oil Field Operations

David Echeverría Ciaurri, Tapan Mukerji, and Louis J. Durlofsky

Abstract. A variety of optimization problems associated with oil production involve cost functions and constraints that require calls to a subsurface flow simulator. In many situations gradient information cannot be obtained efficiently, or a global search is required. This motivates the use of derivative-free (non-invasive, black-box) optimization methods. This chapter describes the use of several derivative-free techniques, including generalized pattern search, Hooke-Jeeves direct search, a genetic algorithm, and particle swarm optimization, for three key problems that arise in oil field management. These problems are the optimization of settings (pressure or flow rate) in existing wells, optimization of the locations of new wells, and data assimilation or history matching. The performance of the derivative-free algorithms is shown to be quite acceptable, especially when they are implemented within a distributed computing environment.

2.1 Introduction

Oil and natural gas account for around 60% of the current worldwide primary energy supply, and the demand for these key resources is expected to increase for several decades. Because the development of new fields is often very expensive and technically challenging, it is essential that these operations are performed as efficiently as possible. In addition, the high expense of discovering and developing new fields provides a substantial economic incentive to maximize production from existing fields. Both of these trends provide strong motivation for the development and application of robust methodologies for the computational optimization of oil field operations.

David Echeverría Ciaurri · Tapan Mukerji · Louis J. Durlofsky
Department of Energy Resources Engineering, Stanford University,
Stanford, CA 94305-2220, USA
e-mail: echeverr@stanford.edu, mukerji@stanford.edu, lou@stanford.edu

X.-S. Yang & S. Koziel (Eds.): Computational Optimization & Applications, SCI 359, pp. 19–55.
springerlink.com

The closed-loop reservoir management paradigm [1] provides a framework for efficiently operating an oil field. This approach relies on the continuous acquisition of field data, which are then used to calibrate the computational reservoir model. This represents a data assimilation or history-matching step. The resulting (history-matched) model is then used for optimizing future production. This can be accomplished by either determining optimal settings/controls (e.g., flow rates, well pressures) for existing wells or by finding the best locations for new wells. Given the fact that many different types of wells can be drilled, such as deviated, horizontal or multi-branched wells, the determination of the appropriate well type can also be viewed as an optimization problem.

In this chapter, we address three of the key optimization problems that arise in reservoir engineering – optimization of well settings, optimization of the placement of new wells, and data assimilation. Although there are inter-relationships between these various problems, they have important differences and are typically addressed in a decoupled manner. Well control optimization usually has real-valued decision variables, and a nonlinear, simulation-based cost function and constraints. The well location (often referred to as field development) problem entails, in general, finding the number, type, location and drilling sequence of new wells. In practice, because wells are associated to cell centers in the underlying simulation grid, the optimization variables are typically integers. The well type is described by categorical variables. Model calibration (data assimilation) can be formulated as an inverse problem where we seek to minimize the discrepancy between measured data and model output. The requisite optimization usually involves a very large number of variables (normally at least one per simulation grid block, and in practical problems there are $O(10^4 - 10^6)$ blocks), so parameter reduction and regularization techniques are commonly applied. The subsurface flow simulations required for all of the aforementioned optimizations entail numerical solutions of sets of discretized partial differential equations. These function evaluations can be very costly, and this is a key consideration when designing the optimization framework.

Although our emphasis in this paper is on the use of derivative-free optimization methods, it is important to recognize that gradient-based approaches are appropriate in many settings. In particular, when gradients are available through an adjoint procedure [2], these techniques can be highly efficient. Successful applications of gradient-based methods to oil field problems have been presented in many papers; see, e.g., [3,4,5,6].

Gradient-based approaches do, however, have some drawbacks. As a result of the nonconvex nature of the optimizations considered here, these problems generally contain multiple optima, and hence, a purely local search, which can get trapped in local solutions, might not be the best approach. In addition, for some problems (particularly well placement), the optimization surface can be very rough, which results in discontinuous gradients. It is also important to recognize that derivative information is often not readily available. Adjoint-based techniques, which are a popular way for computing derivatives efficiently, are invasive with respect to the flow simulator, and are therefore only feasible with full access to, and detailed knowledge of, the simulator source code. Numerical gradients are straightforward to calculate, though this computation is expensive and may be subject to practical difficulties (for example, in finite differencing, the selection of the perturbation size and/or simulation tolerances can be problematic). Thus there is clearly a need for other, derivative-free, techniques for oil reservoir optimization problems.

The derivative-free techniques considered in this work are noninvasive with respect to the flow simulator. They treat the simulator as a black-box – only cost function values are required and no explicit gradient calculations are involved. These methods are therefore much easier to implement than, for example, adjoint-based techniques, though this advantage is counterbalanced by a significant deterioration in computational efficiency compared to adjoint approaches. The computational cost associated with derivative-free methods depends strongly on the number of optimization variables considered (in adjoint-based schemes this dependence is much weaker). However, most of these algorithms parallelize naturally and easily, and therefore their efficiency, measured in terms of elapsed time, is usually satisfactory.

Derivative-free optimization approaches can be divided into deterministic (e.g., generalized pattern search) and stochastic (e.g., particle swarm optimization) techniques. Stochastic approaches can be useful for dealing with rough functions or functions that contain multiple local optima. Based on the computational resources typically available in current practice (e.g., $O(100)$ cores), derivative-free optimization methods are appropriate when the number of optimization variables is at most a few hundred [7, 8].

Although gradient-free methodologies have been in existence for many years, they have become widely used in only the last 20 years or so [9]. This relatively recent uptake can be attributed to several factors, including the wide availability of large numbers of cores (combined with algorithms that parallelize easily), the significant theoretical results achieved in this period, and the successful application of derivative-free techniques in a number of areas. Examples can be found in molecular geometry [10], aircraft design [11, 12], hydrodynamics [13, 14] and medicine [15, 16].

Many derivative-free stochastic schemes have also been applied within the oil industry. The field development problem has often been addressed by means of global stochastic-search techniques; see, e.g., [17, 18, 19, 20, 21]. These stochastic schemes have also been hybridized with deterministic search techniques, as presented in [22, 23, 24]. Both global [25, 26] and local (deterministic) [27, 28] derivative-free search techniques have been applied for well control optimization. The history matching problem has also been approached from both a stochastic point of view [29, 18, 30, 31] and using local methodologies combined with regularization and initial guess selection [32, 33].

Our goal in this chapter is to illustrate the applicability of derivative-free optimization methods for three types of problems arising in oil field operations. The examples presented are taken from [28] (well control optimization), [21] (field development optimization), and [32] (history matching). This chapter is structured as follows. In Section 2.2 we briefly describe the simulation modeling procedures and basic optimizers considered. Examples demonstrating the use of derivative-free techniques for well control optimization, field development optimization, and history matching are presented in Sections 2.3, 2.4 and 2.5, respectively. Enhancements to the basic optimization algorithms required for the target problem are discussed in these three sections. We end the chapter with a summary and recommendations.

2.2 Basic Methodologies

We now discuss the simulation techniques used in the optimizations, and describe the basic optimizers considered in this work.

2.2.1 Simulation Techniques

The optimization problems studied here rely on simulations of fluid flow in subsurface formations. Additionally, in Section 2.5, equations describing wave diffraction tomography must also be solved as part of the inverse modeling process. These simulations require the numerical solution of systems of partial differential equations (PDEs).

In this work we consider oil-water systems. These two components exist in separate phases, both of which reside within the pore space of porous rock. Within the context of oil production, the subsurface formation containing oil (and associated water) is referred to as a reservoir. The flow of oil and water in a reservoir is described by statements of mass conservation combined with constitutive (Darcy's law) relationships that relate phase flow rates to pressure gradient. For single-phase flow, Darcy's law is given by $\mathbf{u} = -(k/\mu)\nabla p$, where $\mathbf{u}$ is the Darcy velocity (volumetric flow rate divided by total area), k is the absolute permeability, which is a key property of the rock, μ is fluid viscosity and p is fluid pressure. For two or three-phase flow, this relationship is modified by the inclusion of the so-called relative permeability function, which is a scalar function of local phase volume fraction. Another key quantity is porosity ϕ, which specifies the fraction of the bulk rock volume that is pore space.

In most reservoir simulators, the governing equations are discretized using a finite volume numerical procedure. The detailed equations and discretizations can be found in, e.g., [34, 35]. In practical applications, simulation models may contain $O(10^5 \sim 10^6)$ grid blocks and may require several hundred time steps (the systems considered here are somewhat smaller). In addition, the discrete system of equations is nonlinear and is solved using a Newton-Raphson procedure. Thus the evaluation of reservoir performance is computationally demanding. In this work we apply Stanford's general purpose research simulator (GPRS; [36, 37]) for two of the cases considered and the commercial streamline simulator 3DSL [38] for the other cases. The streamline simulator shares many similarities with GPRS, though it uses the streamlines from the total velocity field (total velocity is equal to the sum of the water and oil Darcy velocities) to define a coordinate system that is used to solve the water transport equation. This introduces some approximations but it provides a more computationally efficient solution than would typically be achieved using a standard simulator. We note finally that, in the examples presented here, some secondary effects (such as capillary pressure in all cases, compressibility in the streamline simulations) are neglected. These effects could be included if necessary though they would not be expected to impact our basic findings.

Seismic measurements involve first a number of sources, such as dynamite, air guns, or piezoelectric transducers, which send out elastic waves through the reservoir. The transmitted and reflected waves are then recorded on geophones that respond to ground displacement or stresses. The recorded wavefields are processed and analyzed, and by means of a data assimilation process, such as that described in Section 2.5, can be used to infer the rock properties needed in the calculation of oil production forecasts.

In this work diffraction tomography (see e.g., [39, 40, 41]) simulations are used as seismic measurements. The simulations for diffraction tomography require the numerical solution of the elastic wave equation, which describes the propagation of mechanical waves in elastic media. This equation is a statement of conservation of momentum, combined with the constitutive relation for an elastic material relating stresses to strains (Hooke's law). The velocity of the traveling waves depends on the elastic properties of the rock (Young's modulus and Poisson's ratio) and the density, which in turn depend on the rock type, porosity, and the saturations of the pore fluids. Rock physics models relate these rock and fluid properties to the seismic velocities.

The wave equation is solved using the Born approximation [42, 43], which is a perturbation method applied to the scattering of waves in inhomogeneous media. In that approximation, the spatial heterogeneities in elastic properties are divided into a smooth background medium with fluctuations around the background. The wavefield is also divided into an incident wavefield traveling in the background medium along with a scattered wavefield from the heterogeneities. The contributions from the scattered field are expressed in terms of an integral which is computed numerically.

2.2.2 *Optimization Problem Statement*

A general single-objective optimization problem, as is addressed in this chapter, can be stated as:

$$\min_{\mathbf{x}\in\Omega\subset\mathbb{R}^n} f(\mathbf{x}) \quad \text{subject to} \quad \mathbf{g}(\mathbf{x}) \le 0, \tag{2.1}$$

where $f(\mathbf{x})$ is the objective function (e.g., negative of net present value ($-$NPV) or norm of discrepancy between measurements and model output), $\mathbf{x} \in \mathbb{R}^n$ is the vector of control variables (e.g., sequence of well pressures, locations for each well, or calibration parameters), and $\mathbf{g} : \mathbb{R}^n \to \mathbb{R}^m$ represents the nonlinear constraints in the problem. Bound and linear constraints are included in the set $\Omega \subset \mathbb{R}^n$. As indicated above, the objective function (and constraints, in some cases) are computed using the output from a simulator.

Though the optimization problems considered in this work share some commonalities, there are important distinctions between them. Well control optimization is in most cases formulated in terms of continuous variables and includes nonlinear, simulation-based constraints. Previous studies demonstrate that this problem often displays multiple solutions with comparable cost function values [44, 28]. For that reason, this optimization is usually addressed using local search optimization techniques. By contrast, the optimization landscapes found in field development problems can be very rough [20], and this motivates the use of global search approaches.

As is the case with most inverse problems, history matching typically involves more unknowns than informative measurements, which leads to an undetermined optimization problem. Additionally, noise in the measurements can introduce roughness into the cost function. In our application, many of the multiple optima that can result from history matching are not consistent with prior geological information, and should therefore be discarded. Strategies for finding geologically realistic optima include regularization methodologies, performing a global exploration of the search space, and/or selecting a proper initial guess in local optimization schemes. Since the number of optimization parameters in history matching can be comparable to the number of grid blocks in the simulation model, parameter reduction techniques, which can be interpreted in regularization terms, are extremely helpful. These techniques can be used to assure consistency with prior geological information, as described in [45, 46].

Discrete-valued variables are common in optimization problems in the oil and gas industry. Such problems cannot in general be addressed by gradient-based optimizers. In some cases, however, these variables can be treated as real-valued in order to establish a more amenable optimization problem (in this case we say that the discrete-valued variable is relaxed to a real-valued variable).

2.2.3 *Derivative-Free Optimization Methods*

In this section we describe, within an unconstrained real-valued optimization framework, the derivative-free local and global methods applied in this chapter. Most of these procedures can be extended to cases with discrete-valued variables, bound and/or linear constraints and, with slightly more effort, to problems with computationally inexpensive nonlinear constraints (in Section 2.3.1 we provide mathematically sound procedures for handling simulation-based nonlinear constraints). Additional enhancements of these basic methodologies are introduced for the case examples when necessary. It is important to note that the variants devised for discrete optimization are generally based on heuristics. In Sections 2.3 and 2.5, a gradient-based method, sequential quadratic programming (SQP; see [47]), with numerical derivatives is also considered to enable additional comparisons between the various approaches. The SQP implementation used in this work is SNOPT [48].

2.2.3.1 Local Search Algorithms

The local search techniques considered here are two different pattern search methods: generalized pattern search and Hooke-Jeeves direct search. Pattern search optimization has recently become popular as a result of the development of a solid mathematical convergence theory [49, 8, 7] and of the increasing availability of parallel computing resources. Pattern search schemes evaluate iteratively the cost function in a stencil-based manner. This stencil is modified as iterations proceed, and convergence theory requires that the stencil size eventually tends toward zero [8, 7]. By using a relatively large stencil size during the first stages in a pattern search technique, some local minima can be avoided. This strategy may endow pattern search with a degree of robustness against noisy cost functions. We note that pattern search schemes (and, in general, most local as well as global optimizers) can be accelerated by means of computationally inexpensive surrogates. The use of surrogates can be quite useful for reservoir engineering problems given the large number of expensive objective function evaluations that are typically required.

Generalized Pattern Search

Generalized pattern search (GPS; [49, 50]) comprises a family of optimization algorithms. By considering different types of stencils and various strategies for evaluating the stencil points (which is known as polling [50]), multiple GPS-based optimizers can be constructed. For unconstrained optimization, the basic GPS iteration, for a given stencil centered at the intermediate solution $\mathbf{x}_0$, is as follows. First, the objective function is evaluated for a number of stencil points. If some of these points yield cost function improvement, the current solution is updated with either the best point (if the full stencil is evaluated) or the first point that improves the solution (if an opportunistic search is used). The stencil can then be modified, but in most implementations it stays unaltered. If none of the stencil points improves on $\mathbf{x}_0$, then the stencil size is decreased. The search progresses until some stopping criterion is satisfied (typically, a minimum stencil size).

The stencil should contain a generating set for $\mathbb{R}^n$ [8]. A generating set of vectors has the property that, if $\nabla f(\mathbf{x}_0) \neq 0$, then at least one element of the set is a descent direction [8]. Though only $n+1$ points are needed to establish a generating set for $\mathbb{R}^n$, stencils containing $2n$ elements are commonly used in GPS. We illustrate these two types of stencils in Figure 2.1(a) and 2.1(b).

If the stencil polling process is opportunistic then, as soon as a point improving on the current solution is found, the stencil is moved to that new point. Therefore, only a subset of stencil points will be polled at a given iteration. We show an example of opportunistic polling for a two-dimensional compass stencil in Figure 2.1(c). The point in the east direction is assumed to yield improvement over $\mathbf{x}_0$. As a consequence, the other three points are not evaluated.

In GPS the set of directions in the stencil remains the same at each iteration, which typically provides a coordinate or compass search, as depicted in Figure 2.1(a). The approach can be further generalized by iteratively varying the set of directions in the stencil. For example, at a given iteration the stencil for a two-dimensional optimization problem could be as shown in Figure 2.1(a). Upon polling success, the new stencil is rotated arbitrarily, as in Figure 2.1(d). If the stencil is randomly selected from an asymptotically dense set of directions, the resulting

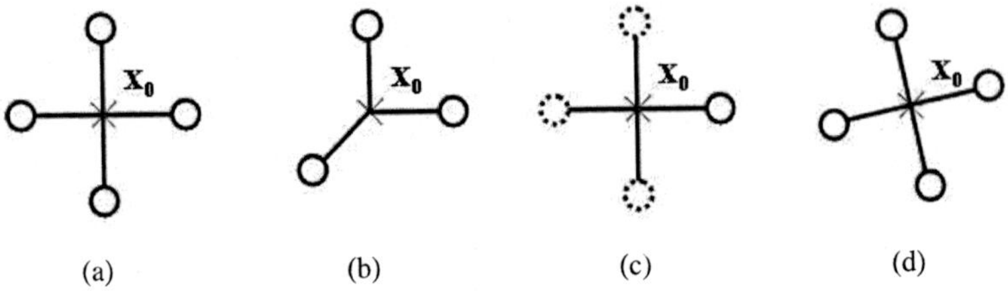

Fig. 2.1 Types of stencil-based search for a two-dimensional space: (a) positive basis with $2n$ directions (compass), (b) positive basis with $n+1$ directions, (c) opportunistic search (the first point tried, the one in the east direction, is assumed to improve on $\mathbf{x}_0$; the other points, for which the cost function is not evaluated, are plotted with dashed lines), and (d) mesh adaptive compass search (the stencil changes randomly at every iteration)

algorithm is the mesh adaptive direct search (MADS; [51]). The MADS approach may be beneficial in situations where the cost function is noisy [51].

If the polling process is not opportunistic (which means the cost function is evaluated for all stencil points), generalized pattern search requires on the order of n function evaluations per iteration. However, the GPS method parallelizes naturally since, at a particular iteration, the objective function evaluations at the polling points are completely independent and can thus be accomplished in a distributed fashion. We note that opportunistic polling is well suited to situations where parallel computing resources are limited or unavailable.

Hooke-Jeeves Direct Search

Hooke-Jeeves direct search (HJDS; [52]) is a compass-based pattern search method. There are two different types of moves in HJDS: exploratory and pattern. In the exploratory move the cost function is evaluated at consecutive perturbations of the stencil center $\mathbf{x}_0$ in the coordinate directions. All directions are polled opportunistically. The exploratory move resembles a numerical gradient estimation (with a perturbation size that may initially be large, but that eventually tends to zero). If no cost function improvement is found in the exploratory step (and this implies $2n$ function evaluations), the stencil size is decreased.

Otherwise, a new point $\mathbf{x}_1$ is obtained, and the next exploratory move is centered at $\mathbf{x}_0 + 2(\mathbf{x}_1 - \mathbf{x}_0)$. This aggressive step in the underlying successful direction is the pattern move, which is somewhat analogous to a line search procedure. The pattern move can be beneficial in situations where an optimum is far from the current solution. If the new exploratory step yields no cost function decrease, another opportunistic compass search is centered at $\mathbf{x}_1$, and if, again, this search yields no improvement, the step size is reduced, keeping the stencil at $\mathbf{x}_1$. Because HJDS is inherently sequential, it is most appropriate for use with serial computing resources.

2.2.3.2 Global Search Algorithms

The global search approaches applied in this work are a genetic algorithm and particle swarm optimization. These techniques share some similarities as they are both based on abstractions of natural processes, have a markedly stochastic nature, and apply sequential updating of a set of solutions (population of individuals in genetic algorithms, swarm of particles in particle swarm optimization).

Genetic Algorithms

Genetic algorithms (GAs) are well known and widely used so our discussion here will be brief (refer to [53] for a detailed description). GAs are inspired by the theory of natural selection. An iteration starts with a population of individuals, which is ranked in terms of cost function (referred to as fitness in the context of GAs). Thereafter, a set of operators, typically selection, crossover and mutation, are applied to generate a new population. The population size, like the swarm size in particle swarm optimization, has a marked impact on the performance of GAs. With a proper population size, a genetic algorithm can be used to explore complex objective function landscapes, and to thus identify promising regions in the search space. A thorough global exploration, even for a moderate number of optimization variables, often requires many function evaluations, and accordingly, a large population size. However, the cost function computation for all of the individuals can be readily performed in a distributed manner.

Particle Swarm Optimization

Particle swarm optimization (PSO; [54, 55]) was introduced by Kennedy and Eberhart in the mid 1990s. The algorithm mimics the social behaviors exhibited by swarms of animals. At each PSO iteration, all particles in the swarm move to a new position in the search space. Let $\mathbf{x}_{i,k} \in \mathbb{R}^n$ be the position of particle i at iteration k, $\mathbf{x}^*_{i,k}$ represent the best position (solution) found by particle i up to iteration k, and $\mathbf{y}^*_{i,k}$ be the best position found by any of the particles in the 'neighborhood' of particle i up to iteration k. The neighborhood can include all of the PSO particles, in which case the algorithm is referred to as global-best PSO. Other neighborhood specifications [56] limit particle communication such that particle i interacts with only a subset of the swarm (this has been observed to be useful in avoiding premature convergence). The new position of particle i at iteration $k+1$, $\mathbf{x}_{i,k+1}$, is computed by adding a so-called velocity term, $\mathbf{v}_{i,k} \in \mathbb{R}^n$, to the current position $\mathbf{x}_{i,k}$ [54, 55, 57]:

$$\mathbf{x}_{i,k+1} = \mathbf{x}_{i,k} + \mathbf{v}_{i,k}. \tag{2.2}$$

The velocity $\mathbf{v}_{i,k}$ is in turn calculated as follows:

$$\mathbf{v}_{i,k} = \omega\,\mathbf{v}_{i,k-1} + c_1\,\mathbf{r}_1 \circ \left(\mathbf{x}^*_{i,k} - \mathbf{x}_{i,k}\right) + c_2\,\mathbf{r}_2 \circ \left(\mathbf{y}^*_{i,k} - \mathbf{x}_{i,k}\right), \tag{2.3}$$

where ω, c_1, and c_2 are weights, $\mathbf{r}_1$ and $\mathbf{r}_2$ are random vectors in $\mathbb{R}^n$ with components uniformly distributed in the interval $(0,1)$, and $\circ$ denotes the Hadamard (component-wise) product. Thus, we see that each particle moves to a new position based on its existing trajectory, its own memory, and the collective experience of neighboring particles. These three velocity contributions are referred to as the inertia, cognitive, and social components [54, 57].

Some constraints can be handled in PSO through use of the 'absorption' technique [56, 58, 59]. With this approach, particles corresponding to infeasible solutions are moved to the nearest constraint boundary, and the corresponding velocity components are set to zero. We should note that this constraint handling procedure should be accompanied by an efficient scheme for projecting infeasible points back into the feasible domain. When this projection algorithm cannot be applied (e.g., for simulation-based constraints), the penalty function approach is a likely viable alternative (though this approach is not exempt from potential issues; see Section 2.3.1).

2.3 Well Control Optimization with Operational Constraints

The optimization of well settings/controls typically entails maximizing either net present value (NPV) or the cumulative volume of oil produced through time by finding the optimal well flow rates or pressures (these pressures are referred to as bottom-hole pressures or BHPs). In many actual scenarios, and in the cases considered here, water is injected to drive the oil toward production wells and to maintain reservoir pressure. Secondary objectives could include minimizing the total volume of water injected or produced, or maximizing the initial oil production rate. The problem is usually solved subject to operational constraints, such as maximum and minimum BHP, maximum water injection rate, maximum well water cut (fraction of water in the produced fluid), etc. The optimization variables are generally real-valued, and the relationships between these variables and both the objective function and constraints are in general nonlinear. Thus, the problem can be addressed by nonlinear programming techniques [47].

The production optimization cases presented here involve the maximization of undiscounted NPV by adjusting the BHPs of water injection and production wells (well flow rates could also have been the optimization variables). The objective function we seek to minimize is

$$f(\mathbf{x}) = -\mathrm{NPV}(\mathbf{x}) = -r_o Q_o(\mathbf{x}) + c_{wp} Q_{wp}(\mathbf{x}) + c_{wi} Q_{wi}(\mathbf{x}), \qquad (2.4)$$

where r_o is the price of oil (\$/STB, where 'STB' stands for stock tank barrel; 1 STB = 0.1590 m^3), c_{wp} and c_{wi} are the costs of produced and injected water (\$/STB), respectively (produced water reduces NPV due to pumping and separation costs), and Q_o, Q_{wp} and Q_{wi} are the cumulative oil production, water production and water injection (STB) obtained from the simulator.

2.3.1 *Constraint Handling Techniques*

The nonlinear programming methods applied here are generalized pattern search (GPS), Hooke-Jeeves direct search (HJDS), and a genetic algorithm (GA), with enhancements introduced to deal with general constraints. Consistent with the derivative-free spirit of this work, the constraint handling techniques considered, namely penalty functions and filter methods, allow us to continue treating the simulator as a black-box. These methodologies are not exclusive to gradient-free optimizers, so they could be implemented with a wide variety of optimization approaches. The description below of constraint handling techniques follows the discussion presented in [28].

Penalty Functions

The penalty function method (see, e.g., [47]) for general optimization constraints entails modification of the objective function with a penalty term that depends on some measure of the constraint violation $h : \mathbb{R}^n \to \mathbb{R}$. The modified optimization problem

$$\min_{\mathbf{x}\in\Omega} f(\mathbf{x}) + \rho\, h(\mathbf{x}), \tag{2.5}$$

where $\rho > 0$ is a penalty parameter, may still have constraints, but they should be straightforward to handle (for example, bound constraints). In this work we apply $h(\mathbf{x}) = ||\mathbf{g}^+(\mathbf{x})||_2^2$, with $\mathbf{g}^+ : \mathbb{R}^n \to \mathbb{R}^m$ defined as $g_i^+(\mathbf{x}) = \max\{0, g_i(\mathbf{x})\}$ (normalizing the constraints can be beneficial since they are all weighted equally in the penalty term). If the penalty parameter is iteratively increased (tending to infinity), the solution of the modified optimization problem (2.5) converges to that of the original nonlinearly constrained problem. However, the sequence of values to use for ρ may require some numerical experimentation and the overall procedure can lead to significant additional computation. In certain cases, a finite (and fixed) value of the penalty parameter also yields the correct solution (this is the so-called *exact* penalty; see [47]). However, for exact penalties, the modified cost function is not smooth around the solution [47], and thus the corresponding optimization problem can be challenging to solve.

Filter Method

The penalty function approach is straightforward to implement but, as discussed above, can introduce some potential difficulties and complications. Filter methods [60, 47] provide an alternate and systematic approach for handling general constraints. A filter is a set of pairs $(h(\mathbf{x}), f(\mathbf{x}))$, such that no pair dominates another pair. The concept of dominance, borrowed from multi-objective optimization, is defined as follows: the point $\mathbf{x}_1 \in \mathbb{R}^n$ dominates $\mathbf{x}_2 \in \mathbb{R}^n$ if and only if either $f(\mathbf{x}_1) \leq f(\mathbf{x}_2)$ and $h(\mathbf{x}_1) < h(\mathbf{x}_2)$, or $f(\mathbf{x}_1) < f(\mathbf{x}_2)$ and $h(\mathbf{x}_1) \leq h(\mathbf{x}_2)$. In this work, the constraint violation h associated to the filter method is computed the same way as described above for the penalty method.

Filters have been combined with a variety of basic optimization algorithms including sequential quadratic programming [60], interior point methods [61], and pattern search techniques [62,63]. They can be understood as essentially an add-on for a basic optimization procedure. Within the context of a pattern search method, a filter acts to modify the standard acceptance criterion which, as discussed in Section 2.2.3, is based only on cost function improvement. At a given iteration, the basic optimization algorithm proposes a number of intermediate solutions. These solutions are accepted if they are not dominated by any point in the filter. Prior to continuing with the next iteration, the filter is updated based on all the points evaluated by the optimizer. Using filters, the original problem (2.1) is thus viewed as a bi-objective optimization: besides minimizing the cost function $f(\mathbf{x})$, we also minimize the constraint violation $h(\mathbf{x})$. Using this multi-objective perspective, the optimization search is enriched by considering infeasible points. We reiterate that the ultimate solution is intended to be feasible (it may however show a very small constraint violation).

2.3.2 *Production Optimization Example*

This example is taken from [28]. The reservoir is a portion of the synthetic SPE 10 model [64]. It is represented on a three-dimensional grid containing $60 \times 60 \times 5$ blocks. The reservoir contains oil and water. The 25 wells (16 water injectors and nine producers) are distributed following a five-spot pattern (see Figure 2.2). This model is similar to models used in practice except it contains fewer grid blocks. The variation in permeability, evident in Figure 2.2, strongly impacts the flow field.

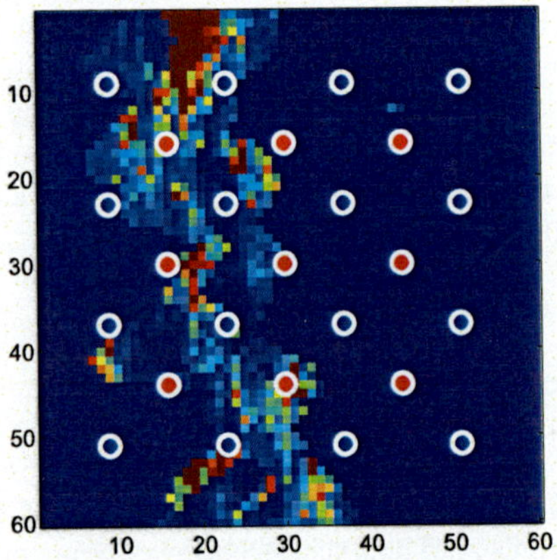

Fig. 2.2 Well configurations and top layer of the geological model considered in the production optimization case in Section 2.3. Grid blocks are colored to indicate value of permeability (red is high permeability, blue is low permeability). Injection and production wells are represented as blue and red circles, respectively (from [28]); see online version for colors.

By optimizing the well settings, we can achieve a more uniform distribution of the injected water, thus increasing the amount of oil produced and maximizing NPV.

Reservoir production proceeds for a total of 1460 days. The BHP of each well is updated every 365 days. There are thus a total of four control intervals. Since there are 25 wells, the number of optimization variables is 100. During each control interval, the BHPs are held constant. Injection well BHPs are specified to be in the range 6500 – 12000 psi and production wells are constrained to the range 500 – 5500 psi.

The additional constraints, which are nonlinear, specify that (1) the maximum field-wide water injection rate not exceed 15000 STB/day, (2) the maximum field-wide liquid (oil+water) production rate not exceed 10000 STB/day, (3) the minimum field-wide oil production rate not fall below 3000 STB/day, and (4) the fraction of water in the produced fluid (water cut) not exceed 0.7 in any of the nine production wells. The oil price considered is \$50/STB, and the costs of produced and injected water are \$10/STB and \$5/STB, respectively. Additional details of the problem specification are provided in [65].

Based on results for another nonlinearly constrained production optimization problem presented in [28], we apply the following four approaches for this case: sequential quadratic programming (SQP) with numerical derivatives and an active set constraint handling method [47], generalized pattern search (GPS) with penalty function, GPS with filter, and Hooke-Jeeves direct search (HJDS) with filter. The gradients required by SQP were computed using second-order finite differencing, with a perturbation size of 0.1 psi (this perturbation size was established through numerical experimentation – we reiterate that this can be an issue when estimating gradients numerically). In all cases, the initial stencil size for GPS and HJDS was 1375 psi. The penalty method relies on some heuristics for increasing the penalty parameter and terminating each corresponding intermediate optimization. Details on the strategy used here can be found in [28]. The two approaches considered with the filter method, GPS and HJDS, do not rely nearly as directly on heuristics.

The initial guess $\mathbf{x}_0$ for all methods was the center of the orthotope given by the bound constraints (i.e., BHP of 9250 psi for all injectors at all times, BHP of 3000 psi for all producers at all times). This reference case has an associated NPV of \$193.43 million and a constraint violation value of 0.3731. The optimization results are summarized in Table 2.1. Consistent with the underdetermined nature of

Table 2.1 Performance summary for the production optimization case (from [28])

Optimization approach	Number of simulations	Max. NPV [\$ MM]	h
SQP + active set	41004	341.32	0.0031
GPS + penalty function	60001	342.95	0.0000
GPS + filter	39201	342.61	0.0001
HJDS + filter	1618	336.28	0.0001

the optimization problem, the solutions computed by the four approaches differ. The NPVs for the first three methods are within 0.5% of one another, though the NPV for the last method (HJDS with filter) is about 1.5% less. Note that all algorithms except GPS with penalty function have nonzero constraint violations. For the filter-based methods, we allowed a constraint violation of 0.0001. Were we to require zero constraint violation, GPS with filter would provide an NPV of $341.12 million, and HJDS with filter would provide an NPV of $332.93 million.

All algorithms other than HJDS were implemented within a distributed computing environment (67 cores were used, which provided a speedup factor of around 50). We therefore observe that, although SQP and GPS with filter required a factor of about 24 times more function evaluations than HJDS, in terms of elapsed time, these two methods required only about half the time as HJDS. The procedure that required the highest number of function evaluations, GPS with penalty function, needed about 3/4 of the time of HJDS. This highlights the impact of the availability of multiple cores on algorithm selection. We note finally that, although the results in Table 2.1 for GPS with penalty function and GPS with filter are similar, the filter method is less heuristic and may, therefore, be preferable for many problems.

We now illustrate the degree of nonlinear constraint satisfaction provided by the various optimization algorithms. Figures 2.3 and 2.4 present the field-wide fluid production rates and the maximum of the water cut in any producer well. The red horizontal lines in these figures indicate the constraint value. It is evident that, at late time, the initial guess settings lead to constraint violations. The constraints are essentially satisfied by the other algorithms, with the exception of SQP. This occurs

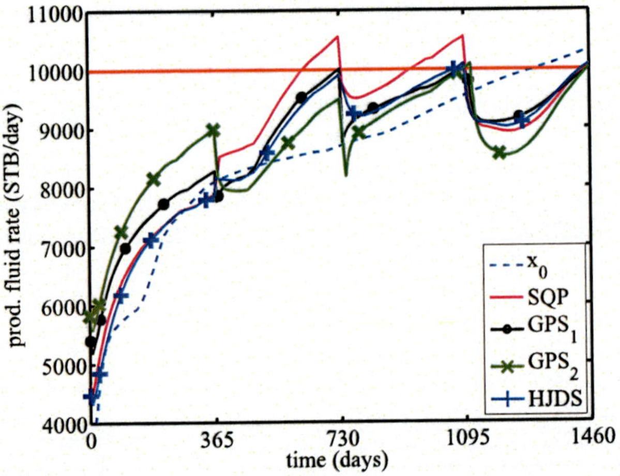

Fig. 2.3 Total field-wide fluid production rate for the initial guess $\mathbf{x}_0$ and the four solutions found for the production optimization case. The red line indicates the maximum total fluid rate allowed. GPS_1 and GPS_2 denote GPS with the penalty function and the filter method, respectively (from [28]); see online version for colors.

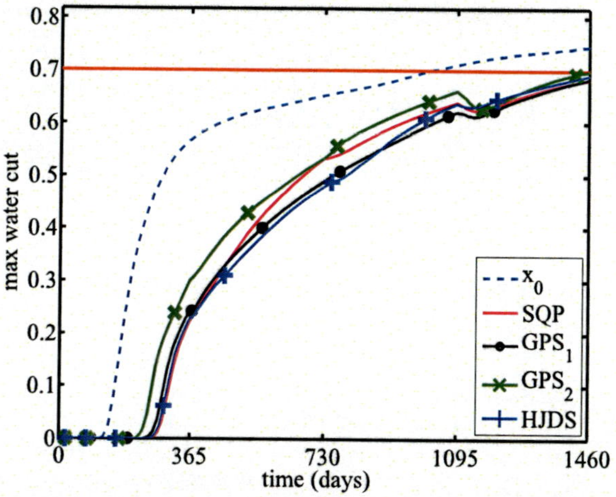

Fig. 2.4 Maximum well water cut for the initial guess $\mathbf{x}_0$ and the four solutions found for the production optimization case. The maximum water cut at a given time is the maximum of the water cut values for all producer wells at that time. The red line indicates the maximum water cut allowed for any producer well. GPS_1 and GPS_2 denote GPS with the penalty function and the filter method, respectively (from [28]); see online version for colors.

because our SQP stopping criterion does not enforce strict feasibility. SQP does, however, encounter solutions during the course of the optimization with lower constraint violations but also with lower NPVs. Thus, it is clear that the SQP results could be improved if it was used with a filter.

The quantities that directly impact NPV are displayed in Figure 2.5, where we show the production and injection profiles for $\mathbf{x}_0$ and for the solution computed by GPS with filter. The peaks in the rates in the optimized solution, evident every 365 days, result from the changes in the well BHPs, which occur at those times. It is evident that, relative to the initial guess, the optimized controls lead to a significant increase in cumulative oil production along with a significant decrease in cumulative water production (note that cumulative oil production corresponds to the integral of the curve shown in Figure 2.5, and similarly for other quantities). The cumulative water injection does not vary significantly between the two cases. This example illustrates the substantial gains that can potentially be achieved in oil field operations through the use of computational optimization.

2.4 Optimal Well Placement with Particle Swarm Optimization

The general problem of field development optimization involves the determination of how many new wells to drill, what type of wells these should be (i.e., injection well or production well; vertical, horizontal or multi-branched well; type of

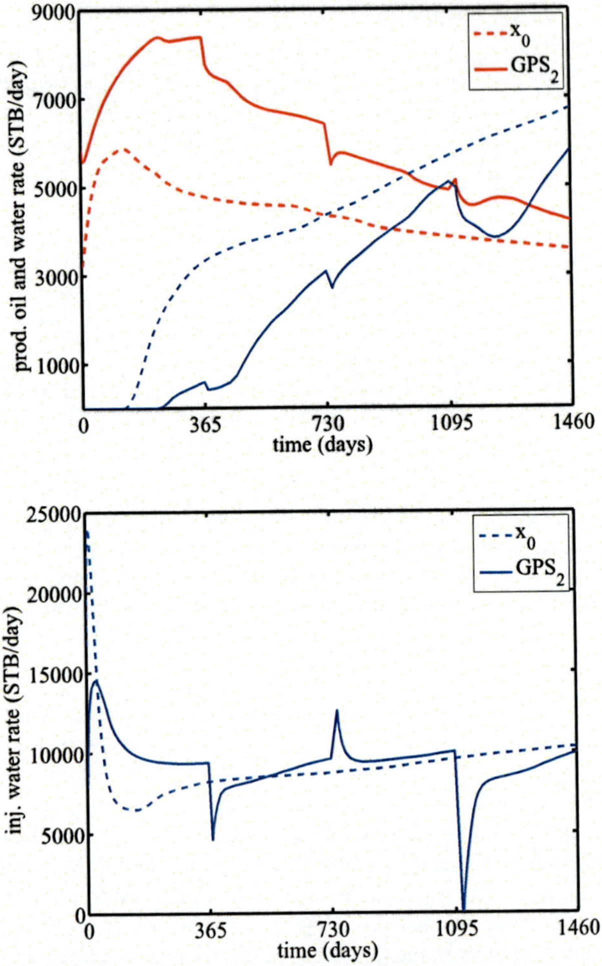

Fig. 2.5 Total field-wide production and injection rates for the initial guess $\mathbf{x}_0$ and solution computed by GPS with filter for the production optimization case. Top: Oil (red) and water (blue) production rates. Bottom: Water injection rate (from [28]); see online version for colors.

downhole instrumentation), and the drilling schedule, in order to maximize a prescribed objective function. In previous work, a number of gradient-based and derivative-free procedures have been developed and applied for this problem (see [20] for a full discussion). Of the stochastic search approaches employed, many researchers have applied genetic algorithms (e.g., [23, 66, 24, 67, 68, 19, 69, 70]), though simultaneous perturbation stochastic approximation algorithms [71], as well as other approaches, have also been explored [68]. Mattot et al. [72] evaluated several optimization algorithms for a groundwater remediation problem and achieved the best results using particle swarm optimization (PSO). This motivated the use of PSO for optimization of oil field development in [20]. Consistent with [72], in [20] PSO was found to outperform GA for several example cases. All of these examples involved relatively few wells (20 or less).

The development of large-scale oil fields, however, often involves drilling many wells. If we restrict ourselves for now to vertical wells (which can be either production or injection wells) that penetrate the entire thickness of the formation, the optimization variables include the areal (x, y) location of each well and a binary variable b defining the well type. Thus there are a total of $n = 3N_w$ optimization variables, where N_w is the number of wells. Even given the restriction of fully-penetrating vertical wells, the optimization problem is challenging. For large-scale problems, N_w can be several hundred, so the number of optimization variables can be large. In addition, for large N_w the imposition of well-to-well distance constraints (which are commonly used in field applications) can lead to a large number of infeasible solutions, and this can negatively impact the performance of a population-based algorithm such as PSO. Another key concern is that the number of wells N_w should itself be an optimization variable. Direct inclusion of N_w as an integer variable in the set of parameters will further complicate the optimization and will lead to much larger computational requirements.

2.4.1 Optimization Methodology

In recent work, a field development optimization procedure that addresses some of the issues raised above was presented [21]. In this implementation, rather than prescribe N_w and optimize $3N_w$ parameters, the wells were constrained to be arranged in repeated patterns (such patterns are commonly used for onshore oil field development). By optimizing the parameters that define the well patterns, a close-to-optimal N_w and the locations and types of all wells can be determined. This method would theoretically be expected to lead to suboptimal results relative to those that could be achieved by optimizing the number of wells and the associated $3N_w$ parameters, but it is much more tractable computationally than the more exhaustive approach.

In this section we describe and then apply this new well pattern optimization procedure and a second-stage optimization that perturbs well locations within the patterns. The core optimizer used is PSO, but the method could be implemented with other derivative-free optimization algorithms including GA.

2.4.1.1 Well Pattern Description

The basic PSO procedure was described in Section 2.2.3.2. In the well pattern description (WPD), the optimization parameters define the target pattern. This pattern is then replicated over the entire domain, with wells that fall outside of the reservoir eliminated. The algorithm considers four different well pattern types, as shown

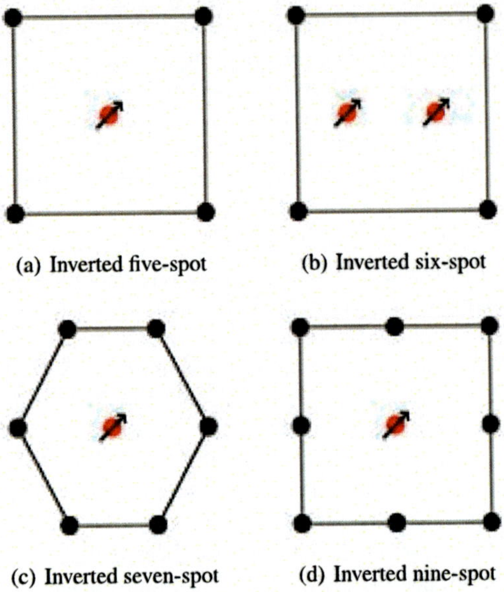

Fig. 2.6 Illustration of the well patterns considered. The solid black circles represent production wells and the circles with arrows represent injection wells. The patterns are referred to as 'inverted' because the injection wells are at the centers of the patterns (from [21]).

in Figure 2.6. Optimization variables include the pattern type (categorical variable I_i^{wp}), the location of one of the wells in the pattern (ξ_i^0, η_i^0), pattern dimensions (a_i, b_i), and parameters associated with a number of pattern operators, which we now describe.

The patterns determined using the representation above will be quite regular and oriented with the $x-y$ coordinate system. It may be advantageous, however, to adjust the orientation of the pattern to better accommodate the reservoir shape or the spatial variation/correlation of rock properties such as permeability. To accomplish this, several different pattern operators were introduced in [21]. These include a rotation operator, a shear operator and a scale operator. Well locations for the target pattern, after application of these operators, can be expressed as:

$$\mathbf{W}_{out}^T = \mathbf{M}\,\mathbf{W}_{in}^T, \tag{2.6}$$

where $\mathbf{W}_{out}$ and $\mathbf{W}_{in}$ are $N_{wp} \times 2$ (relative) well location matrices, where N_{wp} is the number of wells in the pattern, and $\mathbf{M}$ is a 2×2 transformation matrix, defined for each operator. For example, for the rotation operator, we have:

$$\mathbf{M}_\theta = \begin{pmatrix} \cos\theta & \sin\theta \\ -\sin\theta & \cos\theta \end{pmatrix}, \tag{2.7}$$

where θ designates the angle of rotation. **M** matrices are also defined for shear and scale operators; see [21] for details. A fourth operator, referred to as 'switch,' which acts to convert all injection wells to production wells and vice versa, was also introduced. This operator changes the target pattern from the so-called 'normal' form to the 'inverted' form (or back).

The full set of optimization variables for the well pattern description, for PSO particle i, is given by (with the iteration index k omitted for clarity):

$$\mathbf{x}_i = [\underbrace{\{I_i^{wp}, [\xi_i^0, \eta_i^0, a_i, b_i]\}}_{\text{pattern parameters}} \underbrace{\{S_{i,1}, S_{i,2}, \ldots, S_{i,\mathcal{N}_o}\}}_{\text{operator sequence}} \underbrace{\{\mathcal{O}_{i,1}, \mathcal{O}_{i,2}, \ldots, \mathcal{O}_{i,\mathcal{N}_o}\}}_{\text{pattern operators}}]. \qquad (2.8)$$

Here $\{I_i^{wp}, [\xi_i^0, \eta_i^0, a_i, b_i]\}$ are the basic pattern parameters for particle i, $\mathcal{N}_o$ is the number of pattern operators, $\{\mathcal{O}_{i,1}, \mathcal{O}_{i,2}, \ldots, \mathcal{O}_{i,\mathcal{N}_o}\}$ are the parameters associated with the pattern operators, and $\{S_{i,1}, S_{i,2}, \ldots, S_{i,\mathcal{N}_o}\}$ defines the sequence in which the operators are applied. The total number of optimization variables depends on the number and type of operators included, but it is only around 25 when all of the operators noted above are used. All components of $\mathbf{x}_i$ are treated as real numbers in the optimization. Some of these parameters (e.g., I_i^{wp} and $S_{i,j}$) are, however, integers. Where necessary, integer values are determined from real values by simply rounding to the nearest integer.

2.4.1.2 Second-Stage Optimization

Following the determination of the optimum repeated pattern using the well pattern description (WPD) approach described above, a second-stage optimization can be applied to further improve the solution. This procedure is based on a well-by-well perturbation (WWP) and involves the local shifting of wells within patterns. Optimization variables (PSO particles) for WWP optimization are:

$$\mathbf{x}_i = \{\underbrace{\Delta\xi_1, \Delta\eta_1}_{\text{well 1}}, \underbrace{\Delta\xi_2, \Delta\eta_2}_{\text{well 2}}, \ldots, \underbrace{\Delta\xi_j, \Delta\eta_j}_{\text{well } j}, \ldots, \underbrace{\Delta\xi_{N_w}, \Delta\eta_{N_w}}_{\text{well } N_w}\}, \qquad (2.9)$$

where N_w is the number of wells determined in the first-stage (WPD) optimization and $\Delta\xi_j$ and $\Delta\eta_j$ are the perturbations of the spatial locations of well j. The minimum and maximum values of $\Delta\xi_j$ and $\Delta\eta_j$ are constrained to keep wells essentially within their original patterns. The dimension of this optimization problem can be high for large N_w, but the size of the search space is greatly limited by bound constraints on $\Delta\xi_j$ and $\Delta\eta_j$. We note finally that this second-stage optimization could be extended to determine completion intervals (i.e., vertical locations where the well is open to flow), to eliminate particular wells, or to modify individual well types.

2.4.2 Field Development Optimization Example

We now apply the procedures described above to a two-dimensional reservoir model. This example is taken from [21]; refer to that paper for full details. The reservoir domain is irregular, as shown in Figure 2.7, where the dark regions along the boundaries designate non-reservoir zones. Wells that fall outside of the reservoir region are eliminated from the set. The model contains a total of 80×132 grid blocks. The production and injection wells are prescribed to operate at fixed bottom-hole pressures of 1200 psi and 2900 psi, respectively. The total production time is 1825 days. Flow simulations for this case were performed using the streamline simulator 3DSL [38]. Streamline simulators are not as broadly applicable as standard finite-volume based simulators, but when appropriate, as they are in many waterflood simulations, streamline approaches can be considerably more efficient than standard procedures.

The well pattern optimization runs used 40 PSO particles and proceeded for 40 iterations. The optimization was run five times. Following these five runs, the best optimization solution (run 3 in Table 2.2) was used for five subsequent WWP optimizations. Results for NPV for the well pattern optimizations are shown in Table 2.2, while those from the subsequent use of WWP are presented in Table 2.3. It is evident from Table 2.2 that the inverted five-spot was the best pattern in all runs. We see from Table 2.3 that WWP consistently led to improvements of around 20% over the unperturbed patterns. The progress of the overall optimization is displayed in Figure 2.8, where the improvement in NPV during both stages is evident.

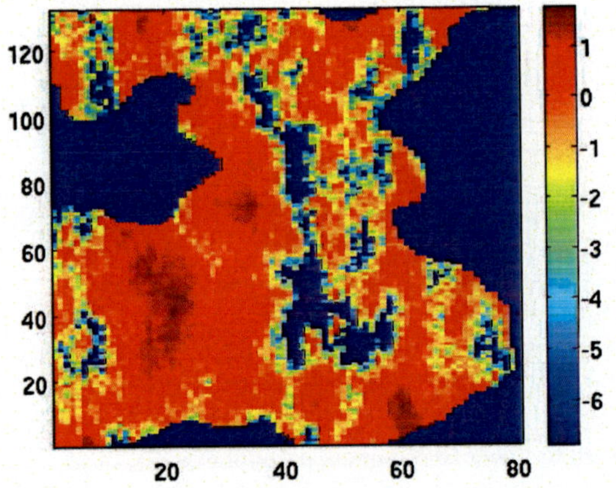

Fig. 2.7 Logarithm of permeability field for field development optimization example (from [21]).

Table 2.2 Optimization results using well pattern description (from [21])

Run	Best pattern	NPV ($MM)	Well count	
			Producers	Injectors
1	inv. 5-spot	1377	16	15
2	inv. 5-spot	1459	15	15
3	inv. 5-spot	1460	15	15
4	inv. 5-spot	1372	15	15
5	inv. 5-spot	1342	13	15
Average		**1402**		

Table 2.3 Optimization results using the second-stage procedure relative to run 3 (from [21])

Run	NPV ($MM)	Increase over well pattern description	
		($MM)	%
1	1777	317	21.7
2	1787	327	22.4
3	1776	316	21.6
4	1801	341	23.4
5	1771	311	21.3
Average	**1782**	**322**	**22.1**

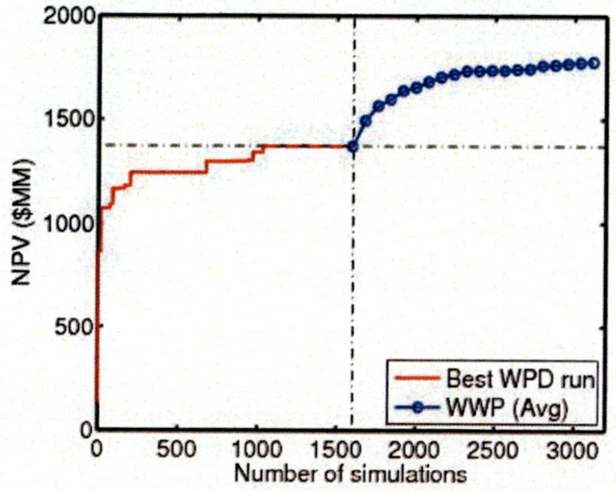

Fig. 2.8 NPV of best result from well pattern description (WPD), and average NPV of the best second-stage well-by-well perturbation (WWP) solutions, versus number of simulations (from [21]).

Figures 2.9(a) and (b) show the optimal well locations from both stages of the optimization. Repeated five-spot patterns are evident in both figures. It is interesting to observe that, although the differences in well locations between the two figures are relatively slight, these perturbations result in an improvement in NPV of 23%.

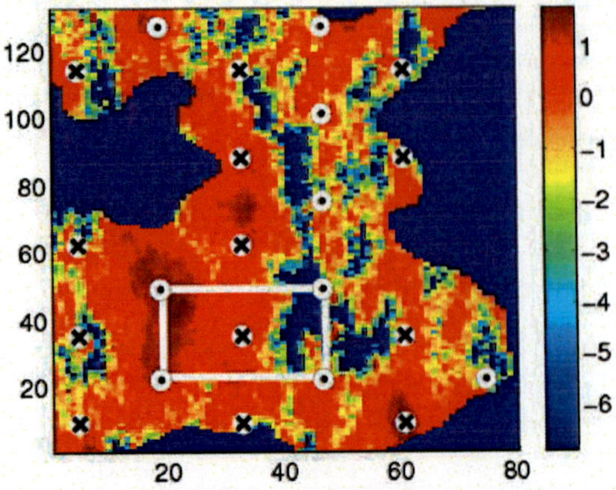

(a) WPD

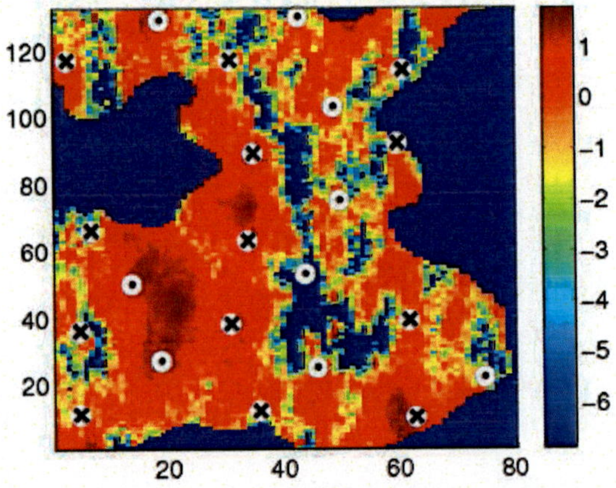

(b) WWP

Fig. 2.9 Well locations for the best well pattern description (WPD) and well-by-well perturbation (WWP) solutions (circles indicate production wells, crosses indicate injection wells). Logarithm of permeability field is shown as background (from [21]).

We note finally that several other examples demonstrating the use of PSO for well placement optimization were presented in [20, 21]. In the examples in [20] the number of wells was always specified, though in some cases the well type was also optimized (e.g., deviated and branched wells were considered in some cases). Comparisons to optimizations using a genetic algorithm (GA) were presented and, as noted above, PSO was shown to consistently outperform the GA considered. In one of the examples in [21], the well pattern optimization followed by well-by-well perturbation was compared to an unconstrained optimization that used $3N_w$ decision variables (the latter is referred to as the 'concatenation' approach). For this case the two-stage optimization consistently outperformed the concatenation approach. Taken in total, the results in [20, 21] display the applicability of PSO for well placement optimization problems, as well as the potential advantages of the well pattern description and the two-stage optimization procedure.

It will clearly be useful to combine the well control optimization described in Section 2.3 with the field development optimization considered here. This coupled optimization problem will be computationally demanding, but the solutions provided can be expected to outperform those determined through the sequential application of the two procedures. Work along these lines is currently underway.

2.5 Assimilation of Reservoir Data (Inverse Modeling)

The reliability of oil production forecasts, and the 'optimal' strategy that is determined based on these predictions, depend strongly on the proper calibration of the reservoir simulation model. In essence, this calibration aims at finding appropriate model parameters given a number of observations. The two model parameters that (in many cases) most directly impact reservoir flow are permeability and porosity. Both of these parameters vary spatially. For a given rock type, which is denoted as facies in this context, porosity and permeability are often correlated, and one can be estimated from the other. In this work, the calibration parameter is taken to be the facies in each grid block, and we assume that each facies corresponds to a particular permeability and porosity.

Historic flow production represents one set of observed data. Such data are crucial because it is precisely the prediction of the reservoir flow response that is the ultimate purpose of the modeling. However, production data provides direct information only at well locations (though of course the flow rates and pressures observed at wells are impacted by reservoir properties outside the well region). In contrast to production data, seismic measurements (such as diffraction tomography) provide more global information and thus can be used to improve estimates of the spatial distribution of rock properties. Here we consider as observable data both flow and seismic measurements.

The use of observational data to infer reservoir properties is an inverse problem. As such, we anticipate that the solution will be non-unique. This is typically the case because there are more parameters to estimate than there are independent measurements, so many combinations of parameters yield similar model responses. In addition to the underspecified nature of the problem, additional complications arise from the approximations used in the forward modeling and from the presence of noise in the data. Uncertainty quantification/assessment involves finding multiple solutions of the inverse problem in order to generate a collection of production forecasts. For more information on data assimilation under uncertainty in this context, refer to, e.g., [73, 45, 74].

2.5.1 Problem Statement

The solutions of a geophysical inverse problem are the set of geological models that, when forward-modeled to provide simulation data, match the observations to within some tolerance. Since the approach here, as shown below, involves formulating the data assimilation process in optimization terms, any model configuration (set of inversion parameters) will be denoted by $\mathbf{x} \in \Omega \subset \mathbb{R}^n$, and Ω is the set of admissible models. The admissibility criteria can be formulated with respect to geological consistency. Geological consistency typically implies a particular spatial correlation of parameters (e.g., a given spatial covariance). The model $\mathbf{x}$ in this work represents the facies type associated with every grid block. Thus, the number of optimization variables n is on the order of the number of grid blocks in the discretized reservoir model (which can be very large in practical models).

From an optimization perspective, the inverse problem can be stated as follows

$$\min_{\mathbf{x}\in\Omega} \|\mathbf{O}(\mathbf{x}) - \mathbf{y}\|^2, \tag{2.10}$$

where $\mathbf{y} \in \mathbb{R}^m$ are the observations and $\mathbf{O}(\mathbf{x}) \in \mathbb{R}^m$ represent the numerically-simulated observations. All the observable data considered are concatenated in $\mathbf{y}$ and $\mathbf{O}(\mathbf{x})$. Thus, if $\mathbf{O}_1(\mathbf{x}) \in \mathbb{R}^{m_1}$ and $\mathbf{O}_2(\mathbf{x}) \in \mathbb{R}^{m_2}$ are the two sets of observable data considered, then $\mathbf{O}(\mathbf{x}) = [\mathbf{O}_1(\mathbf{x}), \mathbf{O}_2(\mathbf{x})]$, with $m_1 + m_2 = m$. In the norm (Euclidean in this work), we can account for data uncertainty and include weights for the different sets of data. Since the observable data in this work are normalized, weights are taken to be unity. We reiterate that there are typically a much larger number of inversion parameters than there are independent measurements ($n \gg m$), and therefore the optimization problem in (2.10) is frequently ill-conditioned.

2.5.2 Methodologies for Data Assimilation

The optimization problem in (2.10) presents a number of challenges in addition to ill-conditioning. The cost function requires costly simulations, and in many cases derivative information is expensive to obtain or not available. The number of optimization variables is often large and the objective function can be non-smooth due to, for example, the presence of noise in the observations. These difficulties can be addressed by means of the following strategies.

The integration of disparate data in reservoir modeling has been suggested in a number of publications (e.g., [75,76,77]) as a means to alleviate the ill-conditioned character of (2.10). In essence, the use of different data types provides a degree of regularization for the inverse problem. Here, as in [32], we use as observable data oil and water production rates and diffraction tomography data. These data sets are complementary since they measure system responses on different spatial and temporal scales.

We can also expect a better conditioned optimization problem if the number of parameters is decreased. Instead of searching in n dimensions, we consider a subspace of dimension n_R. This subspace selection is not arbitrary and essentially aims at reducing the correlation between inversion parameters. The parameter reduction used here is based on principal component analysis (PCA), or the Karhunen-Loève transform, and can also be interpreted from a data compression perspective. The statistical information needed is generally obtained from a prior (rough) knowledge of the reservoir properties, and provides the inversion with geological consistency. We essentially follow the PCA-based parameter reduction technique used in [6], though that approach considers only flow production data and is invasive with respect to the flow simulator (thus it is very efficient but requires source-code access to implement).

In the example in Section 2.5.3, both production and seismic measurements provide the observable data. We reduce the number of optimization variables and introduce geological consistency through principal component analysis, and we approach (2.10) by derivative-free local optimization with an initial guess selected by a heuristic procedure that is based on information obtained by PCA. The use of numerical derivatives and a global procedure (GA) are also considered for comparison. All of these black-box approaches are more demanding computationally than an invasive adjoint-based gradient procedure but, as mentioned above, can be significantly accelerated through distributed computing. We briefly present below the fundamentals of PCA, since that transformation is a key component of our methodology.

2.5.2.1 Parameter Reduction Using Principal Component Analysis

Principal component analysis (PCA) optimally selects a subspace of dimension n_R from a larger space of dimension n. Given N possible models sampled from Ω, the region of the search space where plausible optimal solutions are expected, $\{\mathbf{x}_k\}_{k=1}^{N} \subset \Omega \subset \mathbb{R}^n$, PCA seeks an affine transformation

$$\hat{\mathbf{x}}_k = \sum_{i=1}^{n_R} \left(\mathbf{s}_i^T \left(\mathbf{x}_k - \mu\right)\right) \mathbf{s}_i + \mu \,,$$

with $\mu \in \mathbb{R}^n$ and the set $\{\mathbf{s}_i\}_{i=1}^{n_R} \subset \mathbb{R}^n$ orthonormal. We note that this transformation is essentially an orthogonal projection. PCA is optimal in the sense that the Euclidean reconstruction error $\|\hat{\mathbf{x}}_k - \mathbf{x}_k\|_2$, averaged over $\{\mathbf{x}_k\}_{k=1}^{N}$, is minimized (or, equivalently, that the average reconstruction energy is maximized).

The optimal solution [78] implies that μ is the average of the N models sampled $\{\mathbf{x}_k\}_{k=1}^{N}$, and that each $\mathbf{s}_i$ is an eigenvector for the covariance matrix associated with these models. Additionally, it can be seen that the covariance matrix for the n_R PCA coefficients for $\{\mathbf{x}_k\}_{k=1}^{N}$ is a diagonal matrix, and that the contribution to the average reconstruction error from each of the PCA basis components $\mathbf{s}_i$ is equal to the corresponding eigenvalue.

The selection of the N models $\{\mathbf{x}_k\}_{k=1}^{N}$ is crucial and is done based on prior information. If these models provide an acceptable representation of Ω, a large part of the n_R-dimensional search space will provide solutions that are (in this case, geologically) consistent. Therefore, PCA not only reduces the search space, but also helps to ensure that the solutions obtained are practically acceptable. The value n_R is typically much smaller than n. Low values of n_R yield low-dimensional search spaces that are easier to explore, but the reconstruction error can be unacceptably large. In other words, the optimal search would take place only in a small part of Ω, and thus the solutions obtained in that reduced space may be clearly suboptimal. The determination of the appropriate value for n_R is application specific and is typically done through numerical experimentation.

A ranking for the PCA components can be established based on their respective eigenvalues – the higher the eigenvalue, the higher the rank (and thus the importance) of the associated PCA basis vector. This, together with the fact that the covariance matrix for the PCA coefficients is diagonal, suggests that a sequence of one-dimensional optimizations aimed at computing coefficients for the highest-rank PCA basis vectors may be beneficial in the overall optimization. Based on this observation, a heuristic PCA-based procedure for computing the initial guess in (2.10) can be obtained (please consult [32] for details).

2.5.3 *Data Assimilation Example*

The case study is taken from [32] and is based on a ten-layer synthetic model (with $20 \times 20 \times 10 = 4000$ cells) extracted from the Stanford VI reservoir model [79]. This approach provides a good framework for comparing inversion methodologies since the true model is known. We simulate a five-spot well pattern (four injectors in the corners, and one producer in the center of the domain; see Figure 10(a)). The optimization variable $\mathbf{x}$ is a binary facies indicator in every grid block (designating the block as either sand or shale). Though this variable is binary valued, it can be relaxed to a continuous variable. Thus, a value of 0.5 indicates that in the corresponding grid block, sand and shale are distributed equally.

The observable production data consists of the total field cumulative oil production and water injection, obtained at intervals of ten days up to 90 days (therefore, $m_1 = 10 + 10 = 20$). The production data are computed by solving the (discretized) reservoir flow equations. Here we use Stanford's general purpose research simulator (GPRS; [36,37]). The permeability and porosity fields are functions of the facies parameter $\mathbf{x}$. Given a (real-valued) facies parameter for grid block i, designated x_i, we compute the associated porosity ϕ_i by the following expression

$$\phi_i(x_i) = \phi_0 \exp(x_i \ln(\phi_1/\phi_0)),$$

where the coefficients ϕ_0 and ϕ_1 are the porosity values associated with the shale and sand facies (in practice, these values can be determined through measurements on rock cores or regression). We relate the block permeability k_i to the porosity ϕ_i using the Kozeny-Carman equation (see, e.g., [80])

$$k_i(\phi_i) = \alpha \frac{\phi_i^3}{(1-\phi_i)^2},$$

with the parameter α calculated from measurements or regression.

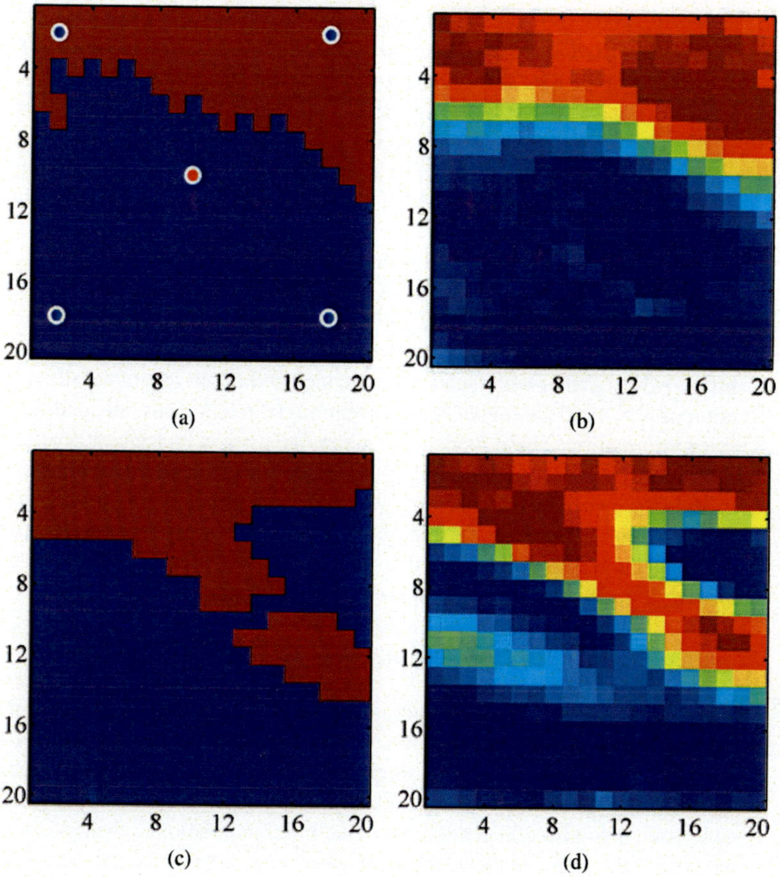

Fig. 2.10 Layer 4 from (a) the true model studied in Section 2.5 (injection and production wells are indicated as blue and red circles, respectively), (b) corresponding reconstruction after PCA with $N = 1000$ realizations and $n_R = 30$, (c) model selected randomly from the set of $N = 1000$ realizations, and (d) corresponding reconstruction after PCA with $n_R = 30$. Red and blue represent sand and shale facies, respectively. The original facies model is binary-valued, but after PCA it becomes continuous (from [32]); see online version for colors.

The second set of observable data is derived from crosswell diffraction tomography. In crosswell tomography, sound wave sources are placed in one (usually vertical) well and recorded and placed in another well (typically some hundred meters away). By recording the waves propagating from one well to another, it is possible to reconstruct approximately the structure of the earth in between the wells. The estimated earth image is sometimes called a crosswell section. In this example we have two crosswell sections obtained by associating diagonally the injectors in the five-spot pattern in Figure 10(a). Each section involves the ten layers in the model and is discretized by a 20 × 20 matrix of velocities (hence, $m_2 = 400 + 400 = 800$). The tomographic data along these two perpendicular crosswell sections are computed only once, after 90 days. The seismic observable data depends on certain rock properties (elastic bulk modulus and density) which in turn are functions of the fluid saturations at each grid block [80]. The input for the seismic tomography simulator thus includes the model **x**, which provides the porosity for each grid block, and fluid saturations. These quantities, together with rock physics models, are used to compute the elastic velocities [80]. In all tomography calculations, both for the observations and during optimization, a simplified geometry for the top of the reservoir is considered, and the associated corrections are not included.

A priori knowledge of the reservoir geology, in the form of a so-called training image [81], together with facies data obtained at the well locations, allow the generation of $N = 1000$ geologically consistent model realizations, all conditioned to the prior information. These models are generated using a multipoint geostatistical algorithm [81], which can represent complex spatial structures. Through application of PCA to these 1000 realizations we reduce the number of inversion parameters from $n = 4000$ to $n_R = 30$. In Figure 2.10 we show two of these models (one of the ten model layers is shown) and their corresponding reconstructions. For our application, these PCA reconstructions are acceptable.

2.5.3.1 Inversion Results and Prediction

We compare here sequential quadratic programming (SQP) using numerical gradients with generalized pattern search (GPS), Hooke-Jeeves direct search (HJDS), and a genetic algorithm (GA). The initial guess for the local optimizers is computed as outlined above (see [32] for details). The GA population is 60 individuals and the algorithm is run for 100 generations. The initial population in the GA does not contain the initial guess taken for the local optimizers. In this way, we can test if GA can be beneficial in cases when useful initial guesses are not available. The distributed computing environment consists of a cluster with 48 nodes, and it is used for the SQP, GPS and GA optimizations. Each observable data value is assigned random noise with an amplitude of 5% of the standard deviation of the corresponding data type.

The models determined through inversion are shown in Figure 2.11. These results are for the same layer as shown in Figure 2.10, though they are generally representative for all ten layers in the model. As noted earlier, after PCA the original binary facies model is continuous (it could be transformed back to binary values using thresholding if necessary). It is evident that all of the methods provide reasonable models. A carefully selected initial guess is crucial for obtaining acceptable inversion results with the SQP, GPS and HJDS methods. Our process for determining the initial guess relies on some heuristics and therefore is not fully general, though it appears adequate for this case. The GA result appears slightly less accurate than the others, though the main model features are captured.

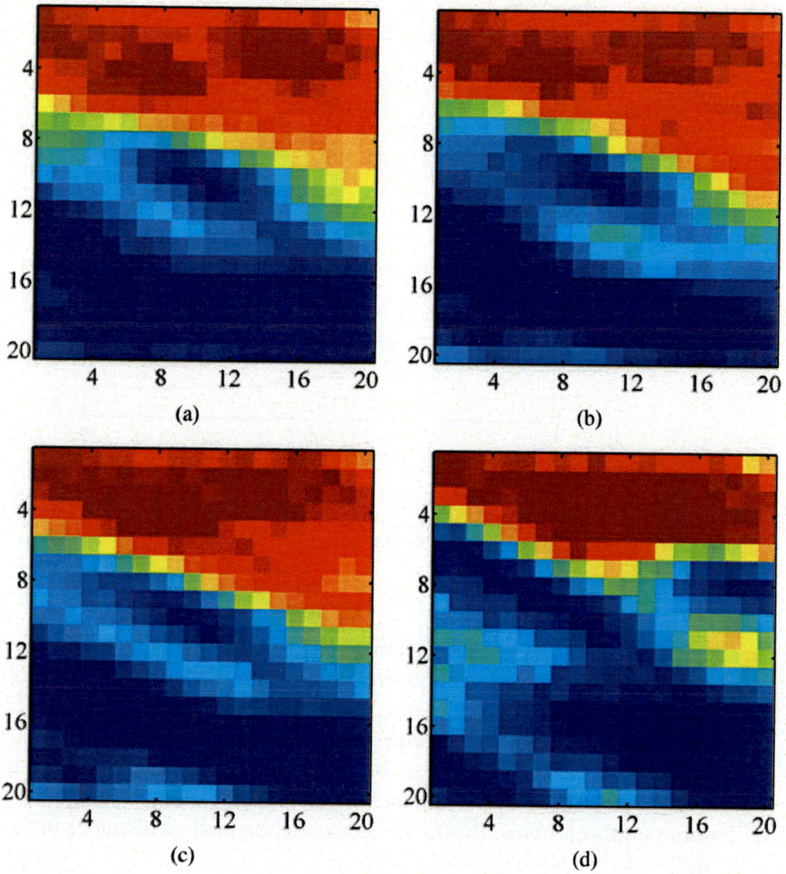

Fig. 2.11 Inverse model results for layer 4 of the reservoir section studied in Section 2.5. Facies distribution obtained by (a) sequential quadratic programming, (b) generalized pattern search, (c) Hooke-Jeeves direct search, and (d) a genetic algorithm. The genetic algorithm, because of its global nature, does not require an initial guess. The true distribution for layer 4 is shown in Figure 10(a). Red and blue represent sand and shale facies, respectively. Though the original facies model is binary-valued, after PCA it becomes continuous (from [32]); see online version for colors.

Figure 2.12 illustrates the performance of the local optimizers used for this problem. In this plot the horizontal axis is the number of equivalent simulations, which is defined as the total number of simulations divided by the speedup obtained by the parallel implementation. The concept of equivalent simulation is used to enable comparisons, in terms of elapsed time (not total computation time), between HJDS and the other (parallel) procedures. Since HJDS is inherently serial, for that algorithm the number of equivalent simulations coincides with the total number of

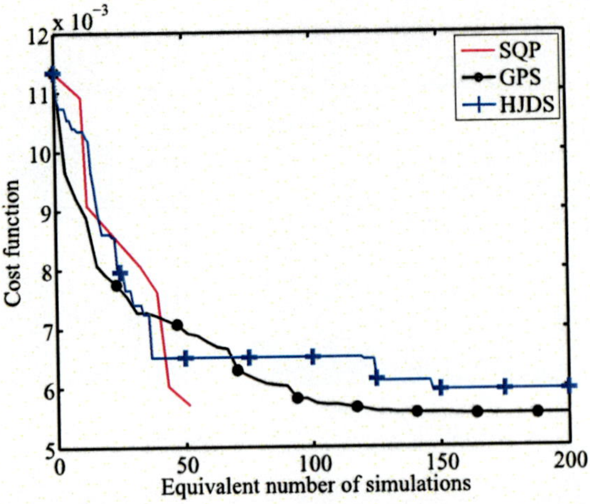

Fig. 2.12 Performance results for the local optimizers studied in the model inversion in Section 2.5 (from [32]); see online version for colors.

simulations. Note that one simulation involves calls to both the flow and seismic tomography simulators and that the initial guess computation for the local optimizers requires roughly five equivalent simulations. It is evident from Figure 2.12 that SQP provides the most efficient performance for this case. However, we expect that SQP performance would degrade if the cost function was less smooth. If the comparison was made in terms of total computation time, HJDS would be the most efficient algorithm for this problem (HJDS would thus be the method of choice in the absence of distributed computing resources).

The best individual in the initial GA population had a cost function of 0.036. After around 200 equivalent function evaluations, the objective function for GA decreased to about 0.006, though more gradually than for the other methods shown in Figure 2.12. This performance is promising since the GA was run without providing any initial guess as input. If a larger population is used, GA can explore the global search space and, as a consequence, potentially identify multiple solutions that are comparable in terms of the cost function. These solutions could then be used for uncertainty assessment.

The oil production and water injection forecasts over 360 days, for the model obtained using SQP, are shown in Figure 13(a) (we note that the inversion involved data over only the first 90 days). Agreement is generally very close, though slight mismatches are evident at later times, and these mismatches grow with time. In order to achieve accuracy over long simulation periods (up to 2000 days), the solution determined by SQP was adjusted as follows. A new data assimilation was performed after the first 1000 days. The observable data considered were the cumulative production of oil and water, together with two new crosswell tomographies at the end

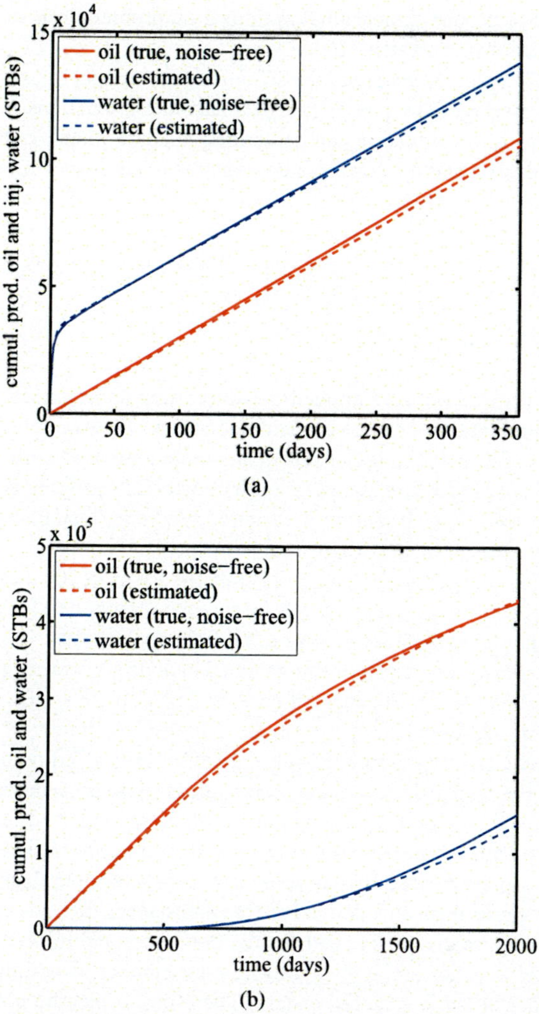

Fig. 2.13 (a) Oil production and water injection forecast (360 days) for the solution obtained by SQP. (b) Oil and water production forecast (2000 days) for the solution recalibrated after 1000 days. In both cases the noise in the observations has been removed (from [32]); see online version for colors.

of the interval. The calibration at 1000 days started with the previously determined model (as shown in Figure 2.11) and it involves only one additional parameter (λ). This parameter simply scales globally the facies distribution; i.e., the new model is given by $\lambda \mathbf{x}$, where $\mathbf{x}$ is the (old) model obtained using data for the first 90 days.

The attendant one-dimensional optimization problem in λ required approximately two additional equivalent simulations.

Figure 13(b) shows predictions from the new model for oil and water production over 2000 days. The model provides accurate predictions over the entire period. This type of recalibration can be done in practice whenever the deviation between the prediction and the corresponding data is larger than some acceptable tolerance. Since this new calibration is performed using a solution calculated previously, the number of parameters considered can be relatively small. Alternative approaches, in which more parameters (or the entire model) are computed, could also be applied.

2.6 Concluding Remarks

In this chapter we have applied derivative-free optimization methods to three different problems relevant to oil field operations. The examples considered are representative of a wide range of practical simulation-based optimization problems and involve oil production optimization with general operating constraints, field development using a well pattern description, and data assimilation based on flow and seismic measurements. These problems involved continuous, integer and categorical variables, and the search spaces contained at most 100 dimensions. The successful use of derivative-free methods for these problems clearly demonstrates that these algorithms are viable for a range of oil field applications.

The derivative-free algorithms studied include generalized pattern search, Hooke-Jeeves direct search, a genetic algorithm, and particle swarm optimization. In order to enable additional comparisons, we also tested a gradient-based method, sequential quadratic programming, with derivatives estimated numerically. With the exception of Hooke-Jeeves direct search, all of these procedures can be readily parallelized and as such benefit immensely when implemented in a distributed manner. When parallel computing resources are limited or nonexistent, Hooke-Jeeves direct search represents a promising serial derivative-free optimization strategy.

The performance of derivative-free approaches depends strongly on the dimension of the search space, and for the computational resources typically available, these approaches are applicable when the number of optimization variables is on the order of a few hundred or less. Therefore, it may be necessary in some occasions to combine these approaches with some type of parameter reduction strategy. In this work, in one case we limited the size of the search space by restricting wells to be located within patterns, while in another case we applied principal component analysis to reduce the number of inversion parameters.

There are still a number of challenges related to the problems considered in this chapter. Though categorical (decision) variables were included in the optimal field development example presented in Section 2.4, a comprehensive study on the use and limitations of derivative-free algorithms for this type of mixed-integer nonlinear optimization problem would be of great interest. In addition, further comparisons between local and global methods, and the development of hybrid procedures, will also be useful. It will be beneficial to jointly address field development optimization and well control optimization, as the optimal well locations will in general depend on how the wells are operated. Multi-objective optimization may be of interest for this and other applications.

The efficient treatment of uncertainty in all of the problems considered is also a topic of great importance. Data assimilation methodologies that generate multiple solutions consistent with observed data are required. Optimization techniques that can efficiently handle multiple models are also needed. Finally, because the forward simulations required for our optimization methods are themselves often very time-consuming, the development of fast and reliable surrogate models will be of great use. Research in many of these areas is currently underway.

Acknowledgements. We are grateful to the industry sponsors of the Stanford Smart Fields Consortium and the Stanford Center for Reservoir Forecasting for partial funding of this work, and to the Stanford Center for Computational Earth and Environmental Science for providing distributed computing resources. We also thank Obiajulu J. Isebor (Stanford University), Jérome E. Onwunalu (now at BP) and Eduardo T. F. Santos (now at CEFET-BA) for their contributions to this work.

References

1. Jansen, J.D., Brouwer, D.R., Naevdal, G., van Kruijsdijk, C.P.J.W.: Closed-loop reservoir management. First Brea 23, 43–48 (2005)
2. Pironneau, O.: On optimum design in fluid mechanics. J. Fluid Mech 64, 97–110 (1974)
3. Ramirez, W.F.: Application of Optimal Control Theory to Enhanced Oil Recovery. Elsevier, Amsterdam (1987)
4. Brouwer, D.R., Jansen, J.D.: Dynamic optimization of waterflooding with smart wells using optimal control theory. SPE Journal 9(4), 391–402 (2004)
5. Zandvliet, M., Handels, M., van Essen, G., Brouwer, R., Jansen, J.D.: Adjoint-based well-placement optimization under production constraints. SPE Journal 13(4), 392–399 (2008)
6. Sarma, P., Durlofsky, L.J., Aziz, K., Chen, W.H.: Efficient real-time reservoir management using adjoint-based optimal control and model updating. Computational Geosciences 10, 3–36 (2006)
7. Conn, A.R., Scheinberg, K., Vicente, L.N.: Introduction to Derivative-Free Optimization. MPSSeries on Optimization, MPS-SIAM (2009)
8. Kolda, T.G., Lewis, R.M., Torczon, V.: Optimization by direct search: new perspectives on some classical and modern methods. SIAM Review 45(3), 385–482 (2003)
9. Wright, M.H.: Direct Search methods: once scorned, now respectable. In: Griffiths, D.F., Watson, G.A. (eds.) Numerical Analysis 1995 (Proceedings of the 1995 Dundee Biennial Conference in Numerical Analysis). Pitman Research Notes in Mathematical Series, pp. 191–208. CRC Press, Boca Raton (1995)
10. Meza, J.C., Martinez, M.L.: On the use of direct search methods for the molecular conformation problem. Journal of Computational Chemistry 15, 627–632 (1994)
11. Booker, A.J., Dennis Jr., J.E., Frank, P.D., Moore, D.W., Serafini, D.B.: Optimization using surrogate objectives on a helicopter test example. In: Borggaard, J.T., Burns, J., Cliff, E., Schreck, S. (eds.) Computational Methods for Optimal Design and Control, pp. 49–58. Birkhäuser, Basel (1998)

12. Marsden, A.L., Wang, M., Dennis Jr., J.E., Moin, P.: Trailing-edge noise reduction using derivative-free optimization and large-eddy simulation. Journal of Fluid Mechanics 572, 13–36 (2003)
13. Duvigneau, R., Visonneau, M.: Hydrodynamic design using a derivative-free method. Structural and Multidisciplinary Optimization 28, 195–205 (2004)
14. Fowler, K.R., Reese, J.P., Kees, C.E., Dennis Jr., J.E., Kelley, C.T., Miller, C.T., Audet, C., Booker, A.J., Couture, G., Darwin, R.W., Farthing, M.W., Finkel, D.E., Gablonsky, J.M., Gray, G., Kolda, T.G.: Comparison of derivative-free optimization methods for groundwater supply and hydraulic capture community problems. Advances in Water Resources 31(5), 743–757 (2008)
15. Oeuvray, R., Bierlaire, M.: A new derivative-free algorithm for the medical image registration problem. International Journal of Modelling and Simulation 27, 115–124 (2007)
16. Marsden, A.L., Feinstein, J.A., Taylor, C.A.: A computational framework for derivative-free optimization of cardiovascular geometries. Computational Methods in Applied Mechanics and Engineering 197, 1890–1905 (2008)
17. Cullick, A.S., Heath, D., Narayanan, K., April, J., Kelly, J.: Optimizing multiple-field scheduling and production strategy with reduced risk. SPE paper 84239 presented at the 2009 SPE Annual Technical Conference and Exhibition, Denver, Colorado, October 5–8 (2009)
18. Velez-Langs, O.: Genetic algorithms in oil industry: an overview. Journal of Petroleum Science and Engineering 47, 15–22 (2005)
19. Artus, V., Durlofsky, L.J., Onwunalu, J., Aziz, K.: Optimization of nonconventional wells under uncertainty using statistical proxies. Computational Geosciences 10, 389–404 (2006)
20. Onwunalu, J., Durlofsky, L.J.: Application of a particle swarm optimization algorithm for determining optimum well location and type. Computational Geosciences 14, 183–198 (2010)
21. Onwunalu, J., Durlofsky, L.J.: A new well pattern optimization procedure for large-scale field development. SPE Journal (in press)
22. Bittencourt, A.: Optimizing Hydrocarbon Field Development Using a Genetic Algorithm Based Approach. PhD thesis. Dept. of Petroleum Engineering, Stanford University (1997)
23. Bittencourt, A.C., Horne, R.N.: Reservoir development and design optimization. SPE paper 38895 presented at the 1997 SPE Annual Technical Conference and Exhibition, San Antonio, Texas, October 5–8 (1997)
24. Yeten, B., Durlofsky, L.J., Aziz, K.: Optimization of nonconventional well type, location and trajectory. SPE Journal 8(3), 200–210 (2003)
25. Harding, T.J., Radcliffe, N.J., King, P.R.: Optimization of production strategies using stochastic search methods. SPE paper 35518 presented at the 1996 European 3-D Reservoir Modeling Confe rence, Stavanger, Norway, April 16–17 (1996)
26. Almeida, L.F., Tupac, Y.J., Lazo Lazo, J.G., Pacheco, M.A., Vellasco, M.M.B.R.: Evolutionary optimization of smart-wells control under technical uncertainties. SPE paper 107872 presented at the, Latin American & Caribbean Petroleum Engineering Conference, Buenos Aires, Argentina, April 15–18 (2007)
27. Carroll III, J.A.: Multivariate production systems optimization. Master's thesis, Dept. of Petroleum Engineering, Stanford University (1990)
28. Echeverría Ciaurri, D., Isebor, O.J., Durlofsky, L.J.: Application of derivative-free methodologies for generally constrained oil production optimization problems. International Journal of Mathematical Modelling and Numerical Optimisation (in press)

29. Schulze-Riegert, R.W., Axmann, J.K., Haase, O., Rian, D.T., You, Y.-L.: Evolutionary algorithms applied to history matching of complex reservoirs. SPE Reservoir Evaluation & Engineering 5(2), 163–173 (2002)
30. Ballester, P.J., Carter, J.N.: A parallel real-coded genetic algorithm for history matching and its application to a real petroleum reservoir. Journal of Petroleum Science and Engineering 59, 157–168 (2007)
31. Maschio, C., Campane Vidal, A., Schiozer, D.J.: A framework to integrate history matching and geostatistical modeling using genetic algorithm and direct search methods. Journal of Petroleum Science and Engineering 63, 34–42 (2008)
32. Echeverria, D., Mukerji, T.: A robust scheme for spatio-temporal inverse modeling of oil reservoirs. In: Anderssen, R.S., Braddock, R.D., Newham, L.T.H. (eds.) Proceedings of the 18th World IMACS Congress and MODSIM 2009 International Congress on Modelling and Simulation, pp. 4206–4212 (2009)
33. Dadashpour, M., Echeverria Ciaurri, D., Mukerji, T., Kleppe, J., Landrø, M.: A derivative-free approach for the estimation of porosity and permeability using time-lapse seismic and production data. Journal of Geophysics and Engineering 7, 351–368 (2010)
34. Aziz, K., Settari, A.: Petroleum Reservoir Simulation. Kluwer Academic Publishers, Dordrecht (1979)
35. Gerritsen, M.G., Durlofsky, L.J.: Modeling fluid flow in oil reservoirs. Annual Review of Fluid Mechanics 37, 211–238 (2005)
36. Cao, H.: Development of Techniques for General Purpose Simulators. PhD thesis, Dept. of Petroleum Engineering, Stanford University (2002)
37. Jiang, Y.: Techniques for Modeling Complex Reservoirs and Advanced Wells. PhD thesis, Dept. of Energy Resources Engineering, Stanford University (2007)
38. Streamsim Technologies Inc., 3DSL v2.30 User Manual (2006)
39. Stewart, R.R.: Exploration Seismic Tomography: Fundamentals. Course Notes Series, Society of Exploration Geophysicists (1991)
40. Devaney, A.J.: Geophysical diffraction tomography. IEEE Transactions on Geoscience and Remote Sensing 22(1), 3–13 (1984)
41. Harris, J.M.: Diffraction tomography with arrays of discrete sources and receivers. IEEE Transactions on Geoscience and Remote Sensing 25(4), 448–455 (1987)
42. Chernov, L.A.: Wave Propagation in a Random Medium. McGraw-Hill, New York (1960)
43. Aki, K., Richards, P.: Quantitative Seismology. W.H. Freeman, New York (1980)
44. van Essen, G.M., van den Hof, P.M.J., Jansen, J.D.: Hierarchical long-term and short-term production optimization. SPE Journal (in press)
45. Oliver, D.S.: Multiple realizations of the permeability field from well test data. SPE Journal 1, 145–154 (1996)
46. Sarma, P., Durlofsky, L.J., Aziz, K.: Kernel principal component analysis for efficient, differentiable parameterization of multipoint geostatistics. Mathematical Geosciences 40, 3–32 (2008)
47. Nocedal, J., Wright, S.J.: Numerical Optimization, 2nd edn. Springer, Heidelberg (2006)
48. Gill, P.E., Murray, W., Saunders, M.A.: SNOPT: an SQP algorithm for large-scale constrained optimization. SIAM Review 47(1), 99–131 (2005)
49. Torczon, V.: On the convergence of pattern search algorithms. SIAM Journal on Optimization 7(1), 1–25 (1997)
50. Audet, C., Dennis Jr., J.E.: Analysis of generalized pattern searches. SIAM Journal on Optimization 13(3), 889–903 (2002)

51. Audet, C., Dennis Jr., J.E.: Mesh adaptive direct search algorithms for constrained optimization. SIAM Journal on Optimization 17(1), 188–217 (2006)
52. Hooke, R., Jeeves, T.A.: Direct search solution of numerical and statistical problems. Journal of the ACM 8(2), 212–229 (1961)
53. Goldberg, D.E.: Genetic Algorithms in Search, Optimization and Machine Learning. Addison-Wesley, Reading (1989)
54. Eberhart, R.C., Kennedy, J.: A new optimizer using particle swarm theory. In: Proceedings of the Sixth International Symposium on Micromachine and Human Science, pp. 39–43 (1995)
55. Kennedy, J., Eberhart, R.C.: Particle swarm optimization. In: Proceedings of IEEE International Joint Conference on Neural Networks, pp. 1942–1948 (1995)
56. Clerc, M.: Particle Swarm Optimization. ISTE Ltd (2006)
57. Shi, Y., Eberhart, R.C.: A modified particle swarm optimizer. In: Proceedings of the 1998 IEEE International Conference on Evolutionary Computation, pp. 69–73 (1998)
58. Helwig, S., Wanka, R.: Theoretical analysis of initial particle swarm behavior. In: Rudolph, G., Jansen, T., Lucas, S., Poloni, C., Beume, N. (eds.) PPSN 2008. LNCS, vol. 5199, pp. 889–898. Springer, Heidelberg (2008)
59. Carlisle, A., Dozier, G.: An off-the-shelf PSO. In: Proceedings of the 2001 Workshop on Particle Swarm Optimization, pp. 1–6 (2001)
60. Fletcher, R., Leyffer, S.: Nonlinear programming without a penalty function. Mathematical Programming 91, 239–269 (2000)
61. Wächter, A., Biegler, T.: On the implementation of an interior-point filter line-search algorithm for large-scale nonlinear programming. Mathematical Programming 106, 25–57 (2006)
62. Audet, C., Dennis Jr., J.E.: A pattern search filter method for nonlinear programming without derivatives. SIAM Journal on Optimization 14(4), 980–1010 (2004)
63. Abramson, M.A.: NOMADm version 4.6 User's Guide. Dept. of Mathematics and Statistics, Air Force Institute of Technology (2007)
64. Christie, M.A., Blunt, M.J.: Tenth SPE comparative solution project: a comparison of upscaling techniques. SPE Reservoir Evaluation & Engineering 4, 308–317 (2001)
65. Isebor, O.J.: Constrained production optimization with an emphasis on derivative-free methods. Master's thesis, Dept. of Energy Resources Engineering, Stanford University (2009)
66. Franstrom, K.L., Litvak, M.L.: Automatic simulation algorithm for appraisal of future infill development potential of Prudhoe Bay. SPE paper 59374 presented at the, SPE/DOE Improved Oil Recovery Symposium, Tulsa, Oklahoma, April 3–5 (2000)
67. Güyagüler, B., Horne, R.N.: Uncertainty assessment of well placement optimization. SPE Reservoir Evaluation & Engineering 7(1), 24–32 (2004)
68. Bangerth, W., Klie, H., Wheeler, M.F., Stoffa, P.L., Sen, M.K.: On optimization algorithms for the reservoir oil well placement problem. Computational Geosciences 10, 303–319 (2006)
69. Litvak, M., Gane, B., Williams, G., Mansfield, M., Angert, P., Macdonald, C., McMurray, L., Skinner, R., Walker, G.J.: Field development optimization technology. SPE paper 106426 presented at the 2007 SPE Reservoir Simulation Symposium, Houston, Texas, February 26–28 (2007)
70. Tupac, Y.J., Faletti, L., Pacheco, M.A.C., Vellasco, M.M.B.R.: Evolutionary optimization of oil field development. SPE paper 107552 presented at the, SPE Digital Energy Conference and Exhibition, Houston, Texas, April 11–12 (2007)
71. Spall, J.C.: An overview of the simultaneous perturbation method for efficient optimization. Johns Hopkins APL Technical Digest 19(4), 482–492 (1998)

72. Mattot, L.S., Rabideau, A.J., Craig, J.R.: Pump-and-treat optimization using analytic element method flow models. Advances in Water Resources 29, 760–775 (2006)
73. Tarantola, A.: Inverse Problem Theory and Methods for Model Parameter Estimation. SIAM, Philadelphia (2005)
74. Caers, J., Hoffman, T.: The probability perturbation method: a new look at Bayesian inverse modeling. Mathematical Geology 38(1), 81–100 (2006)
75. Gosselin, O., van den Berg, S., Cominelli, A.: Integrated history matching of production and 4D seismic data. SPE paper 71599 presented at the 2001 SPE Annual Technical Conference and Exhibition, New Orleans, Louisiana, September–30 October–3 (2001)
76. Waggoner, J.R., Cominelli, A., Seymour, R.H.: Improved reservoir modeling with time-lapse seismic in a Gulf of Mexico gas condensate reservoir. SPE paper 77514 presented at the, SPE Annual Technical Conference and Exhibition, San Antonio, Texas, September–29 October–2 (2002)
77. Aanonsen, S.I., Aavatsmark, I., Barkve, T., Cominelli, A., Gonard, R., Gosselin, O., Kolasinski, M., Reme, H.: Effect of scale dependent data correlations in an integrated history matching loop combining production data and 4D seismic data. SPE paper 79665 presented at the, SPE Reservoir Simulation Symposium, Houston, Texas, February 3–5 (2003)
78. Miranda, A.A., Le Borgne, Y.A., Bontempi, G.: New routes for minimal approximation error to principal components. Neural Processing Letters 27(3), 197–207 (2008)
79. Castro, S.: A Probabilistic Approach to Jointly Integrate 3D/4D Seismic, Production Data and Geological Information for Building Reservoir Models. PhD thesis, Dept. of Energy Resources Engineering, Stanford University (2007)
80. Mavko, G., Mukerji, T., Dvorkin, J.: The Rock Physics Handbook, 2nd edn. Cambridge University Press, Cambridge (2009)
81. Strebelle, S.: Conditional simulation of complex geological structures using multi-point statistics. Mathematical Geology 34, 1–21 (2002)

Chapter 3
Simulation-Driven Design in Microwave Engineering: Application Case Studies

Slawomir Koziel and Stanislav Ogurtsov

Abstract. Application of surrogate-based optimization methods to simulation-driven microwave engineering design is demonstrated. It is essential for the considered techniques that the optimization of the original high-fidelity EM-simulated model is replaced by the iterative optimization of its computationally cheap surrogate. The surrogate is updated using available high-fidelity model data to maintain its prediction capability throughout the optimization process. The surrogate model is constructed from the low-fidelity model which—depending on a particular application case—can be either an equivalent circuit or a coarsely discretized full-wave electromagnetic model. Designs satisfying performance requirements are typically obtained at the cost of just a few evaluations of the high-fidelity model. Here, several surrogate-based design optimization techniques for the use in microwave engineering are discussed. Applications of space mapping, simulation-based tuning, variable-fidelity optimization, as well as various response correction techniques are illustrated. Design examples include planar filters, antennas, and transmission line transitions structures.

Keywords: computer-aided design (CAD), microwave design, simulation-driven optimization, electromagnetic (EM) simulation, surrogate-based optimization, space mapping, tuning, surrogate model, high-fidelity model, coarse model.

3.1 Introduction

In this chapter, first, we describe several simulation-driven design optimization methods exploiting physically-based surrogate models, which can be used to

Slawomir Koziel · Stanislav Ogurtsov
Engineering Optimization & Modeling Center, School of Science and Engineering, Reykjavik University, Menntavegur 1, Reykjavik, 101, Iceland
e-mail: koziel@ru.is; stanislav@ru.is

X.-S. Yang & S. Koziel (Eds.): Computational Optimization & Applications, SCI 359, pp. 57–97.
springerlink.com

design a variety of microwave structures and devices in a computationally efficient way. Second, we illustrate application of these surrogate-based optimization methods for design of microwave components. Examples include a variety of structures such as microstrip filters, ultrawide band (UWB) antenna, planar Yagi antenna, broadband antenna on multilayer substrate, low-loss transition from coplanar waveguide to microstrip and substrate integrated waveguide. All these design problems are computationally expensive so that application of conventional simulation-driven techniques (e.g., gradient-based algorithms) is not practical or even unfeasible. It will be demonstrated that the surrogate based methods exploiting the physically-based low-fidelity models can generate satisfactory designs at the cost corresponding to a few high-fidelity electromagnetic (EM) simulations of the structure of interest.

3.2 Surrogate-Based Design Optimization in Microwave Engineering

Microwave design task can be formulated as a nonlinear minimization problem

$$x^* \in \arg\min_{x \in X_f} U\left(\boldsymbol{R}_f(\boldsymbol{x})\right) \tag{3.1}$$

where $\boldsymbol{R}_f \in R^m$ denotes the response vector of the device of interest, e.g., the modulus of the transmission coefficient $|S_{21}|$ evaluated at m different frequencies. U is a given scalar merit function, e.g., a minimax function with upper and lower specifications [1]. Vector $\boldsymbol{x}^*$ is the optimal design to be determined. Normally, $\boldsymbol{R}_f$ is obtained through computationally expensive electromagnetic simulation. It is referred to as the high-fidelity or fine model.

The conventional way of handling the design problem (3.1) is to employ the EM simulator directly within the optimization loop. This direct approach faces some fundamental difficulties. The most important one is the high computational cost of high-fidelity EM simulation which makes the optimization impractical. Another difficulty is that the responses obtained through EM simulation typically have poor analytical properties. In particular, EM-based objective functions are inherently noisy. Additional problem for direct EM-based optimization is that the sensitivity information may not be available or expensive to compute. Only recently, computationally cheap adjoint sensitivities [2] started to become available in some major commercial EM simulation packages, although for frequency-domain solvers only [3], [4].

Computationally efficient simulation-driven design can be performed using surrogate models. Microwave design through surrogate-based optimization (SBO) [1], [5], [6] is the main focus of this chapter. The primary reason for using SBO approach in microwave engineering is to speed up the design process by shifting the optimization burden to an inexpensive yet reasonably accurate surrogate model of the device.

The generic SBO framework described here that the direct optimization of the computationally expensive EM-simulated high-fidelity model $\boldsymbol{R}_f$ is replaced by an iterative procedure [1], [6]

$$\boldsymbol{x}^{(i+1)} = \arg\min_{\boldsymbol{x}} U\left(\boldsymbol{R}_s^{(i)}(\boldsymbol{x})\right) \tag{3.2}$$

that generates a sequence of points (designs) $\boldsymbol{x}^{(i)} \in X_f$, $i = 0, 1, \ldots$, being approximate solutions to the original design problem (3.1). Each $\boldsymbol{x}^{(i+1)}$ is the optimal design of the surrogate model $\boldsymbol{R}_s^{(i)} : X_s^{(i)} \to R^m$, $X_s^{(i)} \subseteq R^n$, $i = 0, 1, \ldots$. $\boldsymbol{R}_s^{(i)}$ is assumed to be a computationally cheap and sufficiently reliable representation of the fine model $\boldsymbol{R}_f$, particularly in the neighborhood of the current design $\boldsymbol{x}^{(i)}$. Under these assumptions, the algorithm (3.2) is likely to produce a sequence of designs that quickly approach $\boldsymbol{x}_f^*$.

Typically, $\boldsymbol{R}_f$ is only evaluated once per iteration (at every new design $\boldsymbol{x}^{(i+1)}$) for verification purposes and to obtain the data necessary to update the surrogate model. Since the surrogate model is computationally cheap, its optimization cost (cf. (2)) can usually be neglected and the total optimization cost is determined by the evaluation of $\boldsymbol{R}_f$. The key point here is that the number of evaluations of $\boldsymbol{R}_f$ for a well performing surrogate-based algorithm is substantially smaller than for any direct optimization method (e.g., gradient-based one) [7].

In the remaining part of this section we characterize the surrogate models used in microwave engineering (Section 3.2.1) and present several techniques for computationally efficient simulation-driven design of microwave structures (Sections 3.2.2 through 3.2.6). Discussion covers the following methods: space mapping [1], [7], simulation-based tuning [8], shape-preserving response prediction [9], variable-fidelity optimization [10], as well as optimization through adaptively adjusted design specifications [11].

3.2.1 Surrogate Models in Microwave Engineering

There are a number of ways to create surrogate models of microwave and radio-frequency (RF) devices and structures. They can be classified into two groups: functional and physical surrogates. Functional models are constructed from sampled high-fidelity model data using suitable function approximation techniques (e.g., polynomial regression [5] or kriging [5]). Physical surrogates exploit fast but limited-accuracy models that are physically related to the original structure under consideration.

Here, we focus on methods exploiting physical surrogates. Their primary advantage is that they are typically able to ensure good accuracy and generalization capability while using only a few training data points [12]. Physical surrogates are based on underlying physically-based low-fidelity models of the structure of interest (denoted here as $\boldsymbol{R}_c$). Physically-based models describe the same physical phenomena as the high-fidelity model, however, in a simplified manner. In microwave engineering, the high-fidelity model describes behavior of the system in terms of the distributions of the electric and magnetic fields within (and, sometimes in its surrounding) that are calculated by solving the corresponding set of Maxwell equations [13]. Furthermore, the system performance is expressed through certain characteristics related to its input/output ports (such as so-called S-parameters [13]). All of these are obtained as a result of high-resolution electromagnetic simulation where the structure under consideration is finely discretized. In this context, the physically-based low-fidelity model of the microwave device can be obtained through: (i) analytical description of the structure using theory-based or semi-empirical formulas, (ii) different level of physical description of the system. The typical example in microwave engineering is equivalent circuit [1], where the device of interest is represented using lumped components (inductors, capacitors, microstrip line models, etc.), (iii) low-fidelity electromagnetic simulation. This approach allows us to use the same EM solver to evaluate both the high- and low-fidelity models; however, the latter is using much coarser simulation mesh which results in degraded accuracy but much shorter simulation time. The properties of the three groups of models are summarized in Table 3.1.

Table 3.1 Physically-based low-fidelity models in microwave engineering

Model Type	CPU Cost	Accuracy	Availability
Analytical	Very cheap	Low	Rather limited
Equivalent circuit	Cheap	Decent	Limited (mostly filters)
Coarsely-discretized EM simulation	Expensive	Good to very good	Generic: available for all structures

3.2.2 Space Mapping

Space mapping (SM) [1], [7] is probably one of the most recognized SBO techniques using physically-based low-fidelity (or coarse) models in microwave engineering. SM exploits the algorithm (3.2) to generate a sequence of approximate solutions $\boldsymbol{x}^{(i)}$, i = 0, 1, 2, …, to problem (3.1). The surrogate model at iteration i, $\boldsymbol{R}_s^{(i)}$, is constructed from the low-fidelity model so that the misalignment between $\boldsymbol{R}_s^{(i)}$ and the fine model is minimized using so-called parameter extraction process, which is the nonlinear minimization problem by itself [1]. The surrogate is defined as [7]

$$\boldsymbol{R}_s^{(i)}(\boldsymbol{x}) = \boldsymbol{R}_{s.g}(\boldsymbol{x}, \boldsymbol{p}^{(i)}) \tag{3.3}$$

where $\boldsymbol{R}_{s.g}$ is a generic space mapping surrogate model, i.e., the low-fidelity model composed with suitable transformations, whereas

$$\boldsymbol{p}^{(i)} = \arg\min_{\boldsymbol{p}} \sum\nolimits_{k=0}^{i} w_{i.k} \, || \boldsymbol{R}_f(\boldsymbol{x}^{(k)}) - \boldsymbol{R}_{s.g}(\boldsymbol{x}^{(k)}, \boldsymbol{p}) || \tag{3.4}$$

is a vector of model parameters and $w_{i.k}$ are weighting factors; a common choice of $w_{i.k}$ is $w_{i.k} = 1$ for all i and all k.

Various space mapping surrogate models are available [1], [7]. They can be roughly categorized into four groups: (i) Models based on a (usually linear) distortion of coarse model parameter space, e.g., input space mapping of the form $\boldsymbol{R}_{s.g}(\boldsymbol{x}, \boldsymbol{p}) = \boldsymbol{R}_{s.g}(\boldsymbol{x}, \boldsymbol{B}, \boldsymbol{c}) = \boldsymbol{R}_c(\boldsymbol{B}\cdot\boldsymbol{x} + \boldsymbol{c})$ [1]; (ii) Models based on a distortion of the coarse model response, e.g., output space mapping of the form $\boldsymbol{R}_{s.g}(\boldsymbol{x}, \boldsymbol{p}) = \boldsymbol{R}_{s.g}(\boldsymbol{x}, \boldsymbol{d}) = \boldsymbol{R}_c(\boldsymbol{x}) + \boldsymbol{d}$ [7]; (iii) Implicit space mapping, where the parameters used to align the surrogate with the fine model are separate from the design variables, i.e., $\boldsymbol{R}_{s.g}(\boldsymbol{x}, \boldsymbol{p}) = \boldsymbol{R}_{s.g}(\boldsymbol{x}, \boldsymbol{x}_p) = \boldsymbol{R}_{c.i}(\boldsymbol{x}, \boldsymbol{x}_p)$, with $\boldsymbol{R}_{c.i}$ being the coarse model dependent on both the design variables $\boldsymbol{x}$ and so-called preassigned parameters $\boldsymbol{x}_p$ (e.g., dielectric constant, substrate height) that are normally fixed in the fine model but can be freely altered in the coarse model [30]; (iv) Custom models exploiting parameters characteristic to a given design problem; the most characteristic example is the so-called frequency space mapping $\boldsymbol{R}_{s.g}(\boldsymbol{x}, \boldsymbol{p}) = \boldsymbol{R}_{s.g}(\boldsymbol{x}, \boldsymbol{F}) = \boldsymbol{R}_{c.f}(\boldsymbol{x}, \boldsymbol{F})$ [1], where $\boldsymbol{R}_{c.f}$ is a frequency-mapped coarse model, i.e., the coarse model evaluated at frequencies different from the original frequency sweep for the fine model, according to the mapping $\omega \rightarrow f_1 + f_2\omega$, with $\boldsymbol{F} = [f_1\ f_2]^T$.

A though discussion of various issues as well as generalizations of space mapping can be found in the literature [12, 14, 15].

3.2.3 *Simulation-Based Tuning and Tuning Space Mapping*

Tuning space mapping (TSM) [8] combines the concept of tuning, widely used in microwave engineering [16], [17], and space mapping. It is an iterative optimization procedure that assumes the existence of two surrogate models: both are less accurate but computationally much cheaper than the fine model. The first model is a so-called tuning model $\boldsymbol{R}_t$ that contains relevant fine model data (typically a fine model response) at the current iteration point and tuning parameters (typically implemented through circuit elements inserted into tuning ports). The tunable parameters are adjusted so that the model $\boldsymbol{R}_t$ satisfies the design specifications. The second model, $\boldsymbol{R}_c$ is used for calibration purposes: it allows us to translate the change of the tuning parameters into relevant changes of the actual design variables; $\boldsymbol{R}_c$ is dependent on three sets of variables: design parameters, tuning parameters (which are actually the same parameters as the ones used in $\boldsymbol{R}_t$), and SM parameters that are adjusted using the usual parameter extraction process [1] in order to have the model $\boldsymbol{R}_c$ meet certain matching conditions. Typically, the model $\boldsymbol{R}_c$ is a standard SM surrogate (i.e., a coarse model composed with suitable transformations) enhanced by the same or corresponding tuning elements as the model $\boldsymbol{R}_t$. The conceptual illustrations of the fine model, the tuning model and the calibration model are shown in Fig. 3.1.

The iteration of the TSM algorithm consists of two steps: optimization of the tuning model and a calibration procedure. First, the current tuning model $\boldsymbol{R}_t^{(i)}$ is built using fine model data at point $\boldsymbol{x}^{(i)}$. In general, because the fine model with inserted tuning ports is not identical to the original structure, the tuning model response may not agree with the response of the fine model at $\boldsymbol{x}^{(i)}$ even if the values

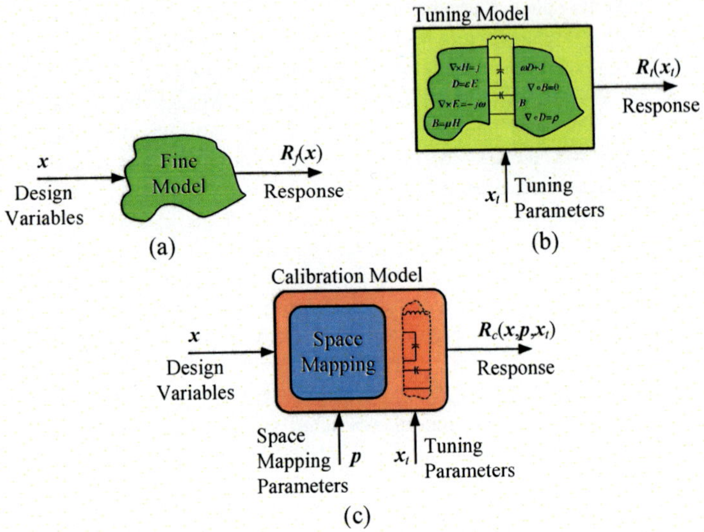

Fig. 3.1 Conceptual illustrations of the fine model, the tuning model and the calibration model: (a) the fine model is typically based on full-wave simulation, (b) the tuning model exploits the fine model "image" (e.g., in the form of *S*-parameters corresponding to the current design imported to the tuning model using suitable data components) and a number of circuit-theory-based tuning elements, (c) the calibration model is usually a circuit equivalent dependent on the same design variables as the fine model, the same tuning parameters as the tuning model and, additionally, a set of space mapping parameters used to align the calibration model with both the fine and the tuning model during the calibration process.

of the tuning parameters $\boldsymbol{x}_t$ are zero, so that these values must be adjusted to, say, $\boldsymbol{x}_{t.0}^{(i)}$, in order to obtain alignment [8]:

$$\boldsymbol{x}_{t.0}^{(i)} = \arg\min_{\boldsymbol{x}_t} \left\| \boldsymbol{R}_f(\boldsymbol{x}^{(i)}) - \boldsymbol{R}_t^{(i)}(\boldsymbol{x}_t) \right\| \tag{3.5}$$

In the next step, one optimizes $\boldsymbol{R}_t^{(i)}$ to have it meet the design specifications. Optimal values of the tuning parameters $\boldsymbol{x}_{t.1}^{(i)}$ are obtained as follows:

$$\boldsymbol{x}_{t.1}^{(i)} = \arg\min_{\boldsymbol{x}_t} U\left(\boldsymbol{R}_t^{(i)}(\boldsymbol{x}_t)\right) \tag{3.6}$$

Having $\boldsymbol{x}_{t.1}^{(i)}$, the calibration procedure is performed to determine changes in the design variables that yield the same change in the calibration model response as that caused by $\boldsymbol{x}_{t.1}^{(i)} - \boldsymbol{x}_{t.0}^{(i)}$ [8]. First one adjusts the SM parameters $\boldsymbol{p}^{(i)}$ of the calibration model to obtain a match with the fine model response at $\boldsymbol{x}^{(i)}$

$$\boldsymbol{p}^{(i)} = \arg\min_{\boldsymbol{p}} \left\| \boldsymbol{R}_f(\boldsymbol{x}^{(i)}) - \boldsymbol{R}_c(\boldsymbol{x}^{(i)}, \boldsymbol{p}, \boldsymbol{x}_{t.0}^{(i)}) \right\|. \tag{3.7}$$

The calibration model is then optimized with respect to the design variables in order to obtain the next iteration point $x^{(i+1)}$

$$x^{(i+1)} = \arg\min_{x} \left\| R_t^{(i)}(x_{t.1}^{(i)}) - R_c(x, p^{(i)}, x_{t.0}^{(i)}) \right\|. \tag{3.8}$$

Note that $x_{t.0}^{(i)}$ is used in (3.7), which corresponds to the state of the tuning model after performing the alignment procedure (3.5), and $x_{t.1}^{(i)}$ in (3.8), which corresponds to the optimized tuning model (cf. (6)). Thus, (3.7) and (3.8) allow finding the change of design variable values $x^{(i+1)} - x^{(i)}$ necessary to compensate the effect of changing the tuning parameters from $x_{t.0}^{(i)}$ to $x_{t.1}^{(i)}$.

Thorough discussion of various variations of tuning space mapping algorithms, calibration procedures, as well as recent development in the TSM technology can be found in the literature [18, 19, 20].

3.2.4 Shape-Preserving Response Prediction

Shape-preserving response prediction (SPRP) [9] is a response correction technique that takes advantage of the similarity between responses of the high- and low-fidelity models in a very straightforward way. SPRP assumes that the change of the high-fidelity model response due to the adjustment of the design variables can be predicted using the actual changes of the low-fidelity model response. Therefore, it is critically important that the low-fidelity model is physically based, which ensures that the effect of the design parameter variations on the model response is similar for both models. In microwave engineering this property is likely to hold, particularly if the low-fidelity model is the coarsely-discretization structure evaluated using the same EM solver as the one used to simulate the high-fidelity model.

The change of the low-fidelity model response is described by the translation vectors corresponding to a certain (finite) number of characteristic points of the model's response. These translation vectors are subsequently used to predict the change of the high-fidelity model response with the actual response of R_f at the current iteration point, $R_f(x^{(i)})$, treated as a reference.

Figure 3.2(a) shows the example low-fidelity model response, $|S_{21}|$ in the frequency range 8 GHz to 18 GHz, at the design $x^{(i)}$, as well as the low-fidelity model response at some other design x. The responses come from the double folded stub bandstop filter example considered in [9]. Circles denote characteristic points of $R_c(x^{(i)})$, selected here to represent $|S_{21}| = -3$ dB, $|S_{21}| = -20$ dB, and the local $|S_{21}|$ maximum (at about 13 GHz). Squares denote corresponding characteristic points for $R_c(x)$, while line segments represent the translation vectors ("shift") of the characteristic points of R_c when changing the design variables from $x^{(i)}$ to x. Since the low-fidelity model is physically based, the high-fidelity model response at the given design, here, x, can be predicted using the same translation vectors applied to the corresponding characteristic points of the high-fidelity model response at $x^{(i)}$, $R_f(x^{(i)})$. This is illustrated in Fig. 3.2(b). Rigorous formulation of SPRP as well as generalizations of the basic algorithm can be found in [9].

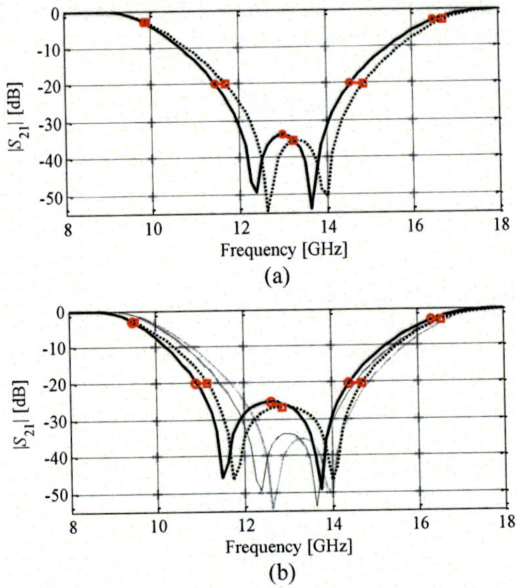

Fig. 3.2 SPRP concept: (a) Example low-fidelity model response at the design $\boldsymbol{x}^{(i)}$, $\boldsymbol{R}_c(\boldsymbol{x}^{(i)})$ (solid line), the low-fidelity model response at $\boldsymbol{x}$, $\boldsymbol{R}_c(\boldsymbol{x})$ (dotted line), characteristic points of $\boldsymbol{R}_c(\boldsymbol{x}^{(i)})$ (circles) and $\boldsymbol{R}_c(\boldsymbol{x})$ (squares), and the translation vectors (short lines); (b) High-fidelity model response at $\boldsymbol{x}^{(i)}$, $\boldsymbol{R}_f(\boldsymbol{x}^{(i)})$ (solid line) and the predicted high-fidelity model response at $\boldsymbol{x}$ (dotted line) obtained using SPRP based on characteristic points of Fig. 3.2(a); characteristic points of $\boldsymbol{R}_f(\boldsymbol{x}^{(i)})$ (circles) and the translation vectors (short lines) were used to find the characteristic points (squares) of the predicted high-fidelity model response; low-fidelity model responses $\boldsymbol{R}_c(\boldsymbol{x}^{(i)})$ and $\boldsymbol{R}_c(\boldsymbol{x})$ are plotted using thin solid and dotted line, respectively [9].

3.2.5 Multi-fidelity Optimization Using Coarse-Discretization EM Models

The most versatile type of physically-based low-fidelity model in microwave engineering is the one obtained through EM simulation of coarsely-discretized structure of interest. The computational cost of the model and its accuracy can be easily controlled by changing the discretization density. This feature has been exploited in the multi-fidelity optimization algorithm introduced in [10].

The design optimization methodology of [10] is based on a family of coarse-discretization models $\{\boldsymbol{R}_{c.j}\}$, $j = 1,\ldots, K$, all evaluated by the same EM solver as the one used for the high-fidelity model. Discretization of the model $\boldsymbol{R}_{c.j+1}$ is finer than that of the model $\boldsymbol{R}_{c.j}$, which results in better accuracy but also longer evaluation time. In practice, the number of coarse-discretization models is two or three.

Having the optimized design $\boldsymbol{x}^{(K)}$ of the last (and finest) coarse-discretization model $\boldsymbol{R}_{c.K}$, the model is evaluated at all perturbed designs around $\boldsymbol{x}^{(K)}$, i.e., at $\boldsymbol{x}_k^{(K)} = [x_1^{(K)} \ldots x_k^{(K)} + \text{sign}(k)\cdot d_k \ldots x_n^{(K)}]^T$, $k = -n, -n+1, \ldots, n-1, n$. A notation of $\boldsymbol{R}^{(k)} =$

$\boldsymbol{R}_{c.K}(\boldsymbol{x}_k^{(K)})$ is adopted here. This data can be used to refine the final design without directly optimizing $\boldsymbol{R}_f$. Instead, an approximation model involving $\boldsymbol{R}^{(k)}$ is set up and optimized in the neighborhood of $\boldsymbol{x}^{(K)}$ defined as $[\boldsymbol{x}^{(K)} - \boldsymbol{d}, \boldsymbol{x}^{(K)} + \boldsymbol{d}]$, where $\boldsymbol{d} = [d_1\ d_2\ \ldots\ d_n]^T$. The size of the neighborhood can be selected based on sensitivity analysis of $\boldsymbol{R}_{c.1}$ (the cheapest of the coarse-discretization models); usually $\boldsymbol{d}$ equals 2 to 5 percent of $\boldsymbol{x}^{(K)}$.

Here, the approximation is performed using a reduced quadratic model $\boldsymbol{q}(\boldsymbol{x}) = [q_1\ q_2\ \ldots\ q_m]^T$, defined as

$$q_j(\boldsymbol{x}) = q_j([x_1 \ldots x_n]^T) = \lambda_{j.0} + \lambda_{j.1}x_1 + \ldots + \lambda_{j.n}x_n + \lambda_{j.n+1}x_1^2 + \ldots + \lambda_{j.2n}x_n^2 \tag{3.9}$$

Coefficients $\lambda_{j.r}$, $j = 1, \ldots, m$, $r = 0, 1, \ldots, 2n$, can be uniquely obtained by solving the linear regression problems

$$\begin{bmatrix} 1 & x_{-n.1}^{(K)} & \cdots & x_{-n.n}^{(K)} & (x_{-n.1}^{(K)})^2 & \cdots & (x_{-n.n}^{(K)})^2 \\ \vdots & \vdots & & \vdots & & \vdots & \vdots \\ 1 & x_{0.1}^{(K)} & \cdots & x_{0.n}^{(K)} & (x_{0.1}^{(K)})^2 & \cdots & (x_{-n.n}^{(K)})^2 \\ \vdots & \vdots & & \vdots & & \vdots & \vdots \\ 1 & x_{n.1}^{(K)} & \cdots & x_{n.n}^{(K)} & (x_{n.1}^{(K)})^2 & \cdots & (x_{-n.n}^{(K)})^2 \end{bmatrix} \cdot \begin{bmatrix} \lambda_{j.0} \\ \lambda_{j.1} \\ \vdots \\ \lambda_{j.2n} \end{bmatrix} = \begin{bmatrix} R_j^{(-n)} \\ \vdots \\ R_j^{(0)} \\ \vdots \\ R_j^{(n)} \end{bmatrix} \tag{3.10}$$

where $x_{k.j}^{(K)}$ is a jth component of the vector $\boldsymbol{x}_k^{(K)}$, and $R_j^{(k)}$ is a jth component of the vector $\boldsymbol{R}^{(k)}$, i.e.,

In order to account for unavoidable misalignment between $\boldsymbol{R}_{c.K}$ and $\boldsymbol{R}_f$, instead of optimizing the quadratic model $\boldsymbol{q}$, it is recommended to optimize a corrected model $\boldsymbol{q}(\boldsymbol{x}) + [\boldsymbol{R}_f(\boldsymbol{x}^{(K)}) - \boldsymbol{R}_{c.K}(\boldsymbol{x}^{(K)})]$ that ensures a zero-order consistency [21] between $\boldsymbol{R}_{c.K}$ and $\boldsymbol{R}_f$. The refined design can be then found as

$$\boldsymbol{x}^* = \arg \min_{\boldsymbol{x}^{(K)} - \boldsymbol{d} \le \boldsymbol{x} \le \boldsymbol{x}^{(K)} + \boldsymbol{d}} U(\boldsymbol{q}(\boldsymbol{x}) + [\boldsymbol{R}_f(\boldsymbol{x}^{(K)}) - \boldsymbol{R}_{c.K}(\boldsymbol{x}^{(K)})]) \tag{3.11}$$

This kind of correction is also known as output space mapping [7]. If necessary, the step (4) can be performed a few times starting from a refined design, i.e., $\boldsymbol{x}^* = \text{argmin}\{\boldsymbol{x}^{(K)} - \boldsymbol{d} \le \boldsymbol{x} \le \boldsymbol{x}^{(K)} + \boldsymbol{d} : U(\boldsymbol{q}(\boldsymbol{x}) + [\boldsymbol{R}_f(\boldsymbol{x}^*) - \boldsymbol{R}_{c.K}(\boldsymbol{x}^*)])\}$ (each iteration requires only one evaluation of $\boldsymbol{R}_f$).

The design optimization procedure can be summarized as follows (input arguments are: initial design $\boldsymbol{x}^{(0)}$ and the number of coarse-discretization models K):

1. Set $j = 1$;
2. Optimize coarse-discretization model $\boldsymbol{R}_{c.j}$ to obtain a new design $\boldsymbol{x}^{(j)}$ using $\boldsymbol{x}^{(j-1)}$ as a starting point;
3. Set $j = j + 1$; if $j < K$ go to 2;
4. Obtain a refined design $\boldsymbol{x}^*$ as in (3.13);
5. END;

Note that the original model $\boldsymbol{R}_f$ is only evaluated at the final stage (step 4) of the optimization process. Operation of the algorithm in illustrated in Fig. 3.3. Coarse-discretization models can be optimized using any available algorithm.

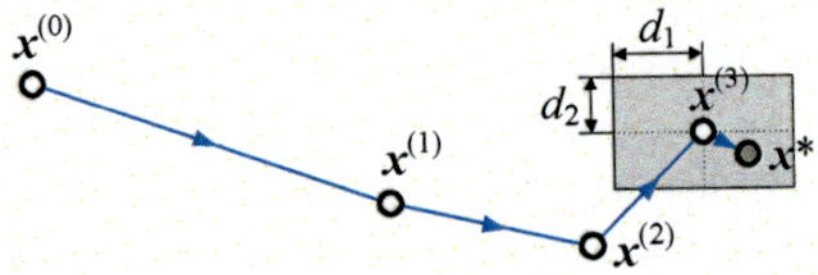

Fig. 3.3 Operation of the multi-fidelity design optimization procedure for $K = 3$ (three coarse-discretization models). The design $\boldsymbol{x}^{(j)}$ is obtained as the optimal solution of the model $\boldsymbol{R}_{c.j}$, $j = 1, 2, 3$. A reduced second-order approximation model $\boldsymbol{q}$ is set up in the neighborhood of $\boldsymbol{x}^{(3)}$ (gray area) and the final design $\boldsymbol{x}^*$ is obtained by optimizing a reduced $\boldsymbol{q}$ as in (3.13).

3.2.6 *Optimization Using Adaptively Adjusted Design Specifications*

The techniques described in Section 3.2.2 to 3.2.5 aimed at correcting the low-fidelity model so that it becomes, at least locally, an accurate representation of the high-fidelity model. An alternative way of exploiting low-fidelity models in simulation-driven design of microwave structures is to modify the design specifications in such a way that the updated specifications reflect the discrepancy between the models. This approach is extremely simple to implement because no changes of the low-fidelity model are necessary.

The adaptively adjusted design specifications optimization procedure introduced in [11] consists of the following two simple steps that can be iterated if necessary:

1. Modify the original design specifications in order to take into account the difference between the responses of $\boldsymbol{R}_f$ and $\boldsymbol{R}_c$ at their characteristic points.
2. Obtain a new design by optimizing the coarse model with respect to the modified specifications.

Characteristic points of the responses should correspond to the design specification levels. They should also include local maxima/minima of the respective responses at which the specifications may not be satisfied. Figure 3.4(a) shows fine and coarse model response at the optimal design of $\boldsymbol{R}_c$, corresponding to the bandstop filter example considered in [11]; design specifications are indicated using horizontal lines. Figure 3.4(b) shows characteristic points of $\boldsymbol{R}_f$ and $\boldsymbol{R}_c$ for the bandstop filter example. The points correspond to –3 dB and –30 dB levels as well to the local maxima of the responses. As one can observe in Fig. 3.4(b) the selection of points is rather straightforward.

In the first step of the optimization procedure, the design specifications are modified (or mapped) so that the level of satisfying/violating the modified specifications by the coarse model response corresponds to the satisfaction/violation levels of the original specifications by the fine model response. Modified design specifications are shown in Fig. 3.4(c).

The coarse model is subsequently optimized with respect to the modified specifications and the new design obtained this way is treated as an approximated solution to the original design problem (i.e., optimization of the fine model with

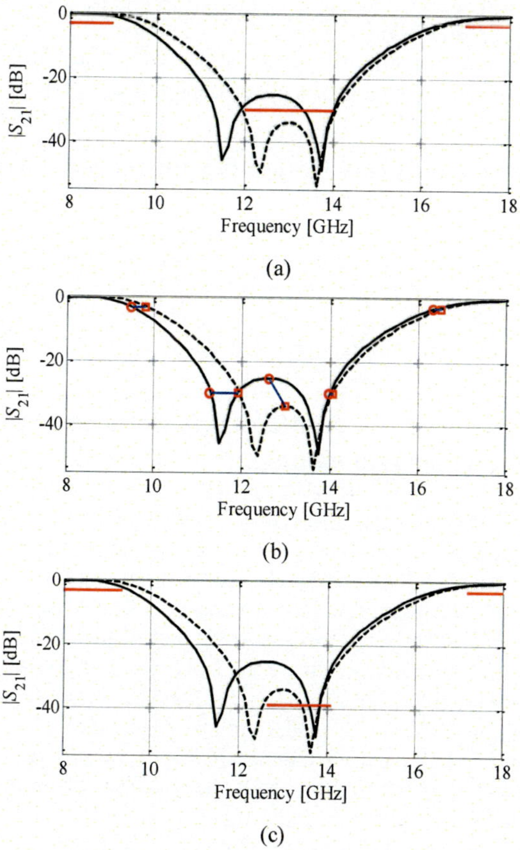

Fig. 3.4 Bandstop filter example (responses of $\boldsymbol{R}_f$ and $\boldsymbol{R}_c$ are marked with solid and dashed line, respectively) [11]: (a) fine and coarse model responses at the initial design (optimum of $\boldsymbol{R}_c$) as well as the original design specifications, (b) characteristic points of the responses corresponding to the specification levels (here, –3 dB and –30 dB) and to the local response maxima, (c) fine and coarse model responses at the initial design and the modified design specifications.

respect to the original specifications). Steps 1 and 2 (listed above) can be repeated if necessary. Substantial design improvement is typically observed after the first iteration, however, additional iterations may bring further enhancement [11].

In the first step of the optimization procedure, the design specifications are modified (or mapped) so that the level of satisfying/violating the modified specifications by the coarse model response corresponds to the satisfaction/violation levels of the original specifications by the fine model response. It is assumed that the coarse model is physically-based, in particular, that the adjustment of the design variables has similar effect on the response for both $\boldsymbol{R}_f$ and $\boldsymbol{R}_c$. In such a case the coarse model design that is obtained in the second stage of the procedure (i.e., optimal with respect to the modified specifications) will be (almost) optimal for $\boldsymbol{R}_f$ with respect to the original specifications. As shown in Fig. 3.4, the absolute matching between the models is not as important as the shape similarity.

3.3 Surrogate-Based Design Optimization of Microwave Filters

In this section, three examples of microwave filter design using various surrogate-based optimization techniques are presented. A common feature of these three cases is that the surrogate model is created exploiting equivalent-circuit coarse model which is computationally much cheaper than the EM-simulated high-fidelity model. This results in a significant speedup of the optimization process.

3.3.1 Optimization of a Microstrip Bandpass Filter Using Space Mapping Technique

Consider the fourth-order ring resonator bandpass filter [22] shown in Fig. 3.5(a). The design parameters are $\boldsymbol{x} = [L_1\ L_2\ L_3\ S_1\ S_2\ W_1\ W_2]^T$ mm. The fine model $\boldsymbol{R}_f$ is simulated in the EM simulator FEKO [23]. The coarse model, Fig. 3.5(b), is an equivalent circuit implemented in Agilent ADS [24]. The design goal is to adjust the design variables so that the modulus of the transmission coefficient of the filter, $|S_{21}|$, satisfies the following requirements: $|S_{21}| \geq -1$ dB for 1.75 GHz $\leq f \leq$ 2.25 GHz, and $|S_{21}| \leq -20$ dB for 1.0 GHz $\leq f \leq$ 1.5GHz and 2.5 GHz $\leq f \leq$ 3.0 GHz, where f stands for frequency. The initial design is the coarse model optimal solution $\boldsymbol{x}^{(0)} = [24.74\ \ 19.51\ \ 24.10\ \ 0.293\ \ 0.173\ 1.232\ 0.802]^T$ mm (minimax specification error +9.0 dB).

Table 3.2 shows the optimization results. The surrogate model is constructed using input and output space mapping of the form $\boldsymbol{R}_s^{(i)}(\boldsymbol{x}) = \boldsymbol{R}_c^{(i)}(\boldsymbol{x} + \boldsymbol{c}^{(i)}) + \boldsymbol{d}^{(i)}$ [7], where $\boldsymbol{c}^{(i)}$ is obtained using the parameter extraction procedure [1], see also Section 3.2.2, eq. (3.4), whereas $\boldsymbol{d}^{(i)} = \boldsymbol{R}_f(\boldsymbol{x}^{(i)}) - \boldsymbol{R}_c(\boldsymbol{x}^{(i)} + \boldsymbol{c}^{(i)})$. Also an enhanced model of the form $\boldsymbol{R}_s^{(i)}(\boldsymbol{x}) = \boldsymbol{R}_c^{(i)}(\boldsymbol{x} + \boldsymbol{c}^{(i)}) + \boldsymbol{d}^{(i)} + \boldsymbol{E}^{(i)}(\boldsymbol{x}\ \ \ \boldsymbol{x}^{(i)})$ is considered, where $\boldsymbol{E}^{(i)}$ is an approximation of the Jacobian of $\boldsymbol{R}_f(\boldsymbol{x}) - \boldsymbol{R}_c(\boldsymbol{x} + \boldsymbol{c}^{(i)})$ obtained using Broyden update [25]. The space mapping algorithm working with the enhanced surrogate uses trust-region convergence safeguard [25].

Figure 3.6 shows the initial fine model response and the optimized fine model response obtained using the algorithm with the enhanced surrogate model. Figure 3.7 shows the convergence plot for the both cases. For this example, the first version of the space mapping algorithm does not converge. Also, the final design is worse than the best one found in the course of optimization. This illustrates one of the difficulties of the standard SM technique: the algorithm does not ensure objective function improvement from iteration to iteration. The algorithm using approximated Jacobian and trust-regions exhibits better performance.

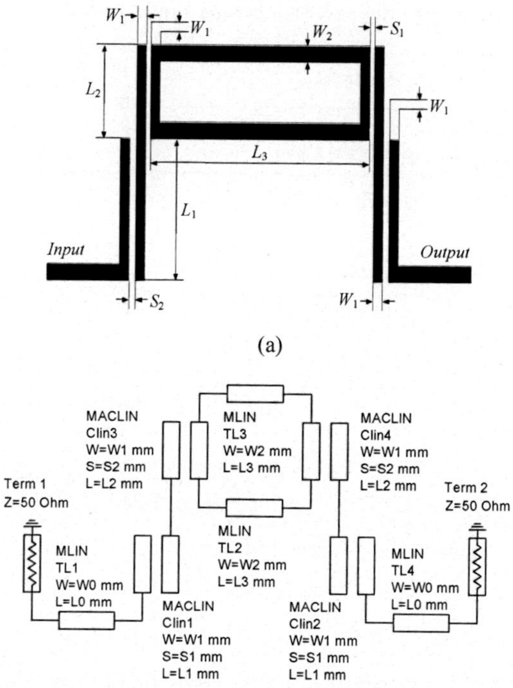

Fig. 3.5 Fourth-order ring resonator bandpass filter: (a) geometry [22], (b) coarse model (Agilent ADS).

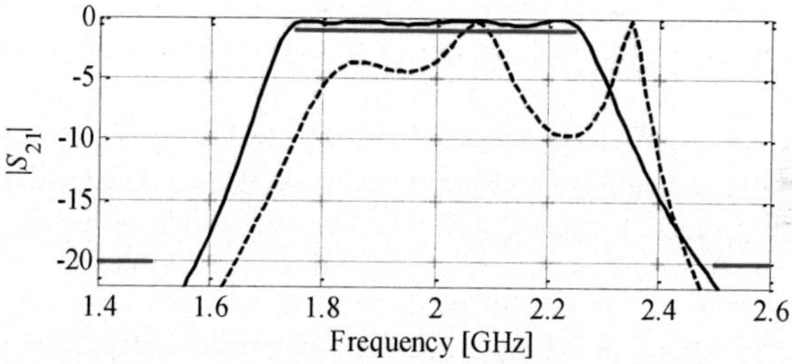

Fig. 3.6 Fourth-order ring resonator filter: Initial (dashed line) and optimized (solid line) $|S_{21}|$ versus frequency; optimization using $SM_{TR\text{-}B2}$ algorithm [25] with the $\boldsymbol{R}_c(\boldsymbol{x}+\boldsymbol{c})$ model: (a) full frequency range, (b) magnification at 1.4 GHz to 2.6 GHz and –22 dB to 0 dB.

Table 3.2 Fourth-order ring-resonator bandpass filter: optimization results

Surrogate Model	Spec. Error [dB] Final	Spec. Error [dB] Best Found	Fine Model Runs [times]
$\boldsymbol{R}_s^{(i)}(\boldsymbol{x}) = \boldsymbol{R}_c^{(i)}(\boldsymbol{x} + \boldsymbol{c}^{(i)}) + \boldsymbol{d}^{(i)}$	–0.2	–0.3	21[#]
$\boldsymbol{R}_s^{(i)}(\boldsymbol{x}) = \boldsymbol{R}_c^{(i)}(\boldsymbol{x} + \boldsymbol{c}^{(i)}) + \boldsymbol{d}^{(i)} + \boldsymbol{E}^{(i)}(\boldsymbol{x} \quad \boldsymbol{x}^{(i)})$	–0.4	–0.4	17

[#] Convergence not obtained; algorithm terminated after 20 iterations.

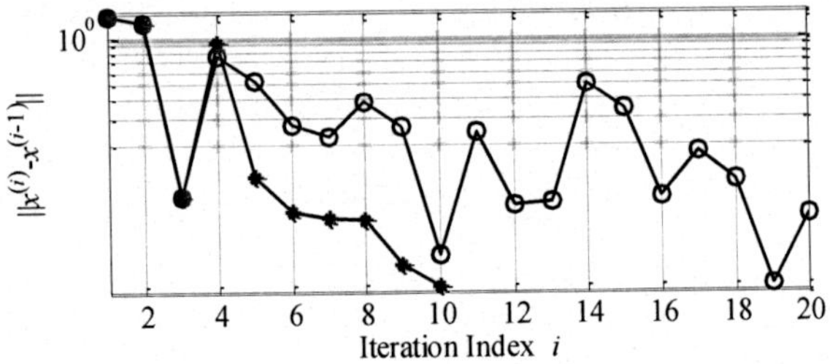

Fig. 3.7 Fourth-order ring resonator filter: convergence plots for the SM algorithm using surrogate model $\boldsymbol{R}_s^{(i)}(\boldsymbol{x}) = \boldsymbol{R}_c^{(i)}(\boldsymbol{x} + \boldsymbol{c}^{(i)}) + \boldsymbol{d}^{(i)}$ (o) and the algorithm using model $\boldsymbol{R}_s^{(i)}(\boldsymbol{x}) = \boldsymbol{R}_c^{(i)}(\boldsymbol{x} + \boldsymbol{c}^{(i)}) + \boldsymbol{d}^{(i)} + \boldsymbol{E}^{(i)}(\boldsymbol{x} \quad \boldsymbol{x}^{(i)})$.

3.3.2 Optimization of a Microstrip Bandpass Filter Using Tuning Space Mapping

Consider the box-section Chebyshev microstrip bandpass filter [26] (Fig. 3.8). The design parameters are $\boldsymbol{x} = [L_1\ L_2\ L_3\ L_4\ L_5\ S_1\ S_2]^T$. The fine model is simulated in Sonnet ***em*** [27] with a grid of 1 mil × 2 mil. The width parameters are $W = 40$ mil and $W_1 = 150$ mil. Substrate parameters are: relative permittivity $\varepsilon_r = 3.63$, and height $H = 20$ mil. The design specifications for the transmission coefficient are $|S_{21}| \le -20$ dB for 1.8 GHz $\le f \le$ 2.15 GHz and 2.65 GHz $\le f \le$ 3.0 GHz, and $|S_{21}| \ge -3$ dB for 2.4 GHz $\le f \le$ 2.5 GHz.

The filter is optimized using the tuning space mapping technology (Section 3.2.3). The tuning model is constructed by dividing the polygons corresponding to parameters L_1 to L_5 in the middle and inserting the tuning ports at the new cut edges. Its S28P data file (i.e., the file generated by the EM solver and containing the fine model *S*-parameter data) is then loaded into the *S*-parameter component in Agilent ADS [24]. The circuit-theory coupled-line components and capacitor components are chosen to be the tuning elements and are inserted into each pair of tuning ports (Fig. 3.9). The lengths of the imposed coupled-lines and the capacitances of the capacitors are assigned to be the tuning parameters, so that one has $\boldsymbol{x}_t = [L_{t1}\ L_{t2}\ L_{t3}\ L_{t4}\ L_{t5}\ C_{t1}\ C_{t2}]^T$ (L_{tk} in mil, C_{tk} in pF).

The calibration model is implemented in ADS and shown in Fig. 3.10. It contains the same tuning elements as the tuning model. It basically mimics the

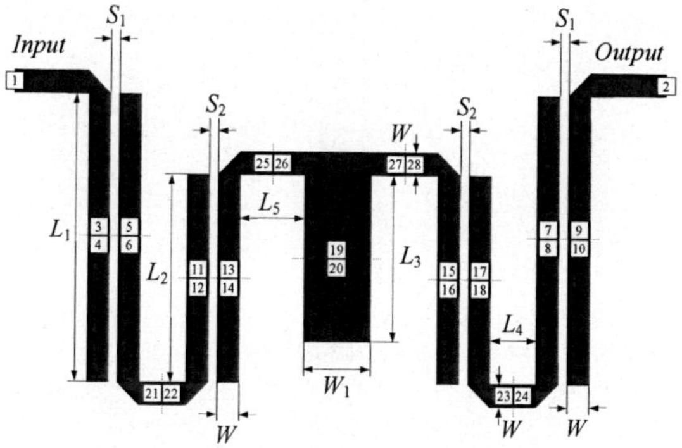

Fig. 3.8 Chebyshev bandpass filter: geometry [26], and the tuning port insertion points.

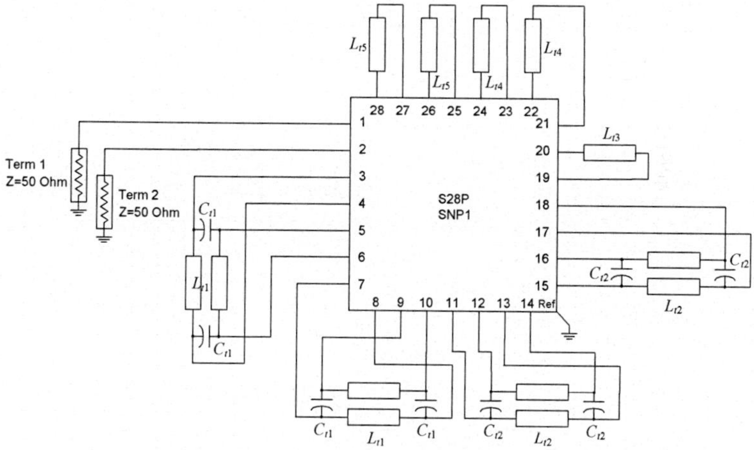

Fig. 3.9 Box-section Chebyshev bandpass filter: tuning model (Agilent ADS).

division of the coupled-lines performed while preparing $\boldsymbol{R}_t$. The calibration model also contains six (implicit) SM parameters that will be used as parameters $\boldsymbol{p}$ in the calibration process [8]. These parameters are $\boldsymbol{p} = [\varepsilon_{r1}\ \varepsilon_{r2}\ \varepsilon_{r3}\ \varepsilon_{r4}\ \varepsilon_{r5}\ H]^T$, where ε_{rk} is dielectric constant of the microstrip line segment of length L_k (Fig. 3.8), and H is the substrate height of the filter. Initial values of these parameters are $[3.63\ 3.63\ 3.63\ 3.63\ 3.63\ 20]^T$.

The initial design, $\boldsymbol{x}^{(0)} = [928\ 508\ 50\ 50\ 201\ 5\ 19]^T$ mil, is the optimal solution of the coarse model, i.e., the calibration model with zero values of the tuning parameters. The specification error is +19 dB.

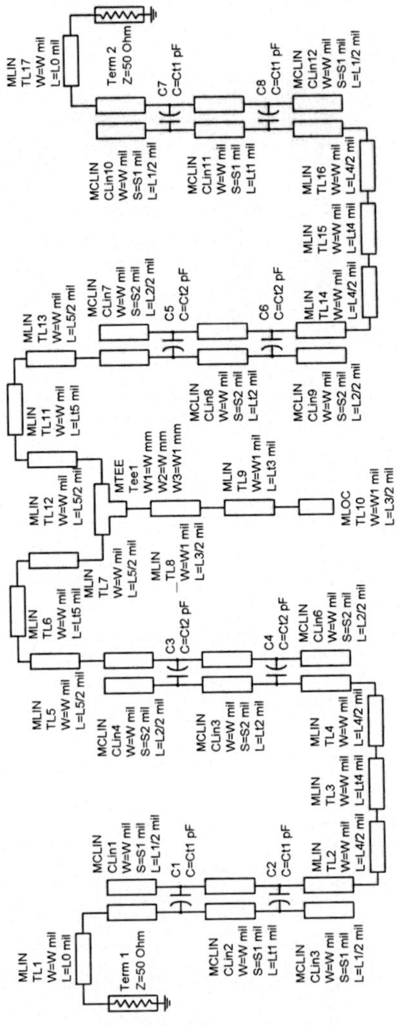

Fig. 3.10 Box-section Chebyshev bandpass filter: calibration model (Agilent ADS) [28].

The misalignment between the fine and the tuning model response with the tuning elements set to zero is negligible (thanks to the co-calibrated port feature in Sonnet ***em*** [8]) so that $\boldsymbol{x}_{t.0}^{(0)} = [0\ 0\ 0\ 0\ 0\ 0\ 0]^T$ was used throughout. The values of the tuning parameters at the optimal design of the tuning model are $\boldsymbol{x}_{t.1}^{(0)} = [-85.2\ 132.5\ 5.24\ 1.13\ -15.24\ 0.169\ -0.290]^T$. Note that some of the parameters take negative values, which is permitted in ADS. The values of preassigned parameters obtained in the first calibration phase [8] are $\boldsymbol{p}^{(0)} = [3.10\ 6.98\ 4.29\ 7.00\ 6.05\ 17.41]^T$.

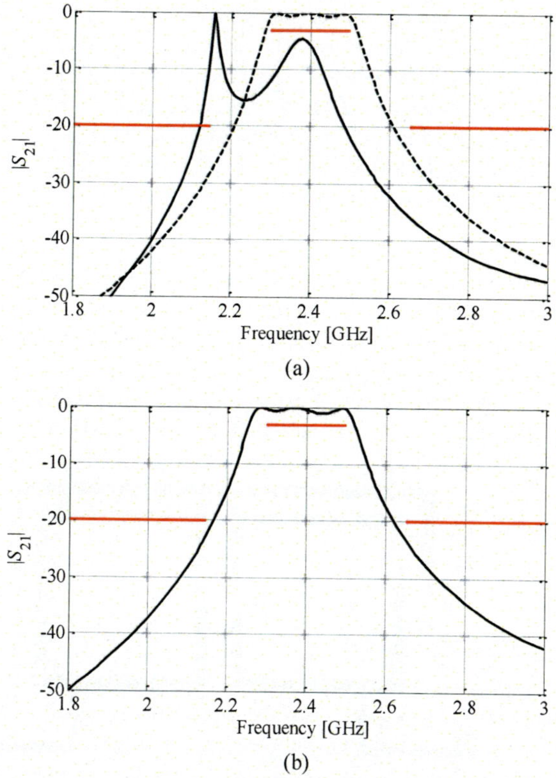

Fig. 3.11 Box-section Chebyshev bandpass filter: (a) the coarse (dashed line) and fine (solid line) model response at the initial design; (b) fine model response at the design found after one iteration of the TSM algorithm.

Figure 3.11 shows the coarse and fine model response at the initial design, as well as the fine model response after just one TSM iteration (two fine model evaluations) with $\boldsymbol{x}^{(1)} = [1022\ 398\ 46\ 56\ 235\ 4\ 10]^T$ mil (specification error –1.8 dB).

It should be emphasized that the evaluation time of both the tuning and the calibration model is very low (a fraction of a second), and, it is negligible compared to the evaluation time of the fine model. Therefore, the computational cost of each tuning space mapping iteration corresponds to two electromagnetic simulations (one for the fine model and one for the "cut" fine model).

3.3.3 Design of Dual-Band Bandpass Filter Using Shape-Preserving Response Prediction

To illustrate the performance of the shape-preserving response prediction (SPRP) algorithm [9] (Section 3.2.4), consider the dual-band bandpass filter [29] shown in

Fig. 3.12. The design parameters are $\boldsymbol{x} = [L_1\ L_2\ S_1\ S_2\ S_3\ d\ g\ W]^T$ mm. The fine model is simulated in Sonnet *em* [27]. The design specifications are $|S_{21}| \geq -3$ dB for 0.85 GHz $\leq f \leq$ 0.95 GHz and 1.75 GHz $\leq f \leq$ 1.85 GHz, and $|S_{21}| \leq -20$ dB for 0.5 GHz $\leq f \leq$ 0.7 GHz, 1.1 GHz $\leq f \leq$ 1.6 GHz and 2.0 GHz $\leq f \leq$ 2.2 GHz. The coarse model is implemented in Agilent ADS [24] (Fig. 3.13). The initial design is $\boldsymbol{x}^{(0)} = [16.14\ 17.28\ 1.16\ 0.38\ 1.18\ 0.98\ 0.98\ 0.20]^T$ mm (the optimal solution of $\boldsymbol{R}_c$). The following characteristic points are selected to set up the SPRP surrogate model [9]: four points for which $|S_{21}| = -20$ dB, four points with $|S_{21}| = -5$ dB, as well as 6 additional points located between –5 dB points. For the purpose of optimization, the coarse model was enhanced by tuning the dielectric constants and the substrate heights of the microstrip models corresponding to the design variables L_1, L_2, d and g (original values of ε_r and H were 10.2 and 0.635 mm, respectively).

Figure 3.13 shows the initial fine model response as well as the fine model response at the design obtained using the stand-alone SPRP. Table 3.3 shows the optimization results. Two variants of the SPRP algorithm were considered [9]:

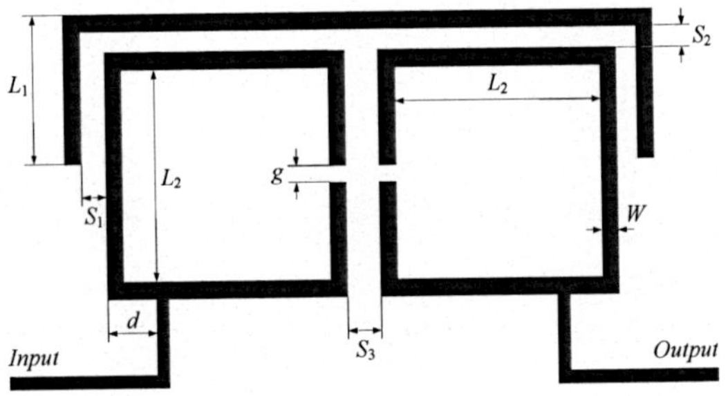

Fig. 3.12 Dual-band bandpass filter: geometry [29].

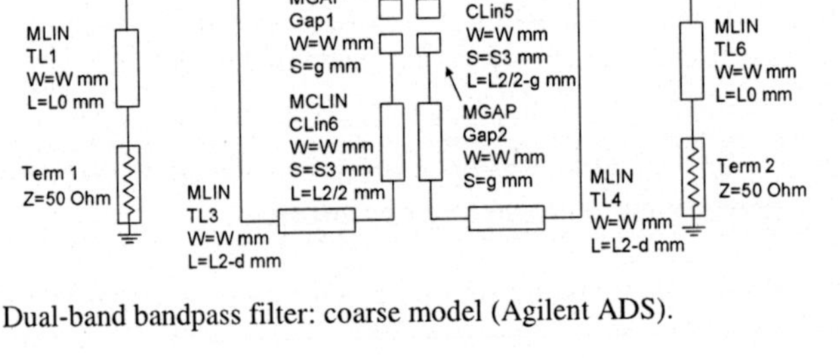

Fig. 3.13 Dual-band bandpass filter: coarse model (Agilent ADS).

Table 3.3 Optimization Results for Dual-Band Bandpass Filter

Algorithm	Final Specification Error [dB]	Fine Model Runs[1] [times]
SPRP	–2.0[3]	3
SPRP + input SM[2]	–1.9[4]	2

[1] Excludes the fine model evaluation at the starting point.
[2] Surrogate model is $\boldsymbol{R}_s^{(i)}(\boldsymbol{x}) = \boldsymbol{R}_c(\boldsymbol{x} + \boldsymbol{c}^{(i)})$; $\boldsymbol{c}^{(i)}$ is found using parameter extraction [9].
[3] Design specifications satisfied after the first iteration (spec. error –1.2 dB).
[4] Design specifications satisfied after the first iteration (spec. error –1.0 dB).

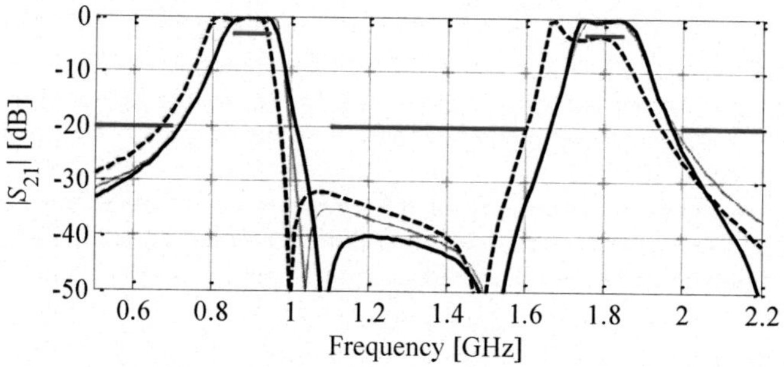

Fig. 3.14 Dual-band bandpass filter: fine model (dashed line) and coarse model (thin dashed line) response at $\boldsymbol{x}^{(0)}$, and the optimized fine model response (solid line) at the design obtained using SPRP.

stand-alone and combined with input SM. Note a very small number of fine model evaluations necessary to yield the optimized design.

3.4 Surrogate-Based Design Optimization of Antennas

Building a surrogate model may not be straightforward for certain types of microwave devices since reliable circuit equivalents, as those used in the previous section for planar microwave filters, may not be available for many types of antennas, e.g., ultrawide band (UWB) antennas, Yagi-type of antennas, or dielectric resonator antennas. Often, the only way to create a surrogate model is to use a coarsely-discretized full-wave EM model which is evaluated using the same EM solver as the one used for the high-fidelity model. However, coarsely-discretized EM simulation is still relatively expensive so that typically only a limited number of such simulations can be afforded. One way to deal with this situation is to generate smooth and computationally inexpensive surrogate by approximating sampled coarse-discretization EM data. The surrogate created this way can be then used in the space mapping optimization process. Another possibility is to exploits techniques that do not require excessive number of coarse-discretization EM simulations. Two of such methods—adaptive design specifications and multi-fidelity optimization algorithm—are also demonstrated in this chapter for antenna design.

3.4.1 Design of UWB Antipodal Vivaldi Antenna Using Coarsely-Discretized EM Models, Kriging and Space Mapping

The example considered here, a UWB antipodal Vivaldi antenna [30] of Fig. 3.15, shows how to combine functional and physical models to build a surrogate. Design variables are $\boldsymbol{x} = [a_1\ a_2\ b_1\ b_3\ h_1\ h_2\ d_1]^T$. The profile of the antipodal metal fins is with arks of ellipses; for the upper fin they are: *BC*, *DE*, and *DB*. The point *A* is the center of two ellipses with the arks of *BC* and *DE,* and the semiaxes of a_1 and b_1 and a_2 and b_2, respectively. The point *F* is the center of the ellipse with the semiaxes of a_3 and b_3. Note that here $a_3 = (a_2\ a_1)/2$, $b_2 = b_1+w_s$, and $d_2 = d_1$. Other parameters are fixed: $w_s = 2.15$, $w_1 = 12.9$, and $h_3 = 5$ (all in mm). Antenna metallization is with 0.05 mm copper. The fins are interfaced with the microstrip input (width of the ground of w_1) through the linear taper of length h_2. Rogers RT5880 (0.787 mm thick) is for the substrate of finite extends, and dielectric losses are maximal at 10 GHz.

The design specifications for reflection are $|S_{11}| \leq -10$ dB for 3.1 GHz to 10.6 GHz. Total lateral and longitudinal dimensions are constrained by 100 mm and 200 mm, respectively. The antenna models include an edge mount SMA connector (AEP part number: 9650-1113-014) [31] and its hex nut since their presence, as it was seen from numerical experiments, can affect the radiation pattern, e.g., tilt the main beam from the end-fire direction, change the gain in the back direction, etc. The connector pin extends 0.5 mm from the flange over the microstrip signal trace. The upper connector tips, the lower connector tips, and the microstrip ground are connected with a pair of vias (1 mm in diameter) going through the substrate.

The mismatch level of the connector-to-input microstrip junction itself is below –28 dB in the bandwidth of interest. The antenna models are excited through the 50 ohm coaxial port which is in the SMA connector.

The initial design is $\boldsymbol{x}^{in} = [30\ 50\ 10\ 10\ 100\ 20\ 2]^T$ mm. The high-fidelity antenna model is evaluated with the CST MWS transient solver [3] (8,954,244 mesh cells at $\boldsymbol{x}^{in}$, simulation time 1h 45 min).

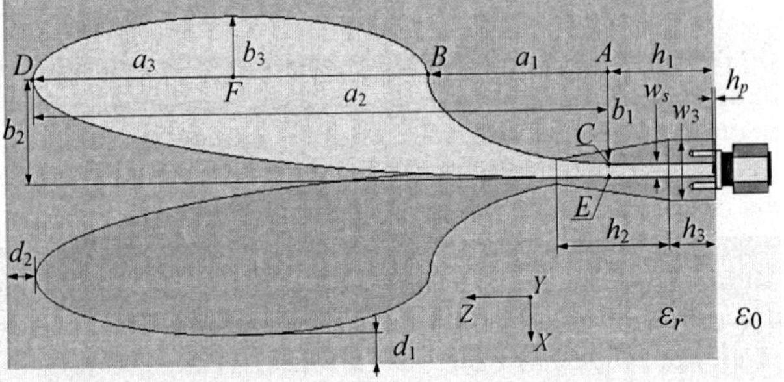

Fig. 3.15 Vivaldi antenna: top view, substrate shown transparent.

Here, a suitable equivalent-circuit coarse model is not available to apply optimization using space mapping. Instead, a coarse-discretization CST model $\boldsymbol{R}_{cd}$ (1,039,008 mesh cells at $\boldsymbol{x}^{in}$, evaluation time 6 minutes) is used. $\boldsymbol{R}_{cd}$ is still computationally too expensive to be used directly as a coarse model, therefore, a coarse model $\boldsymbol{R}_c$ is created in the neighbourhood of the starting point (here, the approximate optimum of $\boldsymbol{R}_{cd}$), using kriging interpolation [5] of the $\boldsymbol{R}_{cd}$ data. The procedure is as follows.

1. Allocate N base designs, $X_B = \{\boldsymbol{x}^1, \ldots, \boldsymbol{x}^N\}$, using Latin Hypercube Sampling [32];
2. Evaluate $\boldsymbol{R}_{cd}$ at each design $\boldsymbol{x}^j$, $j = 1, 2, \ldots, N$;
3. Build $\boldsymbol{R}_c$ as a kriging interpolation of data pairs $\{(\boldsymbol{x}^j, \boldsymbol{R}_{cd}(\boldsymbol{x}^j))\}_{j = 1,\ldots,N}$.

The coarse model created this way is computationally cheap, easy to optimize, and yet retains the features of a physically-based model. The starting point for space mapping optimization, $\boldsymbol{x}^{(0)} = [37.57\ 32.85\ 25.75\ 53.34\ 122.55\ 32.31\ 1.129]^T$ mm, is the approximate optimum of $\boldsymbol{R}_{cd}$. The kriging coarse model $\boldsymbol{R}_c$ is set up in the vicinity of $\boldsymbol{x}^{(0)}$ using $N = 100$ base points.

Figure 3.16 shows the fine model reflection response at the initial design as well as that of the fine and coarse-discretization model $\boldsymbol{R}_{cd}$ at $\boldsymbol{x}^{(0)}$. The final design, $\boldsymbol{x}^{(2)}$=[37.66 33.16 25.21 53.22 122.50 33.06 1.012]^T mm, is obtained after two space mapping iterations (Fig. 3.17). The surrogate model used by the optimization algorithm exploited input and output space mapping of the form $\boldsymbol{R}_s(x) = \boldsymbol{R}_c(\boldsymbol{x} + \boldsymbol{c}) + \boldsymbol{d}$ [7]. Optimization costs are summarized in Table 3.4. The total design time corresponds to about 16 evaluations of the fine model. It should be noted that the design improvement between $\boldsymbol{x}^{(0)}$ and $\boldsymbol{x}^{(2)}$ is somehow limited, which is because of a limited accuracy of the coarse model as shown in Fig. 3.16. The far-field response of the final design at selected frequencies is shown in Fig. 3.18.

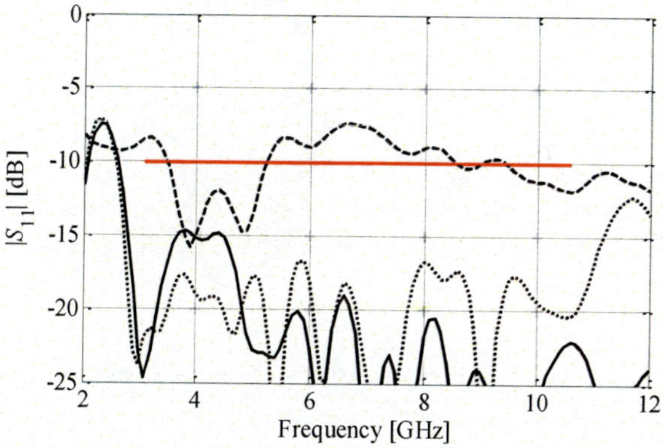

Fig. 3.16 Vivaldi antenna, $|S_{11}|$ versus frequency: Fine model $\boldsymbol{R}_f$ at the initial design (- - -), optimized coarse-discretization model $\boldsymbol{R}_{cd}$ (· · · ·), and $\boldsymbol{R}_f$ at the optimum of $\boldsymbol{R}_{cd}$ (—).

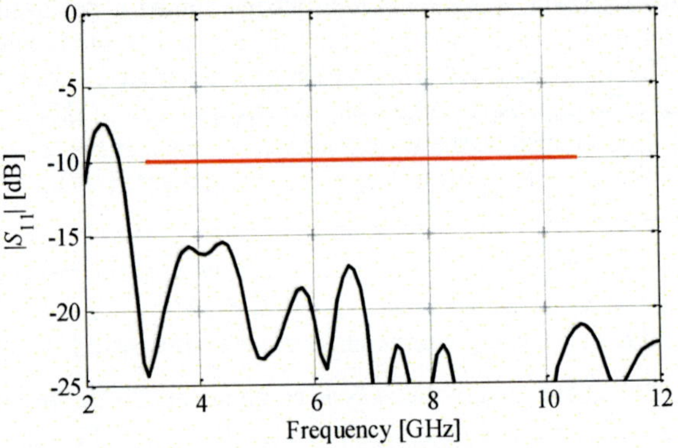

Fig. 3.17 Vivaldi antenna, $|S_{11}|$ versus frequency: $\boldsymbol{R}_f$ at the final design.

Table 3.4 UWB Vivaldi antenna: optimization cost

Algorithm Component	Number of Model Evaluations	CPU Time	
		Absolute	Relative to $\boldsymbol{R}_f$
Optimization of $\boldsymbol{R}_{cd}$	$135 \times \boldsymbol{R}_{cd}$	13.5 hours	7.7
Setting up $\boldsymbol{R}_c$	$100 \times \boldsymbol{R}_{cd}$	10.0 hours	5.7
Evaluation of $\boldsymbol{R}_f$	$3 \times \boldsymbol{R}_f$	5.3 hours	3.0
Total cost	N/A	28.8 hours	16.4

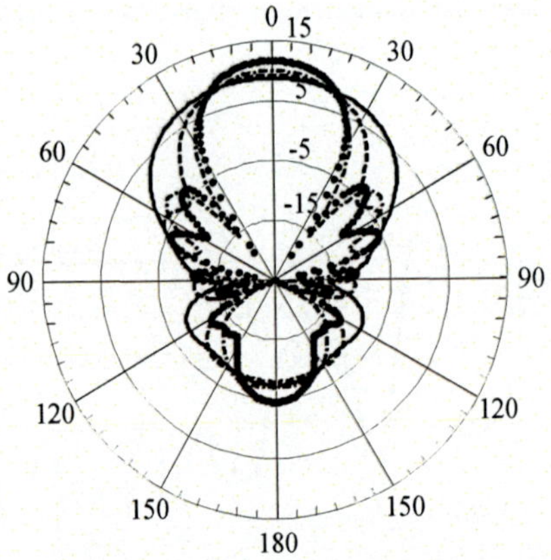

Fig. 3.18 Gain [dBi] of the Vivaldi antenna, x-pol. component: pattern cut in YOZ plane at 4 GHz (—), 6 GHz (- - -), 8 GHz (- · · - -), and 10 GHz (•••). 90^0 on the left, 0^0, and 90^0 on the right are for Y, Z, and Y directions, respectively.

3.4.2 *Design of Planar Yagi Antenna Using Adaptive Design Specifications*

Performance of the adaptive design specifications methodology [11] can be demonstrated with design optimization of a planar Yagi antenna for the 2.4-2.5 GHz band [33]. Optimization of planar Yagi antennas on finite substrate is a challenging task due to the finite substrate and proximity of the feeding circuitry to the radiators both introducing additional degrees of freedom to the design as well as complicate the use of methods developed for Yagi aerials [34, 35] and permits a limited use of existing design techniques [36].

Design geometry. The considered Yagi antenna comprises three directors, one driving element of a modified shape consisting of partially overlapping strips, and the feeding microstrip ground plane serving also as the reflector. The presented antenna can be viewed as a planar realization of the five-element Yagi. An outline of the antenna is given with Fig. 3.19. The antenna components are defined on a single layer of 0.025" thick Rogers RT6010 substrate which has extends of 100 mm × 160 mm. The ground extend is 100 mm × 40 mm. The input 50 ohm microstrip is to be interfaced to the terminals of the driving element through a section of the parallel strip transmission line in a way that provides the balanced input to the antenna. The antenna model is defined with CST MWS, discretized with subgrids, and simulated using the CST transient solver.

Design objectives. Maximum directivity of the principal polarization (E-field is parallel to XOZ plane) in the 2.4-2.5 GHz band is chosen as the main objective. The following antenna figures are treated as constraints (also in the 2.4-2.5 GHz band): the side lobe level relative to maximum (SLL < –10 dB), front-to-back ratio (FBR < –12 dB), direction of maximal radiation θ_m (elevation angle from the

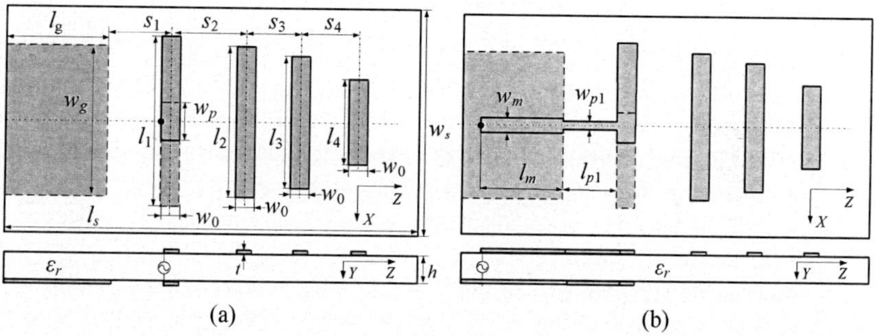

Fig. 3.19 Printed quasi-Yagi antenna: (a) the model used at the optimization stage of design (no feeding section), (b) the model updated with a feeding section starting from the 50 ohm microstrip. Source impedance is not shown at the diagrams. For simplicity, the feeding section at the panel (b) is shown as a simple two section structure: 50 ohm microstrip (dimensions l_m and w_m) and parallel strips (dimensions l_{p1} and w_{p1}). Detailed geometry of the optimized feeding section is given with Fig. 3.21(a).

Z axis, $|\theta_m| < 1°$). The antenna should be interfaced to the 50 ohm environment so that $|S_{11}| < -10$ dB in the 2.4-2.5 GHz band.

Design stages. As the input impedance of Yagi antennas is typically sensitive to variations of antenna dimensions [36], and since its value is not available prior to simulation while it is needed to define the feeding part of the antenna, the design optimization proceeds in two major steps as follows: First, the antenna is optimized for maximal directivity subject to the constraints on SLL, FBR, and θ_m. At this step the excitation is applied directly at the driving element's terminals (Fig. 3.19(a)). Design optimization procedure is based on surrogate-based optimization and involves optimization of a coarse-discretization antenna model. Having the optimal design, the feed interfacing the 50 ohm input and the driving element terminals, is designed.

Optimization methodology. Optimization of the coarse-discretization antenna model is carried out much faster. However, the coarse-discretization model is also less accurate: the figures of interest (e.g., directivity, SLL, FBR) are shifted in frequency with respect to those of the high-fidelity model. The frequency relationship between the two models using characteristic points (e.g., local maxima, points of corresponding response levels) is captured as shown in Fig. 3.20. Using this relationship, the original frequency band of interest is mapped into the corresponding band that is used in the optimization of the coarse-discretization model. This procedure, i.e., mapping of the frequency band, optimization of the coarse-discretization model and evaluation of the high-fidelity model is performed a few times as the frequency dependence between the models' responses may change from one design to another. The high-fidelity model is only evaluated a few times for verification purposes and to set up a new mapping.

Results. The design variables when optimizing the antenna for maximal directivity are $\boldsymbol{x} = [l_1\ l_2\ l_3\ l_4\ s_1\ s_2\ s_3\ s_4\ w_0\ w_p]^T$ (Fig. 3.19(a)). Other parameters are fixed: l_s=160, w_s = 100, l_g = 40, w_g = 100, and h = 0.635 (all in mm). The initial design is $\boldsymbol{x}^{(0)}$ = [40.42 35.7 31.5 27.3 17.85 22.05 22.05 22.05 2.35 1.5]^T. Simulation time of the coarse-discretization model (51,580 cells at $\boldsymbol{x}^{(0)}$) is about 6 minutes, and it is about 2 hours for the original, high-fidelity model (1,096,980 cells at $\boldsymbol{x}^{(0)}$). The optimum is found at $\boldsymbol{x}^*$ = [40.87 37.31 34.33 29.80 17.35 22.55 23.05 24.55 1.55 2.13]^T.

Based on the optimum $\boldsymbol{x}^*$ and the antenna impedance at the driving element terminals Z_t (Fig. 3.19(a)), a feed is designed (Fig. 3.21) using analytical formulas with a microstrip (l_m = 35 mm, w_m = 0.586 mm) and parallel strips ($l_{p1} = s_1 - w_0/2$, w_{p1} = 0.36 mm). The updated antenna model is then simulated. Its reflection does not meet the design specifications for frequency over 2.484 GHz. Therefore, the feed is redesigned with geometry of Fig. 3.21(a) through optimization of its full-wave model and a schematic of Fig. 3.21(b). Dimensions of the simple feed are used as an initial guess. Optimal feed dimensions are found to be $[w_{p1}\ w_{p2}\ w_{p3}\ l_{p1}\ l_{p2}]^T$ = [0.428 0.275 0.245 0.575 8.08]^T. The updated antenna model is then simulated, and responses are shown in Figs. 3.22 through 25 and Table 3.5. Table 3.6 shows the computational cost of the optimization process, which corresponds to only 19 full-wave antenna simulations.

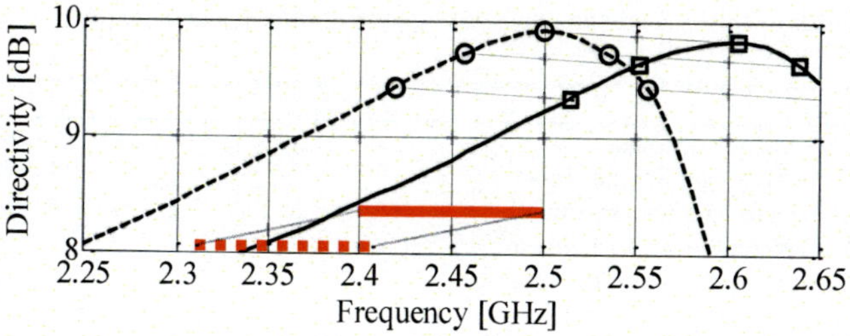

Fig. 3.20 Directivity versus frequency for the antenna structure (solid line) and its coarse-discretization model (dashed line). Characteristic points (squares and circles) are used to establish a frequency relationship between the two responses and to map the original frequency band of interest (2.4 to 2.5 GHz, thick solid line) into the corresponding band used in the coarse-discretization model optimization (thick dashed line).

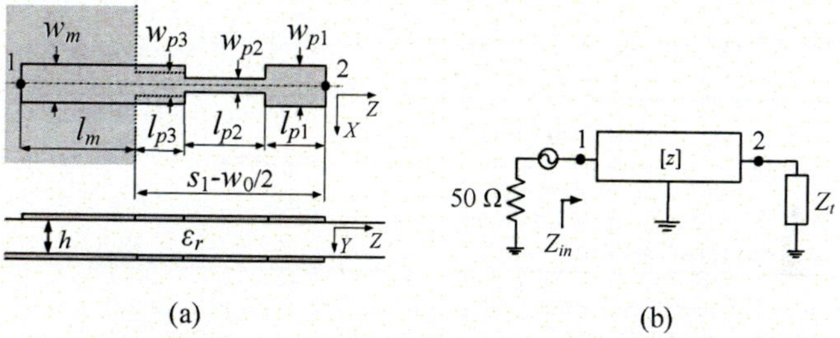

Fig. 3.21 A feed interfacing the 50 ohm input and the driving elements: (a) geometry of its full-wave model; (b) implemented schematic.

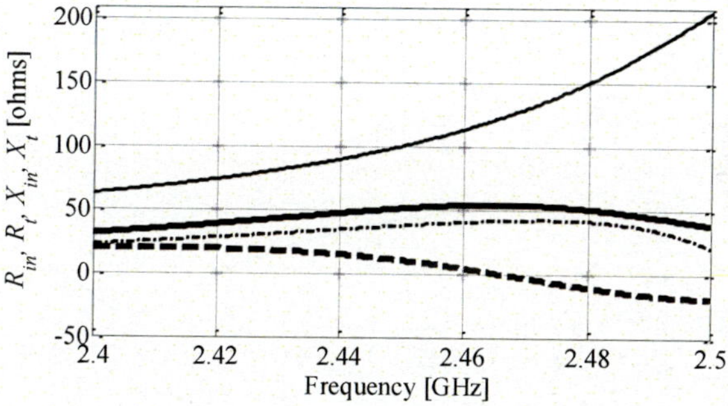

Fig. 3.22 Antenna impedance: resistance R_{in} (thick solid) and reactance X_{in} (thick dash) at the antenna input (Fig. 3.19(b)); R_t (solid) and X_t (dash-dot) at the driving element terminals (Fig. 3.19(a)).

Table 3.5 Printed Yagi antenna: performance summary

Figure	Value
Directivity, maximum*	10 dBi
Directivity, end fire maximum*	9.85 dBi
IEEE gain, end fire maximum*	9.49 dBi
Radiation efficiency, minimum*	92 %
Front to back ratio (FBR), minimum*	15.5 dB
Side lobe level (SLL), maximum*	-10.2 dB
End-fire polarization purity, minimum*	40 dB
3 dB beamwidth at 2.45 GHz	E-plane: 59°, H-plane: 74°
Relative bandwidth ($\|S_{11}\| < -10$ dB)	4.4 %
Input impedance at resonance (2.466 GHz)	54.6 ohms

* maximum/minimum over 2.4 GHz to 2.5 GHz

Table 3.6 Printed Yagi antenna: optimization cost summary

Algorithm Component	# of Model Evaluations	Absolute Time	Relative Time[3]
Coarse-discretization model optimization[1]	316	32 h	16
High-fidelity antenna simulation[2]	3	6 h	3
Total optimization time[4]	-	38 h	19

[1] Total number of evaluations (coarse-discretization model is optimized once per iteration, two iterations were performed in total).
[2] Evaluation at the initial design and after each iteration.
[3] Equivalent number of high-fidelity antenna simulations.
[4] Does not include the time necessary to design the antenna feed, which is negligible compared to the optimization time of the antenna itself.

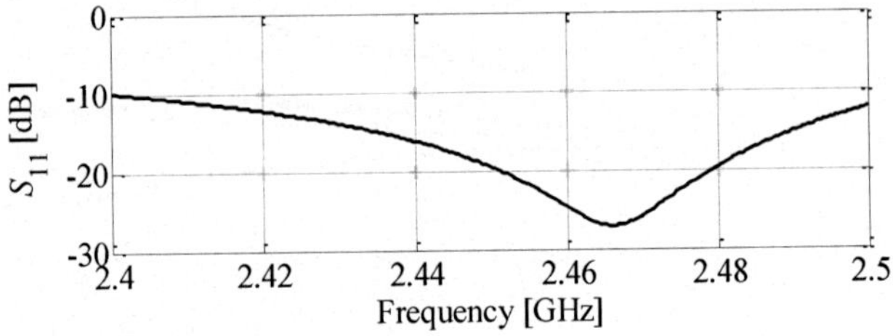

Fig. 3.23 Reflection from the input of the Yagi antenna.

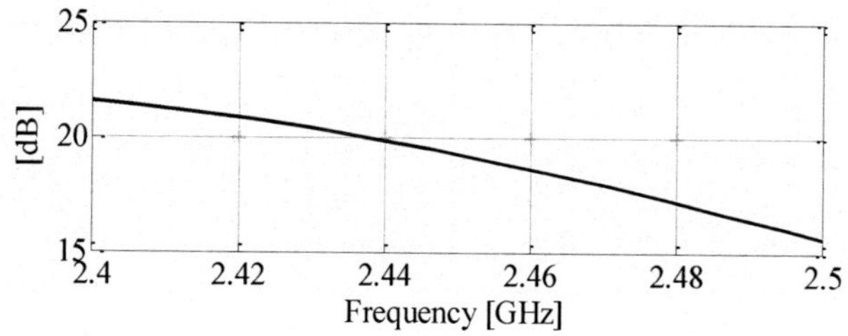

Fig. 3.24 Front-to-back ratio.

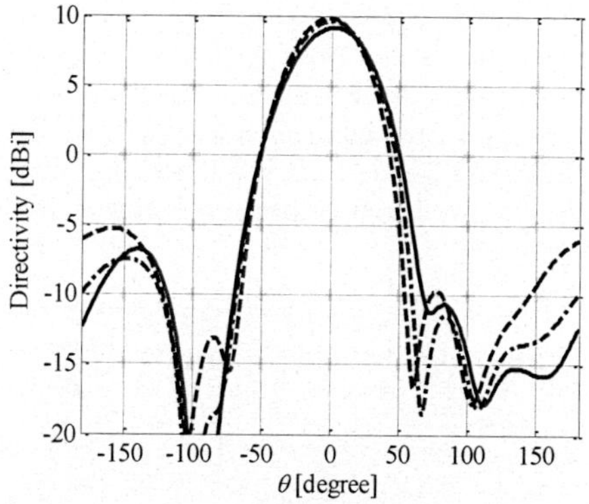

(a)

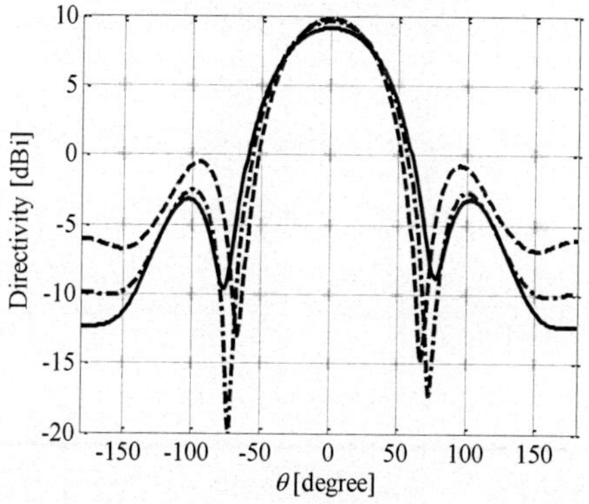

(b)

Fig. 3.25 Directivity pattern at 2.4 GHz (solid), 2.45 GHz (dash-dot), and 2.5 GHz (dash): (a) co-pol. in the E-plane (XOZ); (b) x-pol. in the H-plane (YOZ).

3.4.3 Multi-fidelity Design of Microstrip Broadband Antenna

Application of the multi-fidelity optimization algorithm [10] is demonstrated below using the broadband antenna [37] shown in Fig. 3.26. Here, $\boldsymbol{x} = [l_1\ l_2\ l_3\ l_4\ w_2\ w_3\ d_1\ s]^T$ are the design variables. Multilayer substrate is $l_s{\times}l_s$ (l_s=30 mm). The stack (from bottom-to-top) is: ground, RO4003, signal trace, RO3006 with a through via (trace-to-patch), the driven patch, RO4003, and four patches. Feeding is with 50 ohm SMA connector.

The design objective is $|S_{11}| \leq -10$ dB for 3.1 GHz to 10.6 GHz. IEEE gain not less than 5 dB for the zero elevation angle over band is an optimization constrain. The initial design is $\boldsymbol{x}^{(0)} = [15\ 15\ 15\ 15\ 20\ {-4}\ 2\ 2]^T$ mm. Two coarse-discretization models are used: $\boldsymbol{R}_{c.1}$ (122,713 mesh cells at $\boldsymbol{x}^{(0)}$) and $\boldsymbol{R}_{c.2}$ (777,888 mesh cells). The evaluation times for $\boldsymbol{R}_{c.1}$, $\boldsymbol{R}_{c.2}$ and $\boldsymbol{R}_f$ (2,334,312 mesh cells) are 3 min, 18 min and 160 min at $\boldsymbol{x}^{(0)}$, respectively. Figure 3.27(a) shows the responses of $\boldsymbol{R}_{c.1}$ at $\boldsymbol{x}^{(0)}$ and at its optimal design $\boldsymbol{x}^{(1)}$. Figure 3.27(b) shows the responses of $\boldsymbol{R}_{c.2}$ at $\boldsymbol{x}^{(1)}$ and at its optimized design $\boldsymbol{x}^{(2)}$. Figure 3.26(c) shows the responses of $\boldsymbol{R}_f$ at $\boldsymbol{x}^{(0)}$, at $\boldsymbol{x}^{(2)}$ and at the refined design $\boldsymbol{x}^* = [14.87\ 13.95\ 15.4\ 13.13\ 20.87\ {-5.90}\ 2.88\ 0.68]^T$ mm ($|S_{11}| \leq -11.5$ dB for 3.1 GHz to 4.8 GHz) obtained in two iterations of the refinement step [10], see also Section 3.2.5,

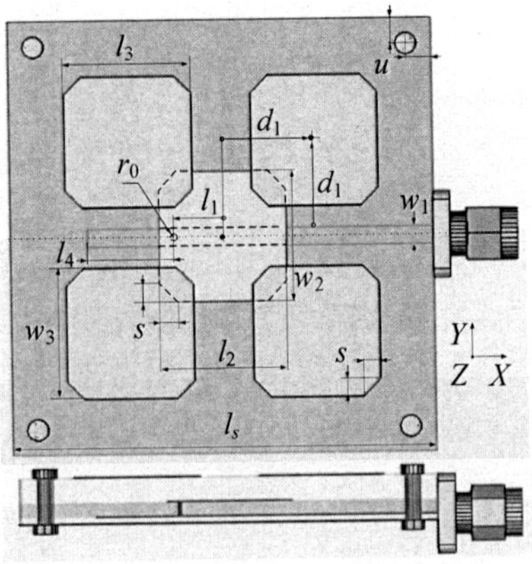

Fig. 3.26 Microstrip broadband antenna: top/side views, substrates shown transparent.

Table 3.7 Microstrip broadband antenna: design cost

Design Step	Model Evaluations	Computational Cost	
		Absolute [hours]	Relative to $\boldsymbol{R}_f$
Optimization of $\boldsymbol{R}_{c.1}$	$125 \times \boldsymbol{R}_{c.1}$	6.3	2.6
Optimization of $\boldsymbol{R}_{c.2}$	$48 \times \boldsymbol{R}_{c.2}$	14.4	5.4
Setup of model $\boldsymbol{q}$	$17 \times \boldsymbol{R}_{c.2}$	5.1	1.9
Evaluation of $\boldsymbol{R}_f$	$2 \times \boldsymbol{R}_f$	5.3	2.0
Total design time	N/A	31.1	11.9

*Excludes $\boldsymbol{R}_f$ evaluation at the initial design.

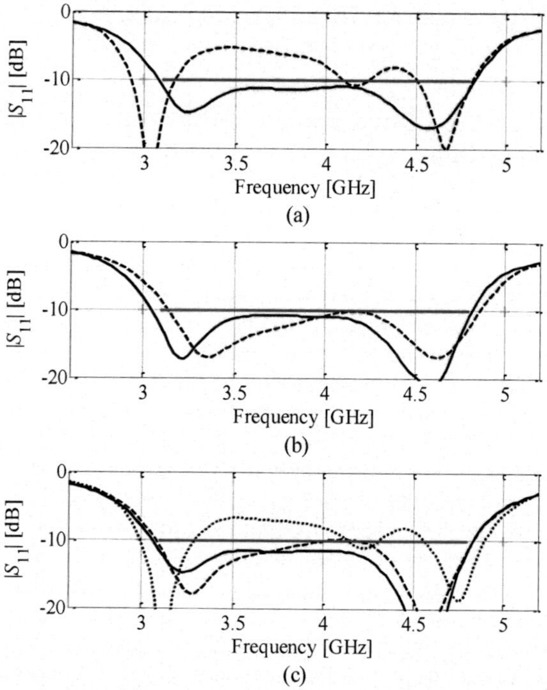

Fig. 3.27 Microstrip broadband antenna: (a) responses of the coarse-discretization model $\boldsymbol{R}_{c.1}$ at the initial design $\boldsymbol{x}^{(0)}$ (- - -) and at the optimized design $\boldsymbol{x}^{(1)}$ (—); (b) responses of the coarse-discretization model $\boldsymbol{R}_{c.2}$ at $\boldsymbol{x}^{(1)}$ (- - -) and at its optimized design $\boldsymbol{x}^{(2)}$ (—); (c) responses of the high-fidelity model $\boldsymbol{R}_f$ at $\boldsymbol{x}^{(0)}$ (····), at $\boldsymbol{x}^{(2)}$ (- - -) and at the refined final design $\boldsymbol{x}^*$ (—).

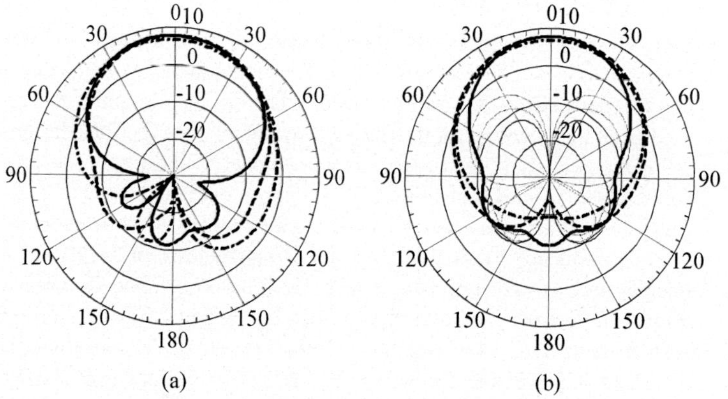

Fig. 3.28 Microstrip antenna, gain [dBi] of the final design at 3.5 GHz (· ·), 4.0 GHz ( ), and 4.5 GHz (—): (a) co-pol. in the *E*-plane (*XOZ*), and connector is at 90^0 on the right; (b) x-pol., primary, (thick lines) and co-pol. (thin lines) in the *H*-plane (*YOZ*).

eq. (3.13). The design cost (Table 3.7) corresponds to about 12 runs of the high-fidelity model $\boldsymbol{R}_f$. Antenna gain at the final design is shown in Fig. 3.28.

3.5 Surrogate-Based Design Optimization of Microwave Transitions

Design of low-loss broad-band transitions interfacing different type of transmission lines at microwave frequencies usually involves full-wave EM simulation to accurately describe the transition responses [38, 39]. Circuit models and analytical formulas, when available, can only be used to get initial designs which should be verified and tuned for required design requirements. Typically, reliable circuit models are either unavailable or require significant amount of development and validation effort. Moreover, additions or modifications introduced in the transition geometry may invalidate existing models, which leads to repeating the model development procedure. On the other hand, optimization techniques exploiting surrogates [40], including those based on coarsely-discretized EM models, may substantially reduce the computational complexity of the conventional optimization methods and, at the same time, be applied to modified/improved geometries without extra effort.

Two examples are presented in this section. The first one illustrates the multi-fidelity design optimization technique to improve performance of a coplanar wavequide-to-microstrip transition based on EM coupling. The second example demonstrates the use of the adaptive design specifications method for design of coplanar waveguide-to-substrate integrated waveguide transition. In both cases the use of coarsely-discretized EM models is essential, since no accurate circuit equivalents are available for the considered structures.

3.5.1 Multi-fidelity Design of Microstrip-to-Coplanar Waveguide Transition

Here, the multi-fidelity optimization algorithm [10] is applied to design optimization of a microstrip-to-CPW transition [41]. The methodology exploits sequential optimization of coarse-discretization EM models. The optimal design of the current model is used as an initial design for the finer-discretization one. The final design is then refined using a polynomial-based approximation model of the responses obtained from the coarse-discretization simulations. The design process is computationally efficient because the optimization burden is shifted to the coarse-discretization models.

Two frequency bands with the center frequencies f_c of 5 GHz and 10 GHz are of interest for this transition [41] (Fig. 3.29). The port-to-port distance is 20 mm. The transition geometry and the input transmission lines (TLs) are on 0.635 mm thick RT6010 substrate. Metallization (5.7e8 S/m) is 0.0254 mm thick. Dimensions of the input TLs are the following: W_m=0.6, W_c=0.8, and s_c=0.3 (all in mm). The ground plane is common to the CPW and microstrip and it is modelled of infinite lateral extend. The low frequency TL impedances are about 50 ohms each. All models are simulated using the CST MWS transient solver.

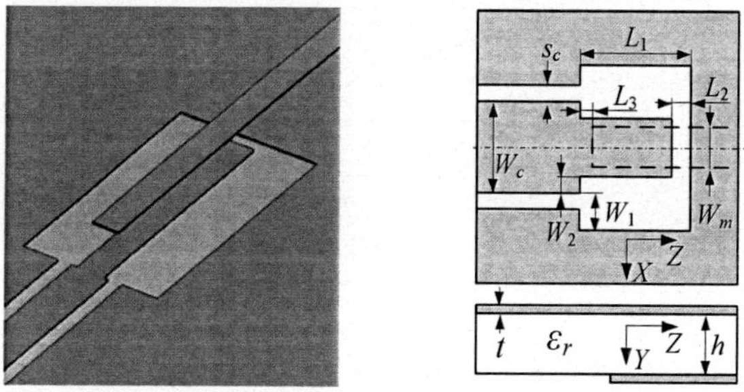

(a) (b)

Fig. 3.29 Coplanar waveguide-to-microstrip transition with EM coupling [41]: (a) 3D view, substrate shown transparent; (b) layout views.

In this example the ground plane is modelled to be of infinite lateral extend. The design objective is the 50% symmetrical bandwidth at the –20 dB level for both $|S_{11}|$ and $|S_{22}|$. The design variables are $\boldsymbol{x} = [L_1\ W_1\ W_2\ L_2\ L_3]^T$ mm. Designs start from $\boldsymbol{x}^{(0)} = [L^{(0)}{}_1\ 0.8\ 0\ 0.3\ 0]^T$, where $L^{(0)}{}_1 = 6$ and 3 for designs of 5 GHz and 10 GHz, respectively. For this example one again uses two coarse-discretization models $\boldsymbol{R}_{c.1}$ and $\boldsymbol{R}_{c.2}$ with the following evaluation times: 60s and 100s (5 GHz) and 71 s and 106 s (10 GHz). The fine model evaluation time is 17 min (5 GHz) and 26 min (10 GHz).

The optimal designs are found to be $\boldsymbol{x}^* = [6.200\ 1.105\ \ 0.113\ 0.319\ –0.033]^T$ for 5 GHz and $[2.877\ 1.017\ 0.038\ \ 0.287\ –0.090]^T$ for 10 GHz. Both final designs meet the specifications completely, which is shown in Fig. 3.30. Figure 3.31 shows the transmission responses $|S_{21}|$ versus frequency at the initial and final designs. Significant improvement in reflection and transmission responses (in level and bandwidth) is achieved: the bandwidth was extended to 53% from initial 0% for the 5 GHz design and to 51% from initial 20% for the 10 GHz design. Design cost are 9.3 evaluations of the fine model for the 5 GHz design and 7.0 evaluations of the fine model for the 10 GHz design (see Table 3.8 for details).

As a comparison with "classical" simulation-driven design, the transition for f_c = 10 GHz has been also designed through direct optimization of the fine model using the pattern search algorithm [42]. The final design obtained this way is almost as good as that produced by the multi-fidelity technique (50% bandwidth), however, the design cost is almost 18 times higher (124 evaluations of $\boldsymbol{R}_f$ versus about 7 for the multi-fidelity algorithm).

The reduced quadratic model [10], see also Section 3.2.5, eqs. (3.11)-(3.12), is also utilized to perform sensitivity analysis of the final designs. For this purpose, however, the quadratic model is set up using the high-fidelity model data. Because sensitivity analysis is performed assuming relatively small deviations around the optimized design (0.0125 and 0.025 mm for geometry variables), the accuracy of the quadratic model is sufficiently good with respect to $\boldsymbol{R}_f$. Results of the sensitivity analysis are shown in Fig. 3.32.

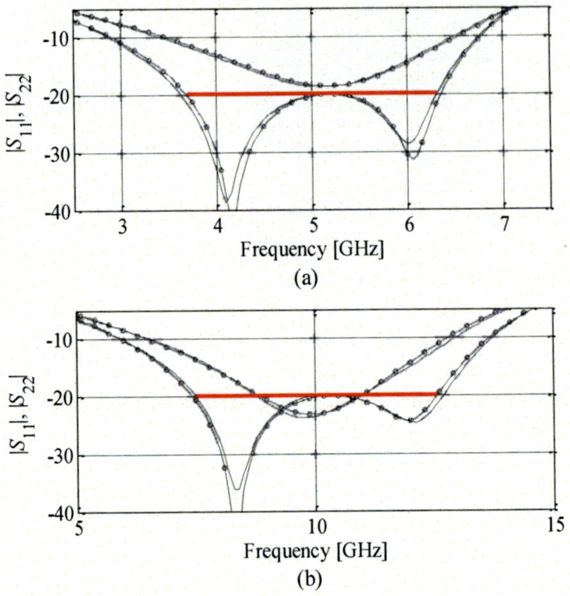

Fig. 3.30 Transition through coupling: fine model responses at initial (dashed line) and final design (solid line) for (a) $f_c = 5$ GHz, and (b) $f_c = 10$ GHz; –20 dB bandwidth at the final design marked with horizontal line.

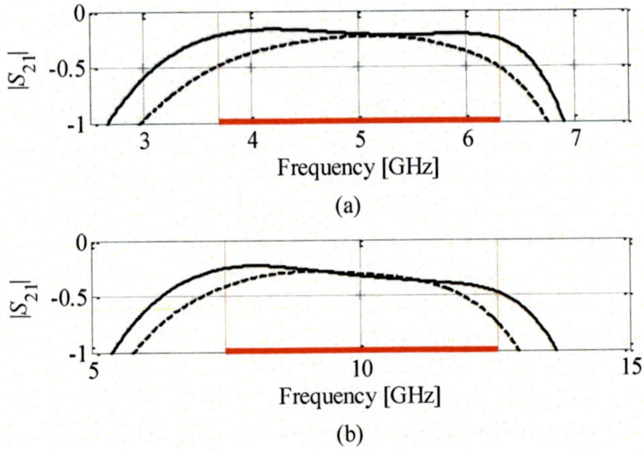

Fig. 3.31 Transition through coupling: transmission response at the initial (dashed line) and the final design (solid line) for (a) $f_c = 5$ GHz and (b) $f_c = 10$ GHz; –20 dB bandwidth at the final design marked with horizontal line.

Table 3.8 CPW-microstrip transition: design cost

Center frequency	Design Procedure Component	Number of Model Evaluations	Evaluation Time Absolute [min]	Evaluation Time Relative to $\boldsymbol{R}_f$
5 GHz	Optimization of $\boldsymbol{R}_{c.1}$	62	62	3.6
	Optimization of $\boldsymbol{R}_{c.2}$	27	45	2.6
	Setup of model $\boldsymbol{q}$	11 ($\boldsymbol{R}_{c.2}$)	18	1.1
	Evaluation of $\boldsymbol{R}_f$	2	34	2.0
	Total design time	N/A	**159**	**9.3**
10 GHz	Optimization of $\boldsymbol{R}_{c.1}$	54	64	2.5
	Optimization of $\boldsymbol{R}_{c.2}$	26	46	1.8
	Setup of model $\boldsymbol{q}$	11 ($\boldsymbol{R}_{c.2}$)	19	0.7
	Evaluation of $\boldsymbol{R}_f$	2	52	2.0
	Total design time	N/A	**181**	**7.0**

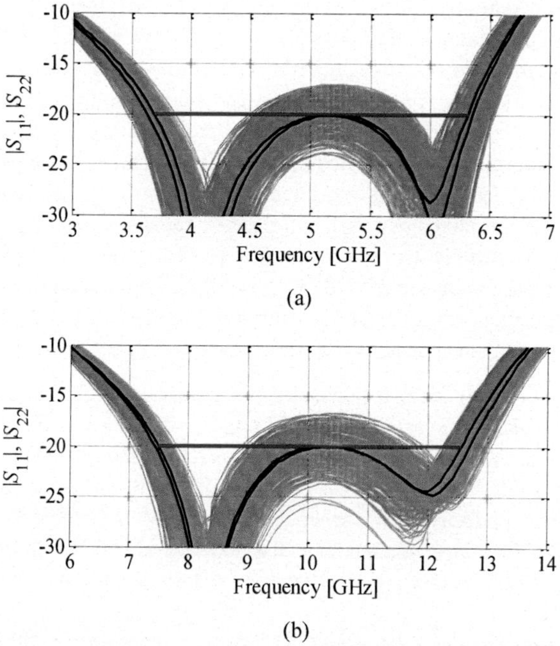

Fig. 3.32 Transition through coupling: sensitivity analysis using 200 random samples allocated in the neighbourhood of the optimized designs: (a) f_c = 5 GHz and (b) f_c = 10 GHz; –20 dB bandwidth at the final design marked with horizontal line. The sensitivity analysis setup is described in the text. Thick solid lines denote transition responses at optimized designs. Thin lines represent the family of responses corresponding to random samples as described above.

3.5.2 Design of Coplanar Waveguide-to-Substrate Integrated Waveguide Transition

Substrate integrated circuits (SICs), and substrate integrated waveguides (SIWs) in particular, find application in modern microwave and millimeter wave engineering due to their capability of low cost realization of waveguide components as well as integration of different components all in the frame of planar technology [43]. One of the major tasks in SIC transition design is the adjustment of geometry parameters so that given design specifications are satisfied. For research in the computer aided design (CAD), this casts into a problem of developing straightforward and reliable procedures to tune geometries of SIC transitions for required performance in a given environment.

Increasing a useable bandwidth of conductor backed coplanar waveguide (CBCPW)-to-SIW transitions is targeted here. Metalized vias partially protruding into substrate in the transition region are used as tuning elements, and the surrogate-based optimization [7] is applied as a design tool. Adjustable metal screws and pins are classical tuning elements in hollow waveguides and cavities [44]; however, they are not used in SICs in the similar way since post-manufacturing adjustment of SICs is hardly possible, and finding optimal position, diameter, and protruding depth of vias represents a challenging task in the case of SIC. Design of SIC transitions can be conducted successfully by means of surrogate-based optimization [7], [45] and coarse-discretization electromagnetic models [46] with the transition dimensions considered as design optimization variables.

Examples include: (i) design optimization of a transition interfacing the conductor backed coplanar waveguide (CBCPW) to SIW without vias (not capable to satisfy the design specifications), and (ii) re-optimized transitions with metalized vias protruding into substrate in the transition region, which improves the usable bandwidth.

Geometry under design. Consider the planar transition interfacing a conductor backed coplanar waveguide (CBCPW) to a SIW shown in Fig. 3.33. The CBCPW, SIW, and transition are on the 3.175 mm RT5880 substrate. The CBCPW upper and lower grounds, the SIW top and bottom walls are of infinite lateral extend. All metal parts have conductivity of copper (5.8e7 S/m). Metallization of the CBCPW signal trace, CBCPW upper ground, and SIW top wall is with 1.5 oz copper (≈ 0.05 mm). Design specifications are $|S_{11}|$, $|S_{22}|$<= 20 dB for the *X*-band (here 8.2 GHz to 11.7 GHz).

The dimensions of the input CBCPW are: signal trace width $w_0 = 2.25$ mm; slot width $s_0 = 0.2$ mm; spacing between the rows of vias $u_0 = 6.95$ mm; spacing between vias in the row $v_0 = 2$ mm; via diameter $d = 1$ mm. The dimensions of the input SIW are: spacing between the rows of vias, $u_2 = 15.95$ mm; spacing between vias in the terminating rows $v_1 = 1.5$ mm; spacing between vias in the row $v_2 = 2$ mm; via diameter is the same as in the CBCPW (1 mm). The cutoff of the SIW's quasi TE_{10} dominant mode is at 6.55 GHz.

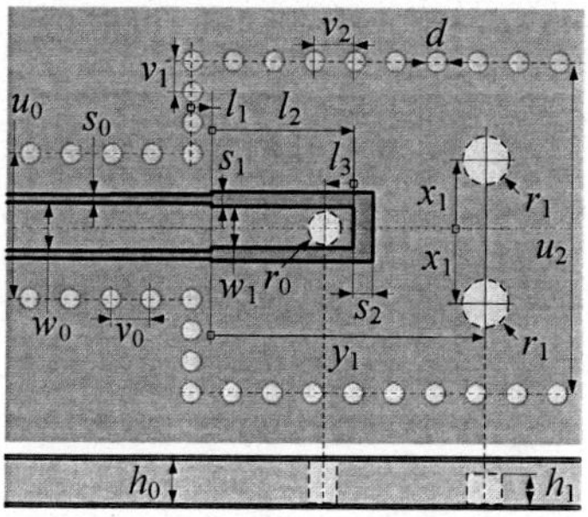

Fig. 3.33 CBCPW-to-SIW transition with two vias added: top and side views. The via walls are not shown on the side view. The dash-dot line shows the location of the symmetry plane (magnetic wall).

The transition comprises a CBCPW section, the probe connecting the CBCPW signal trace to the SIW bottom wall, and pairs of protruding vias. The length of the transition, i.e., $l_1 + l_2$ in the case without protruding vias, and $l_1 + y_1$ in the case of two protruding vias, is constrained to 20 mm ($\approx$ 1.25 of u_2). Figure 3.2 gives a conceptual view of the transition with two extra vias. Via location, x_1, y_1, radius, r_1, and protruding depths, h_1, will be additional (to the dimensions of the CBCPW section) design variables. In the CAD models, the port-to-port distance is 55 mm of which 25 mm is for the SIW section. The SIW is excited through a 2.5 mm section of the equivalent rectangular waveguide [46]. The 50 ohm CBCPW waveguide port has a perfect metal periphery connecting the upper and lower CBCPW grounds. The extra vias protrude into the dielectric from the SIW bottom wall but they are not allowed to touch the SIW top wall.

Design process. The first step of the optimization process is to optimize the coarse-discretization EM model of the transition (low-fidelity model) using pattern search [42]. The design is further improved using adaptively adjusted design specifications technique [46, 47] which consists of the following two steps: (i) Modify the original design specifications to account for the discrepancy between the low- and high-fidelity models; (ii) Obtain a new design by optimizing the low-fidelity model with respect to the modified specifications.

In Step (i), the design specifications are modified so that the level of satisfying/violating the modified specifications by the low-fidelity model response corresponds to the satisfaction/violation levels of the original specifications by the high-fidelity model [46]. The low-fidelity model is then optimized in Step (ii) with respect to the modified specifications and the new design obtained this way is treated as an approximated solution to the original design problem (i.e., optimization of the high-fidelity model with respect to the original specifications). Steps (i) and (ii) can be repeated if necessary. Typically, a substantial design improvement is observed after the first iteration. Additional iterations may bring further enhancement as the discrepancy between the high- and low-fidelity models may change somehow from one design to another. Figure 3.34 illustrates an iteration of this technique used for design of a CBCPW-to-SIW transition.

It should be noted that employing simulation-driven design based on low-fidelity models allows us to find optimal designs that might not be obtainable otherwise.

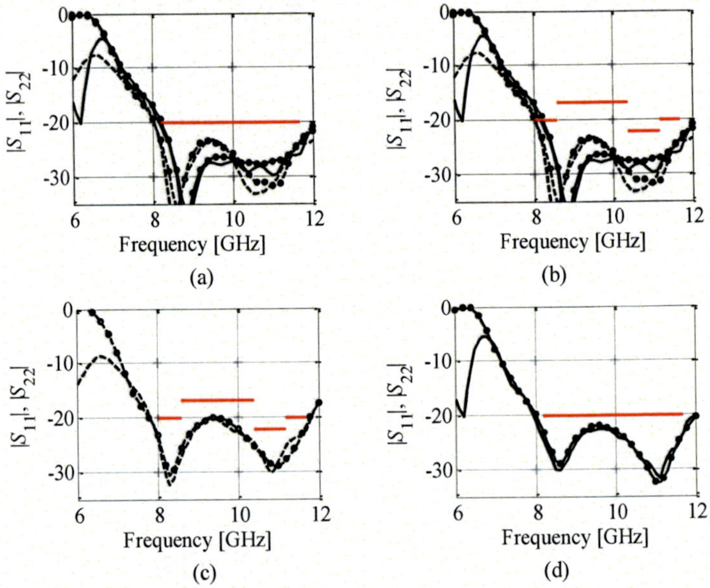

Fig. 3.34 Adaptively adjusted design specification technique applied to optimize CBCPW-to-SIW transitions. High- and low-fidelity model response denoted as solid and dashed lines, respectively. $|S_{22}|$ distinguished from $|S_{11}|$ using circles. Design specifications denoted by thick horizontal lines. (a) High- and low-fidelity model responses at the beginning of the iteration as well as original design specifications; (b) High- and low-fidelity model responses and modified design specifications that reflect the differences between the responses; (c) Low-fidelity model optimized to meet the modified specifications; (d) high-fidelity model at the low-fidelity model optimum shown versus original specifications.

Results. For the case without extra vias the design variables are $\boldsymbol{x}_{0in}=[l_1\ l_2\ l_3\ w_1\ s_1\ s_2\ r_0]^T$. The optimization procedure starts from $\boldsymbol{x}_{0in}=[1.0\ 7.0\ 0.75\ 2.0\ 0.4\ 1.0\ 0.75]^T$. The responses of the initial design are shown in Fig. 3.35 (a) and Fig 3.36(a). The optimal design is found to be $\boldsymbol{x}_{0opt}=[0.695\ 7.451\ 0.323\ 2.387\ 0.764\ 0.235\ 0.250]^T$, its responses are shown at Fig. 3.35(a) and Fig 3.36(a). The optimal design has the improved reflection and transmission responses compared to the initial one; however, it does not meet the design requirements. Therefore, additional vias are introduced (see Fig. 3.33) as tuning elements. The optimum design for the case of two protruding vias is shown in Table 3.9. The response corresponding to this design is shown in Fig. 3.35(b) and Fig 3.36 (b).

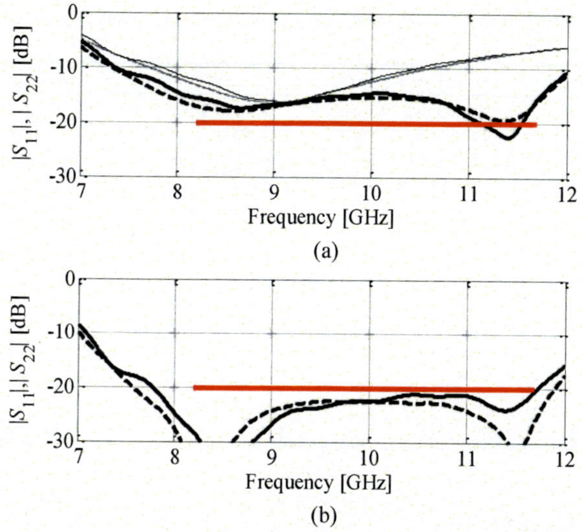

Fig. 3.35 CBCPW-to-SIW transitions, $|S_{11}|$ (solid) and $|S_{22}|$ (dash): (a) the initial (thin) and optimized (thick) design without protruding vias; and (b) optimized design with two vias.

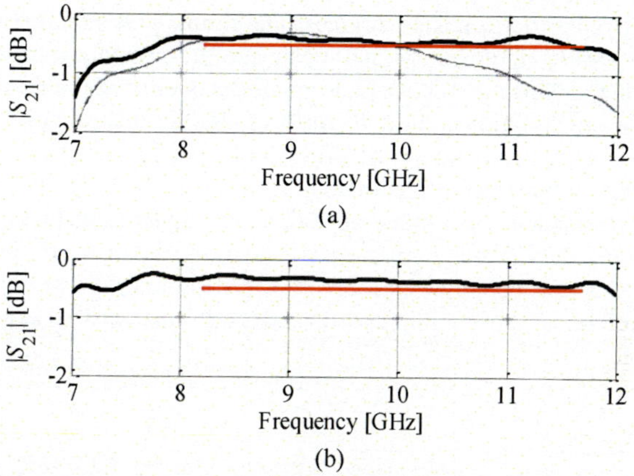

Fig. 3.36 CBCPW-to-SIW transitions, $|S_{21}|$: (a) the initial (thin-dash) and optimized (thick-solid) design without protruding vias; (b) optimized design with two vias.

Table 3.9 CPW-SIW transitions: final designs

Parameter [mm]	Design 1 (no extra vias)	Design 2 (two extra vias)
l_1	0.695	0.570
l_2	7.451	7.389
l_3	0.323	0.323
w_1	2.387	2.387
s_1	0.764	0.839
s_2	0.235	0.235
r_0	0.250	0.250
r_1	-	0.313
x_1	-	1.750
y_1	-	11.375
h_1	-	1.355
$\|S_{11}\|,\|S_{22}\|$ [dB]	<=–14.6	<=–20
Bandwidth [GHz]	8.0–11.75	7.78–11.72

3.6 Conclusion

In this chapter, several techniques for computationally efficient simulation-driven design optimization of microwave structures have been discussed. We also presented a number of design examples concerning various microwave components, including microstrip filters, planar antennas, as well as transition structures. In all cases, the surrogate-based techniques presented in the previous chapter have been employed as optimization engines. The results presented here indicate that the surrogate-based optimization methods make the simulation-driven microwave design feasible and efficient, both in terms of the quality of the final design, and in terms of the computational cost. In most cases, the design cost corresponds to a few high-fidelity electromagnetic simulations of the microwave structure under consideration, typically comparable to the number of design variables. While this kind of performance is definitely appealing, improved robustness and reliability as well as availability through commercial software packages are needed to make the surrogate-based techniques widely accepted by microwave engineering community. Therefore, a substantial research effort in this area is expected in the years to come.

References

1. Bandler, J.W., Cheng, Q.S., Dakroury, S.A., Mohamed, A.S., Bakr, M.H., Madsen, K., Søndergaard, J.: Space mapping: the state of the art. IEEE Trans. Microwave Theory Tech. 52, 337–361 (2004)
2. Director, S.W., Rohrer, R.A.: The generalized adjoint network and network sensitivities. IEEE Trans. Circuit Theory CT-16, 318–323 (1969)
3. CST Microwave Studio, ver. 20109, CST AG. Bad Nauheimer Str.,19, D-64289, Darmstadt, Germany (2010)

4. HFSS, release 13.0, ANSYS (2010), http://www.ansoft.com/products/hf/hfss/
5. Queipo, N.V., Haftka, R.T., Shyy, W., Goel, T., Vaidynathan, R., Tucker, P.K.: Surrogate based analysis and optimization. Progress in Aerospace Sciences 41, 1–28 (2005)
6. Forrester, A.I.J., Keane, A.J.: Recent advances in surrogate-based optimization. Prog. in Aerospace Sciences 45, 50–79 (2009)
7. Koziel, S., Bandler, S.W., Madsen, K.: A space mapping framework for engineering optimization: theory and implementation. IEEE Trans. Microwave Theory Tech. 54, 3721–3730 (2006)
8. Koziel, S., Meng, J., Bandler, J.W., Bakr, M.H., Cheng, Q.S.: Accelerated microwave design optimization with tuning space mapping. IEEE Trans. Microwave Theory and Tech. 57, 383–394 (2009)
9. Koziel, S.: Shape-preserving response prediction for microwave design optimization. IEEE Trans. Microwave Theory and Tech. 58(11), 2829–2837 (2010)
10. Koziel, S., Ogurtsov, S.: Robust multi-fidelity simulation-driven design optimization of microwave structures. In: IEEE MTT-S Int. Microwave Symp. Dig, Anaheim, CA, pp. 201–204 (2010)
11. Koziel, S.: Efficient optimization of microwave structures through design specifications adaptation. In: IEEE Int. Symp. Antennas Propag. Toronto, Canada (2010)
12. Koziel, S., Cheng, Q.S., Bandler, J.W.: Implicit space mapping with adaptive selection of preassigned parameters. IET Microwaves, Antennas & Propagation 4, 361–373 (2010)
13. Pozar, D.M.: Microwave Engineering, 3rd edn. Wiley, Chichester (2004)
14. Koziel, S., Bandler, J.W., Madsen, K.: Quality assessment of coarse models and surrogates for space mapping optimization. Optimization Eng. 9, 375–391 (2008)
15. Koziel, S., Bandler, J.W.: Space-mapping optimization with adaptive surrogate model. IEEE Trans. Microwave Theory Tech. 55, 541–547 (2007)
16. Swanson, D., Macchiarella, G.: Microwave filter design by synthesis and optimization. IEEE Microwave Magazine 8(2), 55–69 (2007)
17. Rautio, J.C.: EM-component-based design of planar circuits. IEEE Microwave Magazine 8(4), 79–90 (2007)
18. Cheng, Q.S., Bandler, J.W., Koziel, S.: Space mapping design framework exploiting tuning elements. IEEE Trans. Microwave Theory and Tech. 58(1), 136–144 (2010)
19. Koziel, S., Bandler, J.W., Cheng, Q.S.: Design optimization of microwave circuits through fast embedded tuning space mapping. In: European Microwave Conference Paris, Septemper 26– October 1 (2010)
20. Cheng, Q.S., Rautio, J.C., Bandler, J.W., Koziel, S.: Progress in simulator-based tuning—the art of tuning space mapping. IEEE Microwave Magazine 11(4), 96–110 (2010)
21. Alexandrov, N.M., Dennis, J.E., Lewis, R.M., Torczon, V.: A trust region framework for managing use of approximation models in optimization. Struct. Multidisciplinary Optim. 15, 16–23 (1998)
22. Salleh, M.H.M., et al.: Quarter-wavelength side-coupled ring resonator for bandpass filters. IEEE Trans. Microwave Theory Tech. 56, 156–162 (2008)
23. FEKO User's Manual. Suite 5.5, EM Software & Systems-S.A (Pty) Ltd, 32 Techno Lane, Technopark, Stellenbosch, 7600, South Africa (2009)
24. Agilent ADS, Version 2009, Agilent Technologies, 395 Page Mill Road, Palo Alto, CA, 94304 (2009)

25. Koziel, S., Bandler, J.W., Cheng, Q.S.: Robust trust-region space-mapping algorithms for microwave design optimization. IEEE Trans. Microwave Theory and Tech. 58, 2166–2174 (2010)
26. Liao, C.-K., Chi, P.-L., Chang, C.-Y.: Microstrip realization of generalized Chebyshev filters with box-like coupling schemes. IEEE Trans. Microwave Theory Tech. 55, 147–153 (2007)
27. Sonnet em. Ver. 12.54, Sonnet Software. North Syracuse, NY (2009)
28. Koziel, S., Bandler, J.W.: Automated tuning space mapping implementation for rapid design optimization of microwave structures. In: Int. Review of Progress in Applied Computational Electromagnetics, ACES 2009, pp. 138–143 (2009)
29. Guan, X., Ma, Z., Cai, P., Anada, T., Hagiwara, G.: A microstrip dual-band bandpass filter with reduced size and improved stopband characteristics. Microwave and Opt. Tech. Lett. 50, 618–620 (2008)
30. Qing, X.M., Chen, Z.N.: Antipodal Vivaldi antenna for UWB applications. In: European Electromagnetics Symposium, UWB SP 7 (2004)
31. SMA Edge Mount P.C. Board Receptacles. Online Catalog, Applied Engineering Products, http://aepconnectors.com/pdf/SMAedg.pdf
32. Beachkofski, B., Grandhi, R.: Improved distributed hypercube sampling. American Institute of Aeronautics and Astronautics, Paper AIAA, 2002–1274 (2002)
33. Koziel, S., Ogurtsov, S.: Computationally Efficient Simulation-Driven Design of a Printed 2.45 GHz Yagi Antenna. Microwave and Optical Technology Letters 52, 1807–1810 (2010)
34. Chen, C.A., Cheng, D.K.: Optimum element lengths for Yagi-Uda arrays. IEEE Trans. Antennas Propag. 23, 8–15 (1975)
35. Cheng, D.K., Chen, C.A.: Optimum element spacings for Yagi-Uda arrays. IEEE Trans. Antennas Propag. 21, 615–623 (1973)
36. Balanis, C.A.: Antenna theory: analysis and design. Wiley-IEEE Press, Chichester (2005)
37. Chen, Z.N.: Wideband microstrip antennas with sandwich substrate. IET Microw. Ant. Prop. 2, 538–546 (2008)
38. Gupta, K.C., Garg, R., Bahl, I., Bhartia, P.: Microstrip lines and slotlines, 2nd edn. Artech House, Norwood (1996)
39. Simons, R.N.: Coplanar waveguide circuits, components, and systems. Wiley, Chichester (2001)
40. Cheng, Q.S., Bandler, J.W., Koziel, S., Bakr, M.H., Ogurtsov, S.: The state of the art of microwave CAD: EM-based optimization and modeling. Int. J. RF and Microwave Computer-Aided Eng. 20, 475–491 (2010)
41. Burke, J.J., Jackson, R.W.: Surface-to-surface transition via electromagnetic coupling of microstrip and coplanar waveguide. IEEE Trans. Microwave Theory Tech. 37, 519–525 (1989)
42. Kolda, T.G., Lewis, R.M., Torczon, V.: Optimization by direct search: new perspectives on some classical and modern methods. In: SIAM Review, vol. 45(3), pp. 385–482 (2003)
43. Wu, K.: Substrate Integrated Circuits (SiCs) – A new paradigm for future Ghz and Thz electronic and photonic systems. IEEE Circuits and Systems Society Newsletter 3(2) (2009)
44. Collin, R.E.: Foundation for Microwave Engineering, 2nd edn. Wiley-IEEE Press, Chichester (2000)
45. Koziel, S., Cheng, Q.S., Bandler, J.W.: Space mapping. IEEE Microwave Magazine 9(6), 105–122 (2008)

46. Ogurtsov, S., Koziel, S., Rayas-Sánchez, J.E.: Design optimization of a broadband microstrip-to-SIW transition using surrogate modeling and adaptive design specifications. In: International Review of Progress in Applied Computational Electromagnetics, Tampere, Finland. ACES, pp. 878–883 (2010)
47. Koziel, S.: Adaptively adjusted design specifications for efficient optimization of microwave structures. Progress in Electromagnetic Research B (PIER B) 21, 219–234 (2010)

Chapter 4
Airfoil Shape Optimization Using Variable-Fidelity Modeling and Shape-Preserving Response Prediction

Slawomir Koziel and Leifur Leifsson

Abstract. Shape optimization of airfoils is of primary importance in the design of aircraft and turbomachinery with computational fluid dynamic (CFD) being the major design tool. However, as CFD simulation of the fluid flow past airfoils is computationally expensive, and numerical optimization often requires a large number of simulations with several design variables, direct optimization may not be practical. This chapter describes a computationally efficient and robust methodology for airfoil design. The presented approach replaces the direct optimization of an accurate but computationally expensive high-fidelity airfoil model by an iterative re-optimization of a corrected low-fidelity model. The shape-preserving response prediction technique is utilized to correct the low-fidelity model by aligning the pressure and skin friction distributions of the low-fidelity model with the corresponding distributions of the high-fidelity model. The algorithm requires one evaluation of the high-fidelity CFD model per design iteration. The algorithm is applied to several example case studies at both transonic and high-lift flow conditions.

4.1 Introduction

The use of optimization methods in the design process, as a design support tool or for design automation, has now become commonplace. In aircraft design, the development of numerical optimization techniques started in the mid 1970's when Hicks and Henne [1] used gradient-based optimization methods coupled with computational fluid dynamic (CFD) codes to design airfoils and wings at both

Slawomir Koziel · Leifur Leifsson
Engineering Optimization & Modeling Center,
School of Science and Engineering, Reykjavik University,
Menntavegur 1, Reykjavik, Iceland
email: koziel@ru.is, leifurth@ru.is

X.-S. Yang & S. Koziel (Eds.): Computational Optimization & Applications, SCI 359, pp. 99–124.
springerlink.com

subsonic and transonic conditions. Substantial progress in gradient-based methods for aerodynamic design has been made since then. Jameson [2] introduced control theory and continuous adjoint methods to the optimal aerodynamic design for two-dimensional airfoils and three-dimensional wings. Initially, using inviscid flow solvers [3, 4], and later using viscous flow solvers [5, 6].

The use of higher fidelity methods, coupled with optimization techniques, has led to improved design efficiency. However, simulation-driven aerodynamic design optimization involves numerous challenges. In particular, the high-fidelity CFD simulations are computationally expensive (e.g., three-dimensional simulations of turbulent flows can take many days on a parallel computer), the design optimization normally requires a large number of simulations, and a large number of design variables are often involved. Therefore, direct optimization of the high-fidelity CFD model may be impractical, especially when using traditional gradient-based techniques.

Computationally feasible design exploiting CFD simulations can be realized using surrogate-based optimization (SBO) techniques [7, 8]. One of the objectives of SBO is to reduce the number of evaluations of the high-fidelity models, and thereby making the optimization process more efficient. This is achieved by using computationally cheap surrogate functions in lieu of the CPU-intensive high-fidelity models. The surrogate models can be created either by approximating the sampled high-fidelity model data using regression (so-called function-approximation surrogates), or by correcting physics-based low-fidelity models which are less accurate but computationally cheap representations of the high-fidelity models.

A variety of techniques are available to create the function-approximation surrogate model, such as polynomial regression [7] and kriging [9]. Function-approximation models are versatile, however, they normally require substantial amount of data samples to ensure good accuracy. The physics-based surrogates are constructed by correcting the underlying low-fidelity models, which can be obtained through simplified physics models [10], coarse-discretization CFD simulation [11], or relaxed convergence criteria [12]. Popular correction methods include response correction [13] and space mapping [14].

The physics-based surrogate models are typically more expensive to evaluate than the function-approximation surrogates, but less high-fidelity model data is needed to obtain a given accuracy level. In many cases, SBO algorithms that utilize physics-based low-fidelity models—so-called variable- or multi-fidelity SBO—typically require only a single high-fidelity model evaluation per algorithm iteration. Due to this, the variable-fidelity SBO method is more scalable to larger numbers of design variables (assuming that no derivative information is required). A review of SBO methods popular in aerospace design can be found in [7] and [8].

In this chapter we describe a computationally efficient variable-fidelity airfoil shape optimization methodology [15, 16, 17], which employs physics-based low-fidelity surrogate models created by means of the shape-preserving response prediction (SPRP) technique [18]. Section 4.2 describes briefly the problem formulation for airfoil shape optimization. The optimization methodology is described in detail in Section 4.3. Application of the method to transonic and high-lift airfoil design is given in Sections 4.4 and 4.5, respectively. Section 4.6 summarizes the chapter.

4.2 Airfoil Shape Optimization

Out of a variety of design problems in aerodynamics, we focus here on airfoil shape optimization (ASO). Material concerning general aerodynamic optimization, CFD analysis, shape parameterization, and other relevant issues can be found in [26], Chapter 9.

An airfoil is a streamlined aerodynamic surface such as the one shown in Fig. 4.1. The function of the airfoil is to generate a lift force $\boldsymbol{l}$ at a range of operating conditions (Mach number M_∞, Reynolds number Re, angle of attack α). The drag force increases quadratically with increasing lift. Normally, the drag force is to be minimized for a given lift. These forces are non-dimensionalized by dividing them by $q_\infty S$, where $q_\infty \equiv (1/2)\rho_\infty V_\infty^2$ is the dynamic pressure, ρ_∞ is the air density, V_∞ is the free-stream velocity, and S is a reference surface. After non-dimensionalization they are called the lift coefficient, denoted by C_l, and the drag coefficient, denoted by C_d.

In direct ASO, the objective is to determine an airfoil shape that maximizes a performance criterion for a given set of constraints at a particular operating condition. Usually, the lift coefficient is maximized, the drag coefficient is minimized, or the lift-to-drag ratio is maximized. For example, if the lift coefficient is maximized, then a constraint is necessary on the maximum allowable drag coefficient. Further constraints are often included, e.g., to account for the wing structural components inside the airfoil one sets a constraint on the airfoil cross-sectional area. In inverse ASO, the airfoil shape is designed to attain a specific flow behavior which is defined a priori. Typically, a target airfoil surface pressure distribution is prescribed.

The optimization methodology described in this chapter is illustrated using the direct ASO approach. However, the method itself is more general and can also be applied to inverse airfoil design.

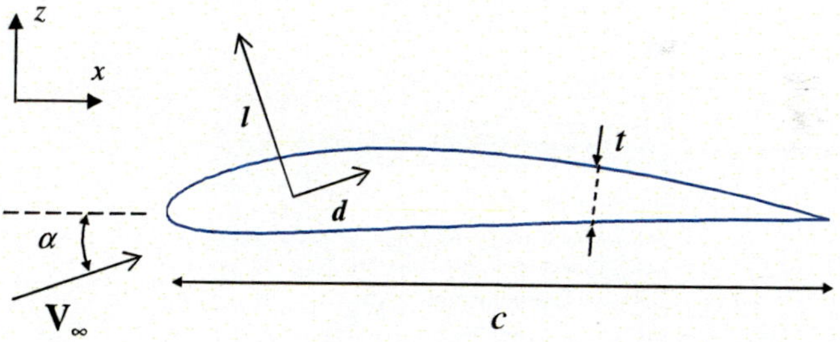

Fig. 4.1 A single-element airfoil section of chord length c and thickness t. $\mathbf{V}_\infty$ is the free-stream velocity at an angle of attack α relative to the x-axis. $\boldsymbol{l}$ is the lift force (perpendicular to $\mathbf{V}_\infty$) and $\boldsymbol{d}$ is the drag force (parallel to $\mathbf{V}_\infty$).

4.3 Optimization Methodology

In this section, we formulate the variable-fidelity airfoil optimization methodology that exploits the CFD-based low-fidelity models and the shape-preserving response prediction [18] methodology as the model correction tool. The NACA four-digit airfoil parameterization method is used due to its simplicity. The details of this parameterization, as well as the details of the CFD modeling methodology using the grid generator ICEM CFD [19] and the flow solver FLUENT [20] can be found in [26], Chapter 9.

4.3.1 General Description

The method follows the general principles of SBO [7, 8], as shown in Fig. 4.2, where the optimization burden is shifted to the low-cost surrogate model (referred to as s), whereas the high-fidelity model (referred to as f) is referenced occasionally for verification purposes and to obtain data necessary to update the surrogate. The surrogate is a corrected physics-based low-fidelity model (referred to as c).

The low-fidelity model is corrected to become a reliable representation of the high-fidelity model. Normally, the figures of interest in the optimization, i.e., the objectives and constraints, are aligned between the high-fidelity and low-fidelity models using a correction procedure, e.g., space mapping [14]. However, in the

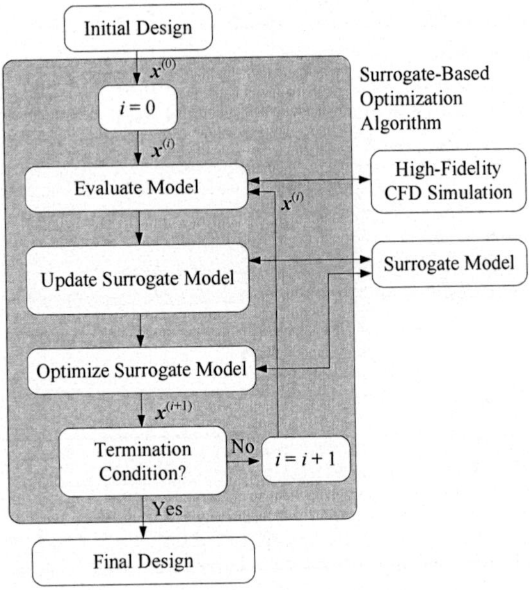

Fig. 4.2 A flowchart of the surrogate-based optimization algorithm

case of aerodynamic shape optimization, the figures of interest, such as the lift and drag coefficients, are scalars for a given operating condition and a given design vector **x**, which results in non-uniqueness of any alignment procedure that could be applied in order to match the low-fidelity model with the high-fidelity one, unless a sufficiently large amount of high-fidelity data is used in the model matching process.

Here, the model alignment is performed using intermediate simulation results, more specifically, the pressure and skin friction distributions, whose dimensionality can be made as large as necessary by selecting sufficient number of control points along the airfoil chord. As the objectives and constraints are uniquely determined by the pressure and skin friction distributions, alignment of the corresponding distributions for the low- and high-fidelity models will result in an (unique) alignment of the figures of interest. The SPRP methodology [18] is adopted here for the alignment procedure.

4.3.2 *Surrogate Modeling Using Shape-Preserving Response Prediction*

The SPRP model is formulated here using the pressure distribution. The formulation for the skin friction part is analogous. We denote the pressure distributions for the high- and low-fidelity models as $C_{p.f}$ and $C_{p.c}$, respectively. The surrogate model is constructed assuming that the change of $C_{p.f}$ due to the adjustment of the design variables **x** can be predicted using the actual changes of $C_{p.c}$. The change of $C_{p.c}$ is described by the translation vectors corresponding to certain (finite) number of its characteristic points on the pressure distribution. These translation vectors are subsequently used to predict the change of $C_{p.f}$, whereas the actual $C_{p.f}$ at the current design, $C_{p.f}(\mathbf{x}^{(i)})$, is treated as a reference.

Figure 4.3(a) shows the pressure distribution $C_{p.c}$ of the low-fidelity model at $\mathbf{x}^{(i)} = [0.02\ 0.4\ 0.12]^T$ (NACA 2412 airfoil) for $M_\infty = 0.7$ and $\alpha = 1$ deg, as well as $C_{p.c}$ at $\mathbf{x} = [0.025\ 0.56\ 0.122]^T$; $\mathbf{x}^{(i)}$ will denote a current design (at the ith iteration of the optimization algorithm; the initial design will be denoted as $\mathbf{x}^{(0)}$ accordingly). Circles denote characteristic points of $C_{p.c}(\mathbf{x}^{(i)})$, here, representing, among others, x/c equal to 0 and 1 (leading and trailing airfoil edges, respectively), the maxima of $C_{p.c}$ for the lower and upper airfoil surfaces, as well as the local minimum of $C_{p.c}$ for the upper surface. The last two points are useful to locate the pressure shock. Squares denote corresponding characteristic points for $C_{p.c}(\mathbf{x})$, while small line segments represent the translation vectors that determine the "shift" of the characteristic points of $C_{p.c}$ when changing the design variables from $\mathbf{x}^{(i)}$ to **x**.

In order to obtain a reliable prediction, the number of characteristic points has to be larger than illustrated in Fig. 4.3(a). Additional points are inserted in between initial points either uniformly with respect to x/c (for those parts of the pressure distribution that are almost flat) or based on the relative pressure value with respect to corresponding initial points (for those parts of the pressure distribution that are "steep"). Figure 4.3(b) shows the full set of characteristic points (initial points are distinguished using larger markers).

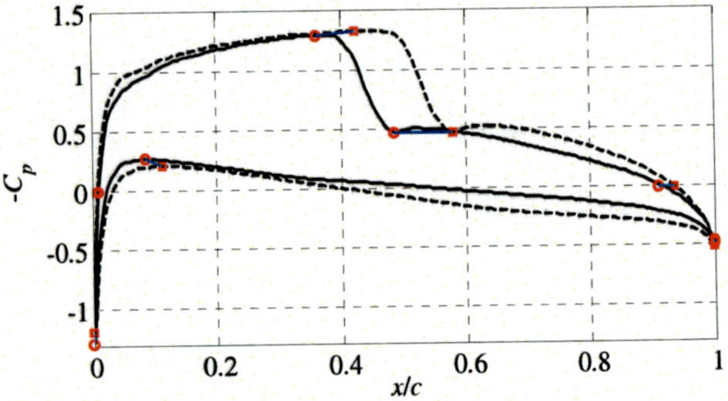

(a)

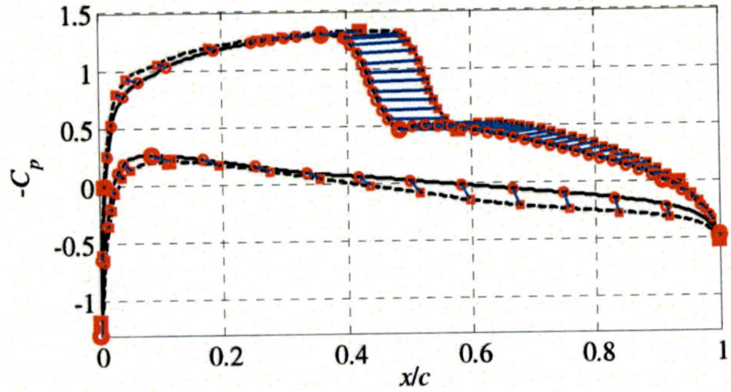

(b)

Fig. 4.3 (a) Example low-fidelity model pressure distribution at the design $\mathbf{x}^{(i)}$, $C_{p.c}(\mathbf{x}^{(i)})$ (solid line), the low-fidelity model pressure distribution at other design $\mathbf{x}$, $C_{p.c}(\mathbf{x})$ (dotted line), characteristic points of $C_{p.c}(\mathbf{x}^{(i)})$ (circles) and $C_{p.c}(\mathbf{x})$ (squares), and the translation vectors (short lines); (b) low-fidelity model pressure distributions, initial characteristic points (large markers) and translation vectors from Fig. 4.3(a) as well as additional points (small markers) inserted in between the initial points either uniformly with respect to x/c (for the "flat" parts of the pressure distribution) or based on the relative pressure value with respect to corresponding initial points (for the "steep" parts of the pressure distribution)

The pressure distribution of the high-fidelity model at the given design, here, $\mathbf{x}$, can be predicted using the translation vectors applied to the corresponding characteristic points of the pressure distribution of the high-fidelity model at $\mathbf{x}^{(i)}$, $C_{p.f}(\mathbf{x}^{(i)})$. This is illustrated in Fig. 4.4(a) where only initial characteristic points and translation vectors are shown for clarity. Figure 4.4(b) shows the predicted pressure distribution of the high-fidelity model at $\mathbf{x}$ as well as the actual $C_{p.f}(\mathbf{x})$. The agreement between both curves is very good.

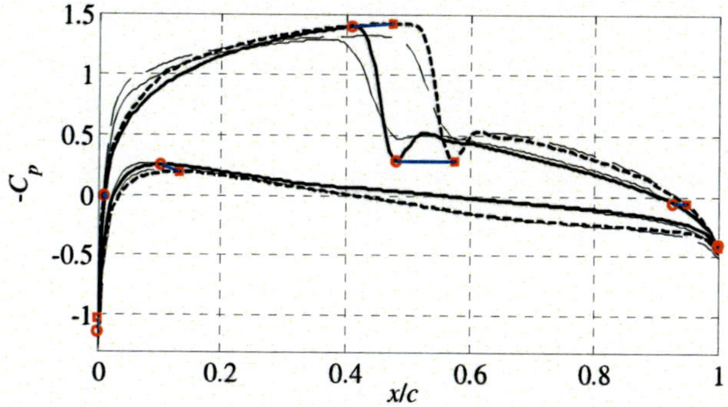

(a)

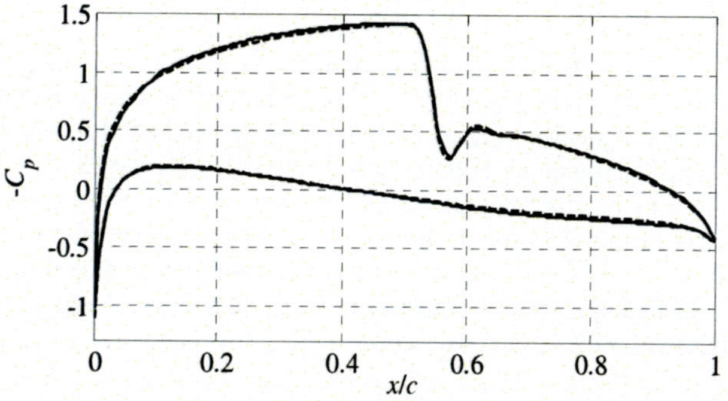

(b)

Fig. 4.4 (a) High-fidelity model pressure distribution at $\mathbf{x}^{(i)}$, $C_{p.f}(\mathbf{x}^{(i)})$ (solid line) and the predicted high-fidelity model C_p at $\mathbf{x}$ (dotted line) obtained using SPRP based on characteristic points of Fig. 4.3(b); characteristic points of $C_{p.f}(\mathbf{x}^{(i)})$ (circles) and the translation vectors (short lines) were used to find the characteristic points (squares) of the predicted high-fidelity model pressure distribution (only initial points are shown for clarity); low-fidelity model distributions $C_{p.c}(\mathbf{x}^{(i)})$ and $C_{p.c}(\mathbf{x})$ are plotted using thin solid and dotted line, respectively; (b) high-fidelity model pressure distribution at $\mathbf{x}$, $C_{p.f}(\mathbf{x})$ (solid line), and the predicted high-fidelity model pressure distribution at $\mathbf{x}$ obtained using SPRP (dotted line)

SPRP can be rigorously formulated as follows. Let $C_{p.f}(\mathbf{x}) = [c_{p.f}(\mathbf{x},y_1) \ \ldots \ c_{p.f}(\mathbf{x},y_m)]^T$ and $C_{p.c}(\mathbf{x}) = [c_{p.c}(\mathbf{x},y_1) \ \ldots \ c_{p.c}(\mathbf{x},y_m)]^T$, where y_j, $j = 1, \ldots, m$, are control points on the x/c axis (we assume that $y_{j+1} > y_j$ and $0 \le y_j \le 1$ for all j). To simplify the notation we assume that $C_{p.f}$ ($C_{p.c}$) is the pressure distribution for the upper surface only. Formulation for the lower surface is identical. Let $p_j^f = [y_j^f \ r_j^f]^T$, $p_j^{c0} = [y_j^{c0} \ r_j^{c0}]^T$, and $p_j^c = [y_{jc} \ r_j^c]^T$, $j = 1, \ldots, K$, denote the sets of characteristic points of $C_{p.f}(\mathbf{x}^{(i)})$, $C_{p.c}(\mathbf{x}^{(i)})$ and $C_{p.f}(\mathbf{x})$, respectively. Here, y and r denote the x/c and magnitude components of the respective point. The translation vectors of the low-fidelity model pressure distribution are defined as $T_i = [y_j^t \ r_j^t]^T$, $j = 1,\ldots, K$, where $y_j^t = y_j^c - y_j^{c0}$ and $r_j^t = r_j^c - r_j^{c0}$.

The SPRP surrogate model is defined as follows

$$C_{p.s}^{(i)} = \left[c_{p.s}^{(i)}(\mathbf{x}, y_1) \ldots c_{p.s}^{(i)}(\mathbf{x}, y_m)\right]^T \tag{4.1}$$

where

$$c_{p.s}^{(i)}(\mathbf{x}, y_j) = \overline{c}_{p.f}(\mathbf{x}^{(i)}, F(y_j, \{-y_k^t\}_{k=1}^K)) + r(y_j, \{r_k^t\}_{k=1}^K) \tag{4.2}$$

for $j = 1, \ldots, m$. $\overline{c}_{p.f}(\mathbf{x}, y)$ is an interpolation of $\{c_{p.f}(\mathbf{x},y_1), \ldots, c_{p.f}(\mathbf{x},y_m)\}$ onto the interval [0,1]. The scaling function F interpolates the data pairs $\{y_1,y_1\}$, $\{y_1^f,y_1^f-y_1^t\}$, …, $\{y_K^f,y_K^f-y_K^t\}$, $\{y_m,y_m\}$, onto the interval [0,1]. The function r does a similar interpolation for data pairs $\{y_1,r_1\}$, $\{y_1^f,r_1^f-r_1^t\}$, …, $\{y_K^f,r_K^f-r_K^t\}$, $\{y_m,r_m\}$; here $r_1 = c_{p.c}(\mathbf{x},y_1) - c_{p.c}(\mathbf{x}^r,y_1)$ and $r_m = c_{p.c}(\mathbf{x},y_m) - c_{p.c}(\mathbf{x}^r,y_m)$. Note that $C_{p.s}^{(i)}(\mathbf{x}^{(i)}) = C_{p.f}^{(i)}(\mathbf{x}^{(i)})$ as all translation vectors are zero at $\mathbf{x} = \mathbf{x}^{(i)}$.

The prediction method assumes that the high- and low-fidelity model pressure distributions have corresponding sets of characteristic points. This is usually the case for the practical ranges of design variables because the overall shape of the distributions is similar for both models. In case of a lack of correspondence, original definitions of characteristic points are replaced by their closest counterparts. The typical example would be non-existence of the local minimum of the pressure distribution for the upper surface for the high- and/or low-fidelity model at certain designs. In this case, the original point (local minimum) is replaced by the points characterized by the largest curvature.

4.3.3 Objective Function

Due to unavoidable misalignment between the pressure distributions of the high-fidelity model and its SPRP surrogate at the designs other than the one at which the model is determined, i.e., $\mathbf{x}^{(i)}$, it is not convenient to handle constraints (e.g., drag) directly, because the design that is feasible for the surrogate model, may not be feasible for the high-fidelity model. In particular, the design obtained as a result of optimizing the surrogate model $C_{p.s}^{(i)}$, i.e., $\mathbf{x}^{(i+1)}$, will be feasible for $C_{p.s}^{(i)}$. However, if $\mathbf{x}^{(i+1)}$ is not feasible for the high-fidelity model, it will not be feasible for $C_{p.s}^{(i+1)}$ because we have $C_{p.s}^{(i+1)}(\mathbf{x}^{(i+1)}) = C_{p.f}(\mathbf{x}^{(i+1)})$ by the definition of the surrogate model. In order to alleviate this problem, we shall use the penalty function approach to handle the constraints.

More specifically, if the figure of interest is the lift coefficient, while the drag and the airfoil cross-sectional area are constraints, the objective function is defined as

$$H(C_p(\mathbf{x})) = -C_{l.s}(C_p(\mathbf{x})) + \beta[\Delta C_{d.s}(C_p(\mathbf{x}))]^2 + \gamma[\Delta A(\mathbf{x})]^2 \tag{4.3}$$

where $\Delta C_{d.s} = 0$ if $C_{d.s} \le C_{d.s.max}$ and $\Delta C_{d.s} = C_{d.s} - C_{d.s.max}$ otherwise, and $\Delta A = 0$ if $A \ge A_{min}$ and $\Delta A = A - A_{min}$ otherwise. In the numerical experiments, presented in the next section, we use $\beta = \gamma = 1000$. Here the pressure distribution for the surrogate model is $C_p = C_{p.s}$, and for the high-fidelity model $C_p = C_{p.f}$. Also, $C_{l.s}$ and $C_{d.s}$ denote the lift and drag coefficients (both being functions of the pressure distribution).

4.3.4 Optimization with SPRP Surrogate

The efficiency of variable-fidelity optimization with SPRP model is illustrated by airfoil design at $M_\infty = 0.75$ and $\alpha = 0°$. The initial design is set as NACA 2412 and the objective function is defined by Eq. (4.3), with $C_{d.s.max}$ =0.0040 and A_{min} = 0.075. The side constraints on the design variables are $0 \le m \le 0.03$, $0.3 \le p \le 0.6$, and $0.09 \le t \le 0.13$. Constraint tolerance bands are set to 5%.

The high-fidelity model is based on the Euler equations and it is solved as described in [26], Chapter 9. The low-fidelity model is the transonic small-disturbance equation (TSDE) and it is solved using the computer code TSFOIL [21], which was developed at NASA in the 1970s. The code is capable of solving the TSDE for flow past lifting airfoils in both free air and various wind-tunnel environments by using a finite-difference method and an iterative successive line over-relaxation (SLOR) algorithm. The computational grid is a simple, fixed Cartesian grid.

Five iterations of the SPRP-based design methodology were executed. The computational cost is 5 high-fidelity and 161 surrogate model evaluations. The surrogate model optimization is performed using the pattern-search algorithm [22]. The results are given in Table 4.1. As the surrogate model evaluates quite fast (about 1 to 3 seconds depending on the design) and the high-fidelity model evaluation takes a few minutes, the total cost of evaluating the low-fidelity model in the whole optimization run corresponds to roughly 1-2 evaluations of the high-fidelity model. The equivalent number of high-fidelity model evaluations is less than 7 for this particular case.

Table 4.1 Numerical results of the design optimization. All the numerical values are from the high-fidelity model. N_c is number of low-fidelity model evaluations and N_f is the number high-fidelity model evaluations.

Variable	Initial	Direct[#]	VF-SPRP[$]
m	0.0200	0.0160	0.0173
p	0.4000	0.5999	0.5930
t	0.1200	0.1199	0.1163
C_l	0.4732	0.4770	0.5085
C_d	0.0100	0.0040	0.0041
A	0.0808	0.0808	0.0783
N_c	N/A	0	161
N_f	N/A	130	5
Total cost[*]	N/A	130	< 7

[#] Direct optimization of the high-fidelity model using the pattern-search algorithm [22].
[$] Design obtained using the methodology described here and the pattern-search algorithm [22].
[*] The total optimization cost is expressed in the equivalent number of high-fidelity model evaluations.

The optimizer achives this design by reducing maximum ordinate of the mean camber line (m) from 2% to 1.6%, and moving the location of the maximum camber (p) is rearward from 40% to 59% (which is close to the side constraint upper limit).

By reducing the camber, the flow velocity decreases on the upper surface and the shock strength is reduced. This can be seen in Fig. 4.5. By moving the maximum camber rearward, the aft camber increases and the pressure distribution opens up behind the shock, where flow is subsonic, and lift is increased by 3.5 lift counts (one lift count is $\Delta C_l = 0.01$). This can be seen in Fig. 4.6. The drag is reduced to satisfy the constraint limit. This is achived by reducing the thickness from 12% to 11.6%.

The case was also performed by direct optimization of the high-fidelity model using the pattern-search algortihm. Direct optimization obtained a very simlar optimal design, but required 130 high-fidelity model evaluations.

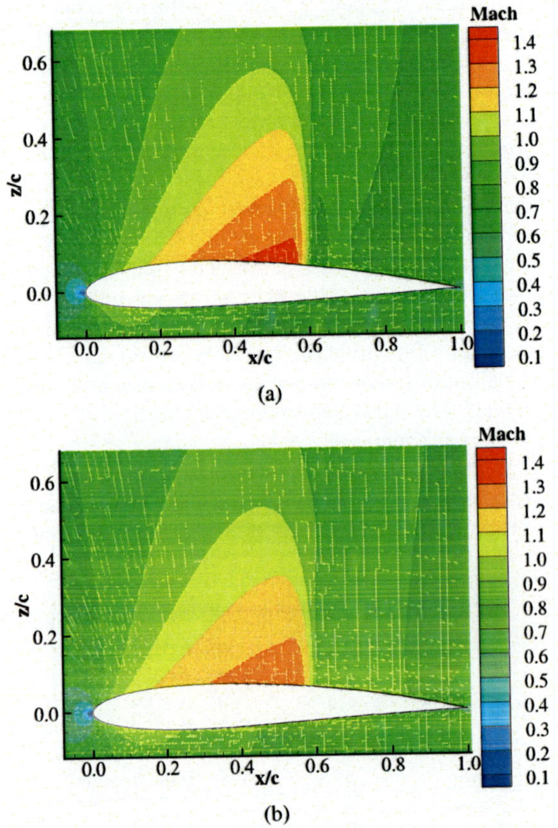

Fig. 4.5 (a) Mach number contours for the initial design (NACA 2412) at $M_\infty = 0.75$ and $\alpha = 0°$, (b) Mach number contours of the optimum design at the operating condition as in (a)

It should be emphasized that despite the encouraging results obtained for the considered test case, the use of the TSFOIL code may be problematic in general. In particular, TSFOIL would not converge to a valid flow solution for all designs in the solution domain. This may results in the convergence problems of the optimization algorithm. Furthermore, it turns out that the TSDE-based surrogate does not give a sufficiently reliable prediction of the drag coefficient. In particular, for small (local) changes of the design variables the low-fidelity model does not follow closely of the high-fidelity model.

The conclusions drawn from this study are:

- a proper low-fidelity model needs to be selected to provide a reliable prediction of the high-fidelity model, especially for small changes in the design variables,
- the low-fidelity model has to be reliable in terms of execution, and
- the optimization algorithm should be endowed with suitable convergence safeguards.

4.3.5 Optimization Algorithm

In this section, we formulate the optimization algorithm exploiting the SPRP-based surrogate model and a trust-region convergence safeguard [23]. This algorithm is used to solve the transonic and high-lift airfoil design cases, presented in Sections 4.4 and 4.5. The algorithm flow can be summarized as follows:

1. Set $i = 0$; Select λ (initial trust region radius); Evaluate $C_{p.f}(\mathbf{x}^{(0)})$;
2. Setup SPRP model;
3. Obtain $\mathbf{x}^{(i+1)} = \text{argmin}\{l \le \mathbf{x} \le u,\ \|\mathbf{x} - \mathbf{x}^{(i)}\| \le \lambda : H(C_{p.s}{}^{(i)}(\mathbf{x}))\}$;
4. Evaluate high-fidelity model to get $C_{p.f}(\mathbf{x}^{(i)})$;
5. If $H(C_{p.f}{}^{(i)}(\mathbf{x}^{(i+1)})) < H(C_{p.f}{}^{(i)}(\mathbf{x}^{(i)}))$ accept $\mathbf{x}^{(i+1)}$; Otherwise $\mathbf{x}^{(i+1)} = \mathbf{x}^{(i)}$;
6. Update λ;
7. Set $i = i + 1$;
8. If termination condition is not satisfied, go to 2.

The SPRP surrogate model is updated before each iteration of the optimization algorithm using the high-fidelity model data at the design obtained in the previous iteration. The trust-region parameter λ is updated after each iteration, i.e., decreased if the new design was rejected or the improvement of the high-fidelity model objective function was too small compared to the prediction given by the SPRP surrogate, or increased otherwise. Classical updating rules are used (see, e.g., [23, 24]). The algorithm is terminated if $\|\mathbf{x}^{(i+1)} - \mathbf{x}^{(i)}\| < 0.001$ or $\lambda < 0.001$.

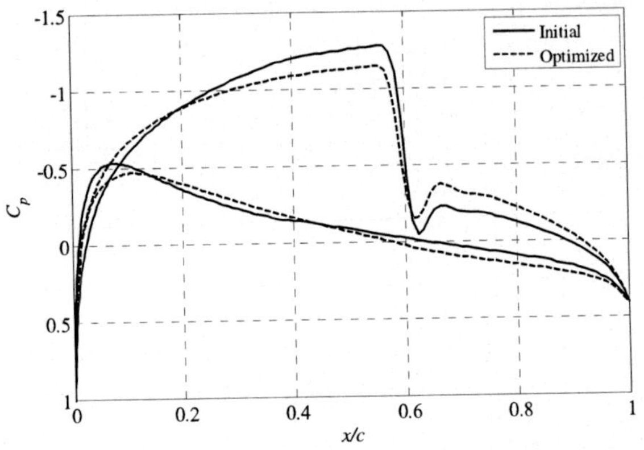

(a)

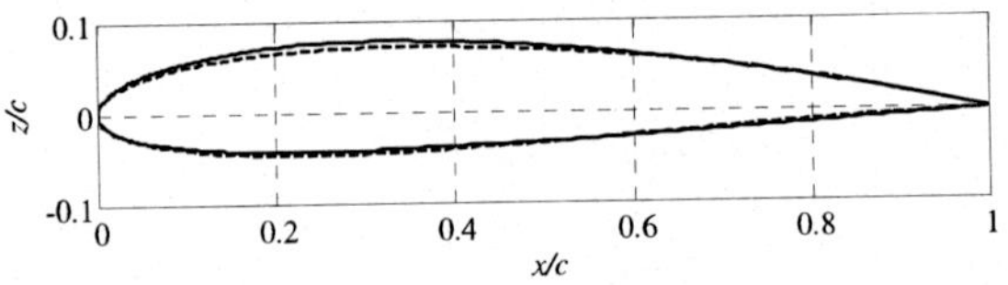

(b)

Fig. 4.6 (a) Pressure distributions of the initial (solid) and optimized (dashed) airfoil shapes, (b) initial (solid) and optimized (dashed) airfoil shapes

4.4 Transonic Airfoil Design

In this section, the variable-fidelity optimization algorithm is applied to airfoil design at a steady transonic flow condition. The surrogate model optimization is performed using the pattern-search algorithm [22]. The results of the design methodology are compared to the results obtained through direct optimization of the high-fidelity model.

4.4.1 Case Setup

The initial design is set as NACA 3210 and the the objective function is defined by Eq. (4.3), with $C_{d.s.max}$ =0.0041 and A_{min} = 0.065. The operating condition is M_∞ = 0.75 and α = 1°. The side constraints on the design variables are $0 \leq m \leq 0.1$, $0.2 \leq p \leq 0.8$, and $0.05 \leq t \leq 0.20$. Constraint tolerance bands are set to 5%.

4.4.2 Model Setup

In this case, variable-resolution modeling is employed. The high- and low-fidelity models solve the Euler equations, but the low-fidelity model uses a coarser computational mesh and relaxed convergence criteria. The particulars of the low-fidelity model mesh and convergence criteria are found by performing a parametric study on a typical airfoil section.

The NACA 2412 was selected for the parametric study. The Mach number is taken to be $M_\infty = 0.75$ and the angle of attack is set to $\alpha = 1$ deg. First a fine mesh is developed with a total of 320 points in the y-direction, 180 points on the airfoil surface and 160 points in the wake behind the airfoil, with a total of 106 thousand cells. Then, the flow is solved to full convergence to get the reference values. The convergence history is shown in Fig. 4.7(a). The solver needed 216 iterations to reach a converged solution based on the residuals. However, the lift and drag

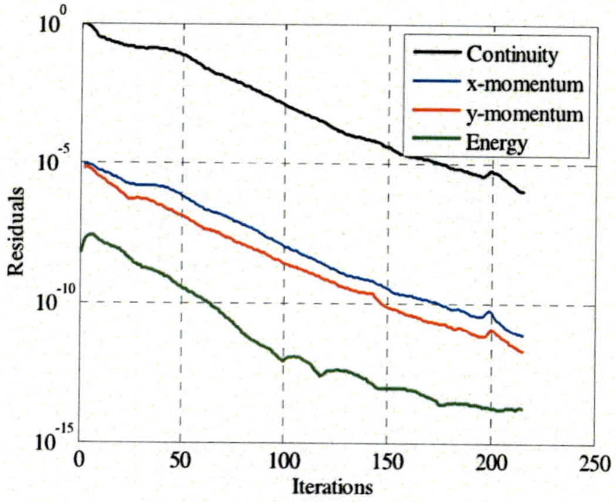

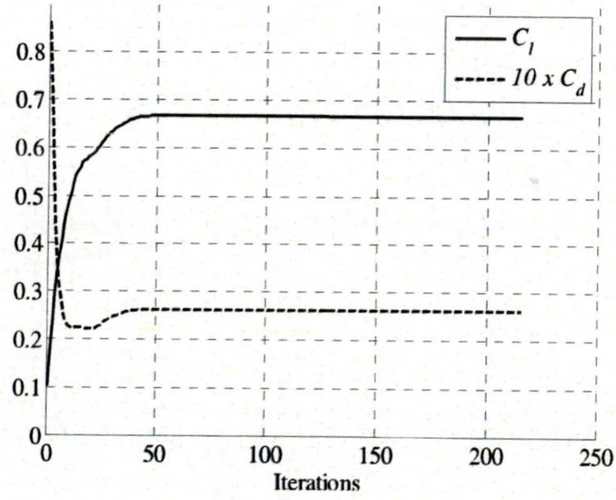

Fig. 4.7 (a) Convergence history of the simulation of the flow past the NACA 2412 at $M_\infty = 0.75$ and $\alpha = 1$ deg., (b) convergence of the lift and drag coefficients. The converged values of the lift coefficient is $C_l = 0.67$ and the drag coefficient is $C_d = 0.0261$.

coefficient values have reached a converged value after approximately 50 iterations, as can be seen in Fig. 4.7(b). Therefore, the number of iterations limit is set to 100 iterations in the subsequent steps.

Subsequently, the number of mesh points was reduced. This was done in two steps. First, the number of mesh points in the y-direction and the number of mesh points behind the airfoil were halved in each step. Then, the number of mesh points on the airfoil surface was reduced incrementally. In each step, the pressure distribution was plotted. This was done so the overall number mesh points could be reduced as much as possible, without reducing the mesh density on the airfoil surface, so that the shock could be resolved adequately.

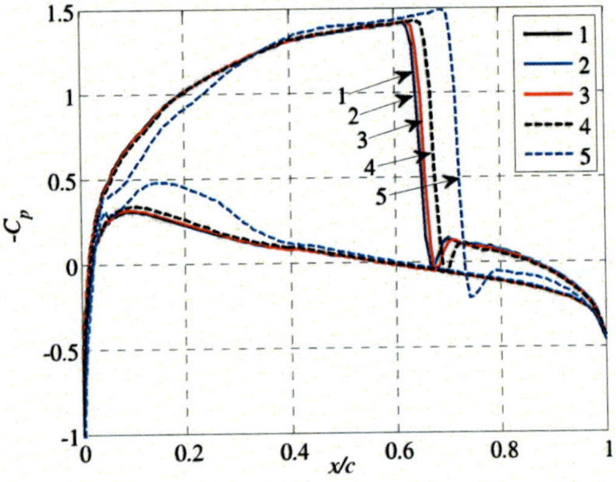

(a)

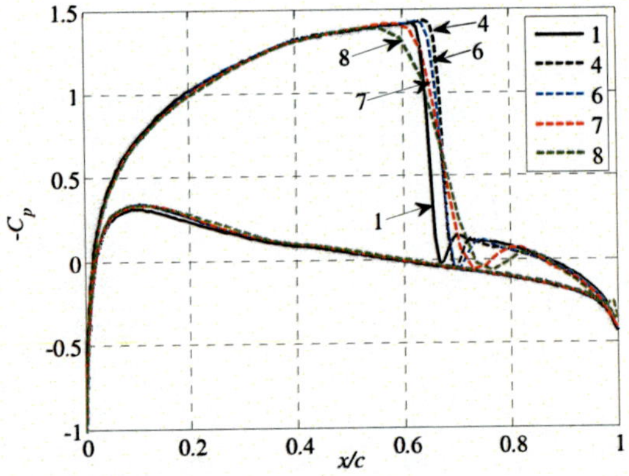

(b)

Fig. 4.8 (a) Pressure distributions for the first part of the parametric mesh study where the mesh points are reduced in the y-direction and in the wake behind the airfoil, (b) pressure distributions for the second part where the mesh points are reduced on the airfoil surface. The Mach number is $M_\infty = 0.75$ and the angle of attack is $\alpha = 1$ deg.

The results of the first mesh reduction are shown in Fig. 4.8(a). In the first four steps the number of cells is reduced from 106 thousand to 8295, but the pressure distribution does not change significantly, aside in the region of the shock, where the shock has strengthened and moved aft by less than 2.5% of the chord length.

This has led, however, to a significant increase in the estimation of the drag coefficient (+23.7%), as can be seen in Fig. 4.9(a), and a moderate increase in the lift coefficient (+2.7%). The evaluation time has been reduced from 470 s to 40 s (Fig. 4.9(b)). In the last step the number of mesh points in the y-direction is reduced to only 12 and the total number of cells is 3750. Now, there is a large change in the shock strength and location, but the pressure distribution is also

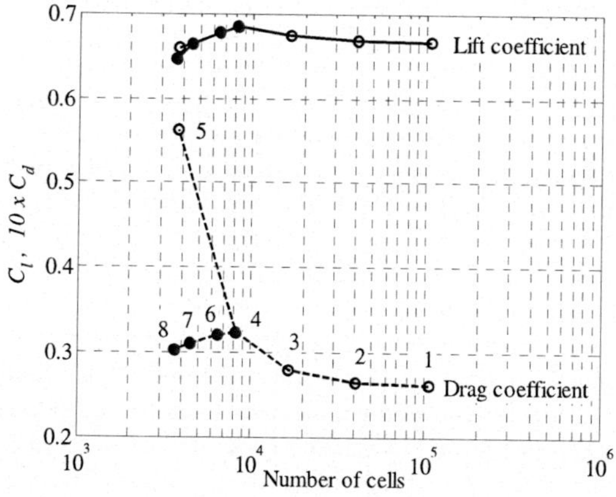

(a)

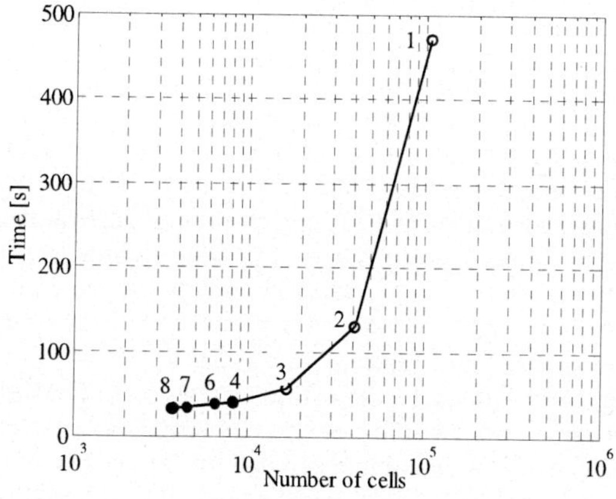

(b)

Fig. 4.9 (a) Lift and drag coefficients as a function of the number of cells for the parametric mesh study, (b) evaluation time of the CFD simulation model as a function of the number of cells. The Mach number is $M_\infty = 0.75$ and the angle of attack is $\alpha = 1$ deg.

altered in the front part the airfoil, leading to a large increase in the drag coefficient and a reduction in the lift coefficient.

The fourth mesh was selected for the second mesh reduction. The results are shown in Fig. 4.8(b). The number of mesh points on the airfoil surface was reduced by 50 in the first two steps (meshes 6 and 7) and then by 20 (mesh 8). It is clear, as the mesh gets coarser on the airfoil surface, the shock is smeared over a larger area and the estimated shock strength is reduced. As can be seen from Fig. 4.9 (a), both the drag and lift coefficients are reduced in this process. The overall evaluation time is reduced to about 34 s in the last step (Fig. 4.9(b)).

The second but last mesh (number 7) was selected as a basis to construct the low-fidelity model. The mesh has 48 points in the y-direction, 115 points on the airfoil surface, and 20 points in the wake behind the airfoil, with a total of 8295 thousand cells. The reason for selecting this particular mesh is that the difference in evaluation time is insignificant between the last two meshes (7 and 8), but the difference in the shock is quite substantial: it is easier to correct the low-fidelity model if the difference between it and the high-fidelity model is smaller.

For the airfoil considered in this parametric study, the overall evaluation time for the low-fidelity model using the above mentioned mesh, and an iteration limit of 100, is about 35 s, which is approximately 13.5 times faster than the high-fidelity model using the fine mesh and traditional convergence criteria. The criteria used in this work for the high-fidelity model is a maximum residual of 10^{-6}, or a maximum number of iterations of 1000. The overall evaluation time of the high-fidelity model in this parametric study is 471 s with a total of 216 iterations. In many cases the solver does not fully converge with respect to the residuals and goes on up to 1000 iterations. Then the overall evaluation time goes up to 2500 s, and the low-fidelity model is approximately 73 times faster. Note that the evaluation times reported here includes the time required for connecting to twice to the license server, once for the grid generator, ICEM CFD [19], and once for the flow solver, FLUENT [20].

4.4.3 Results and Discussion

The optimization method presented here was able to meet the design goals and yield the optimized design—within the given constraint bands—using 330 low-fidelity model evaluations and 11 high-fidelity model evaluations (Table 4.2). The equivalent number of high-fidelity model evaluations is less than 18 (using the ratio of the high-fidelity model evaluation time to the corrected low-fidelity model as 50). The direct method obtained a similar optimized design, but required 120 high-fidelity model evaluations.

To meet the design goals, the optimizer does three fundamental shape changes: (i) the maximum ordinate of the mean camber line (m) is reduced or kept constant, (ii) the location of the maximum ordinate of the mean camber line (p) is moved aft, thus increasing the trailing-edge camber, and (iii) the thickness (t) is reduced. Shape changes (i) and (iii) reduce the shock strength and, thus, reduce the drag coefficient. The associated change in the pressure distribution reduces the lift coefficient. However, shape change (ii) improves (or recovers a part of) the lift by opening up the pressure distribution behind the shock. These effects are clearly demonstrated in the pressure distribution plot in Fig. 4.10(a), the airfoil shape plots in Fig. 4.10(b), and the Mach contour plots in Figs. 4.11(a) and 4.11(b).

Table 4.2 Numerical results for lift maximization while keeping drag below a desired value at $M_\infty = 0.75$ and $\alpha = 1$ deg. All the numerical values are from the high-fidelity model. N_c and N_f are the numbers of low- and high-fidelity model evaluations, respectively

Variable	Initial	Direct[#]	VF-SPRP[$]
m	0.0300	0.0080	0.0090
p	0.2000	0.6859	0.6732
t	0.1000	0.1044	0.1010
C_l	0.8035	0.4641	0.4872
C_d	0.0410	0.0041	0.0040
A	0.0675	0.0703	0.0680
N_c	N/A	0	330
N_f	N/A	120	11
Total cost[*]	N/A	120	< 18

[#] Direct optimization of the high-fidelity model using the pattern-search algorithm [22].
[$] Design obtained using the algorithm described in Section 4.3; surrogate model optimization performed using the pattern-search algorithm [22].
[*] The total optimization cost is expressed in terms of the equivalent number of high-fidelity model evaluations. The ratio of the high-fidelity model evaluation time to the corrected low-fidelity model evaluation time varies between 13.5 to 73 depending on the design. We use a fixed value of 50 here.

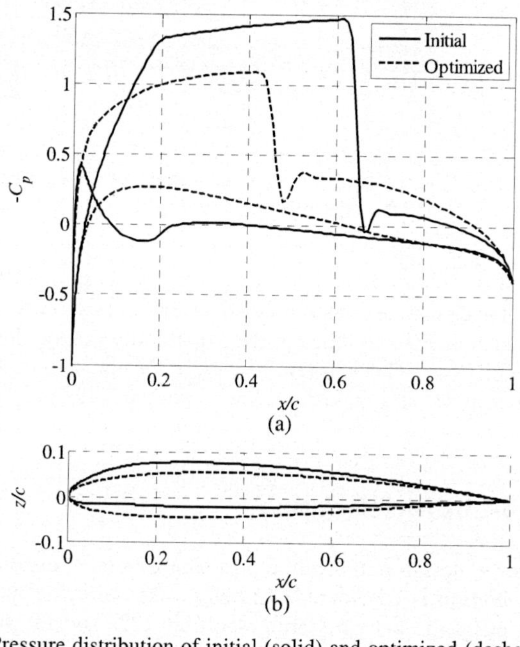

Fig. 4.10 (a) Pressure distribution of initial (solid) and optimized (dashed) airfoils, (b) initial (solid) and optimized (dashed) airfoil shapes. The Mach number is $M_\infty = 0.75$ and the angle of attack is $\alpha = 1$ deg.

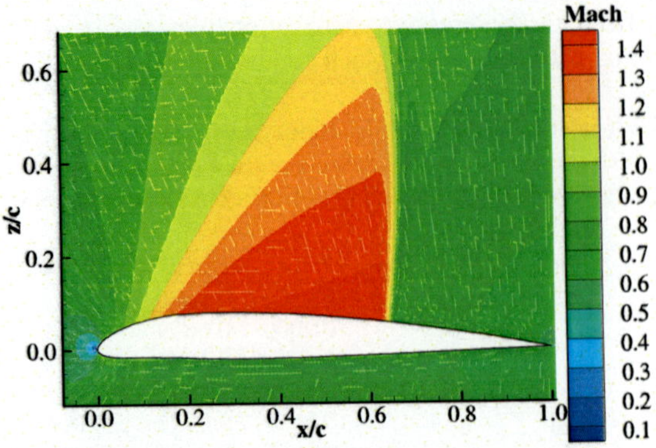

(a)

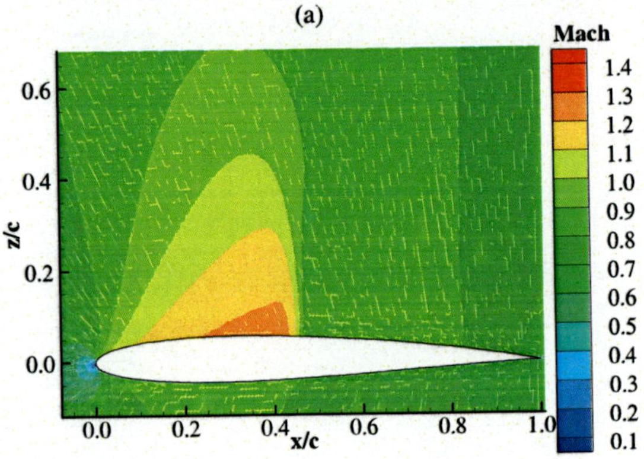

(b)

Fig. 4.11 Mach contour plots of (a) the initial airfoil, (b) the optimized airfoil. The Mach number is $M_\infty = 0.75$ and the angle of attack is $\alpha = 1$ deg.

The variable-resolution modeling exploited in this study exhibits consistent behavior, i.e., the changes of the pressure distribution (and, consequently, the figures of interest such as lift and drag) of the low-fidelity model closely follows that of the high-fidelity one. This was not the case in the example in Section 4.3.4, where variable-fidelity physics modeling was exploited with the Euler equations and TSDE.

4.5 High-Lift Airfoil Design

In this section, design optimization of a single-element airfoil at steady subsonic high-lift condition is considered. As before, the surrogate model optimization is performed using the pattern-search algorithm [22] and the results of the design methodology are compared to the results obtained through direct optimization of the high-fidelity model.

4.5.1 Case Setup

The Mach number is set to $M_\infty = 0.2$, the angle of attack is $\alpha = 12$ degree, and the Reynolds number is $Re = 2.3$ million. The initial design is set as NACA 0012 and the the objective function is defined by Eq. (3), but with the skin friction distribution C_f included, as viscous effects are important at this condition. The maximum allowable drag is set to $C_{d.s.max} = 0.0212$. The area constraint is not used here. The side constraints on the design variables are $0 \le m \le 0.08$, $0.3 \le p \le 0.6$, and $0.08 \le t \le 0.14$. Constraint tolerance bands are set to 5%.

4.5.2 Model Setup

The high-fidelity model f solves the RANS equations with the Spalart-Allmaras one equation turbulence model [25]. The details of the CFD model are as described in [26], Chapter 9. The low-fidelity model c is constructed as a low-order polynomial approximation of the high-fidelity model data, i.e., both the pressure distribution $C_{p.f}$ and the skin friction distribution $C_{f.f}$. The low-fidelity model is established in the entire design using evaluations of f at the following seven designs: $\mathbf{x}^0 = [0.04\ 0.45\ 0.11]^T$ (center point), and $\mathbf{x}^1 = [0.0\ 0.45\ 0.11]^T$, $\mathbf{x}^2 = [0.08\ 0.45\ 0.11]^T$, $\mathbf{x}^3 = [0.04\ 0.3\ 0.11]^T$, $\mathbf{x}^4 = [0.04\ 0.6\ 0.11]^T$, $\mathbf{x}^5 = [0.04\ 0.45\ 0.08]^T$, $\mathbf{x}^6 = [0.044\ 0.45\ 0.14]^T$ (single-variable perturbations for all design variables). The low-fidelity model is defined as a reduced quadratic model (no mixed terms)

$$c(\mathbf{x}) = \lambda_0 + \lambda_1 m + \lambda_2 p + \lambda_3 t + \lambda_4 m^2 + \lambda_5 p^2 + \lambda_6 t^2 \tag{4.4}$$

where the coefficients λ are found by solving the linear system $c(\mathbf{x}^j) = f(\mathbf{x}^j)$, $j = 0, 1, \ldots, 6$.

The reason for choosing the approximation-based model c is that the pressure distribution does not change significantly for the design space considered in this case. The simple model (4.4) is a reasonable compromise between the accuracy and the computational cost of creating the response surface. Still, the low-fidelity model has to be corrected in order to become a reliable representation of the high-fidelity one in the optimization process.

Figure 4.12 shows the construction of the SPRP model for the high-lift airfoil design. When compared to the transonic case, the pressure distribution is simpler (no pressure shock). However, the figures of interest (particularly drag) are very sensitive to the changes of the distribution, so that much attention has to be put to detailed "description" of the distribution through SPRP characteristic points, particularly for x/c close to zero, where the pressure gradients with respect to x/c are large. The pressure distributions of the low-fidelity model are illustrated in Fig. 4.12, at $\mathbf{x}^{(i)} = [0.01\ 0.40\ 0.09]^T$ for $M_\infty = 0.2$ and $\alpha = 10°$, as well as $C_{p.c}$ at $\mathbf{x} = [0.02\ 0.35\ 0.10]^T$. The pressure distribution of the high-fidelity model at the given design, here, $\boldsymbol{x}$, is predicted using the translation vectors applied to the corresponding characteristic points of the pressure distribution of the high-fidelity model at $\mathbf{x}^{(i)}$, $C_{p.f}(\boldsymbol{x}^{(i)})$. This is illustrated in Fig. 4.13. The predicted pressure distribution (magnified parts only) of the high-fidelity model at $\boldsymbol{x}$ as well as the actual $C_{p.f}(\mathbf{x})$ is shown in Fig. 4.14.

4.5.3 Results and Discussion

The optimization method of Section 4.3 with the low-fidelity model (4.4) improves the lift coefficient from 1.235 to 1.491 (+25.6 lift counts) by increasing camber by 2.34% and moving the location of maximum camber more aft, from 0.45 to 0.60, which is the upper bound (Table 4.3). The thickness is increased from 12% to 14% (which is the upper bound). A comparison of the initial and optimized airfoil shapes is given in Fig. 4.15(c). The optimized design is achieved by using only 11 high-fidelity CFD evaluations. The direct optimization method required 65 high-fidelity CFD evaluations and improves the lift only by 11.3%.

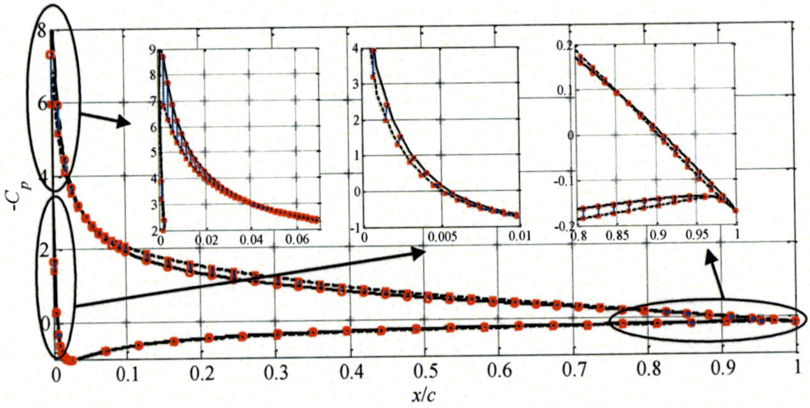

Fig. 4.12 Example low-fidelity model pressure distribution at the design $\mathbf{x}^{(i)}$, $C_{p.c}(\mathbf{x}^{(i)})$ (solid line), the low-fidelity model pressure distribution at other design $\mathbf{x}$, $C_{p.c}(\mathbf{x})$ (dotted line), characteristic points of $C_{p.c}(\mathbf{x}^{(i)})$ (circles) and $C_{p.c}(\mathbf{x})$ (squares), and the translation vectors (short lines). Only selected points and vectors are shown for the sake of clarity of the picture. Selected parts of the distributions are magnified.

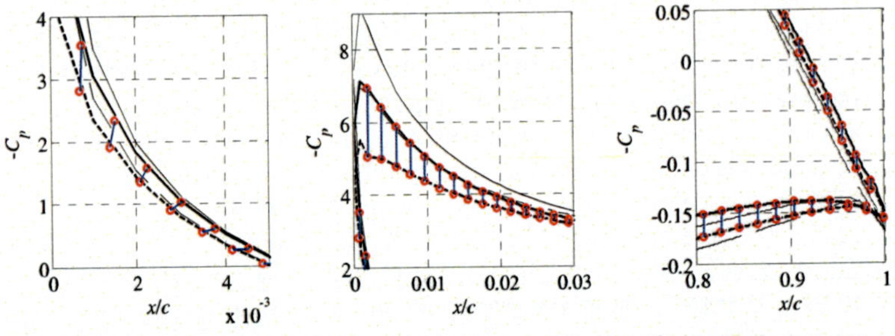

Fig. 4.13 High-fidelity model pressure distribution at $\mathbf{x}^{(i)}$, $C_{p.f}(\mathbf{x}^{(i)})$ (solid line) and the predicted high-fidelity model C_p at $\mathbf{x}$ (dotted line) obtained using SPRP based on characteristic points of Fig. 4.12; characteristic points of $C_{p.f}(\mathbf{x}^{(i)})$ (circles) and the translation vectors (short lines) were used to find the characteristic points (squares) of the predicted high-fidelity model pressure distribution

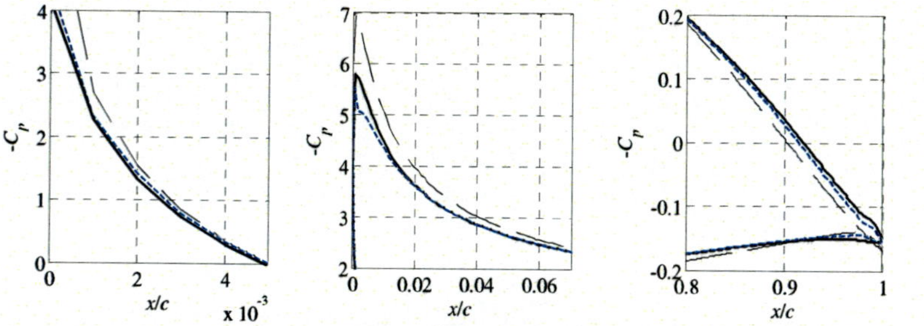

Fig. 4.14 High-fidelity model pressure distribution at **x**, $C_{p.f}(\mathbf{x})$ (solid line), and the predicted high-fidelity model pressure distribution at **x** obtained using SPRP (thick dotted line). Low-fidelity model pressure distribution at **x** is shown using a thin dashed line. SPRP model ensures better accuracy than the low-fidelity model.

Table 4.3 Numerical results for lift maximization while keeping drag below a desired value at $M_\infty = 0.2$, $\alpha = 12$ deg, and $Re = 2.3$ million

Variable	Initial	Direct[#]	VF-SPRP[$]
m	0.0000	0.0150	0.0234
p	0.4500	0.4840	0.6000
t	0.1200	0.1247	0.1400
C_l	1.235	1.392	1.491
C_d	0.0212	0.0212	0.0210
Design cost	N/A	65	11

[#] Direct optimization of the high-fidelity CFD model using the pattern-search algorithm [22].
[$] Design obtained using the algorithm described in Section 4.3; surrogate model optimization performed using the pattern-search algorithm [22].

There are three major changes in the optimized design when compared to the initial one. First of all, the increased camber opens up the pressure distribution over the whole airfoil, as can be seen in Fig. 4.15(a), and thus the lift increases.

Also, the aft camber opens the pressure distribution up near the trailing-edge, also increasing lift. Finally, the increased thickness reduces the pressure peak near the leading-edge, thus creating a milder expansion around the leading-edge, and thereby reducing pressure drag. The result is an optimized airfoil with improved lift coefficient at the same drag coefficient. A comparison of the lift and drag curves is given in Fig. 4.16. Although the airfoil was optimized at $\alpha = 12°$, the entire lift curve is shifted upwards. In this case, the angle of attack at maximum lift increases slightly (approximately by 1°).

Figure 4.17 shows the optimization history. In particular, one can observe a convergence plot (Fig. 4.17(a)), as well as the evolution of the objective function (Fig. 4.17(b)), the lift coefficient (Fig. 4.17(c)) and the drag coefficient (Fig. 4.17(d)). It follows that the algorithm exhibits a good convergence pattern and that the mechanisms introduced in the algorithm (in particular the trust region approach and the penalty function) enforce the drag limitation to be satisfied while increasing the lift coefficient as much as possible.

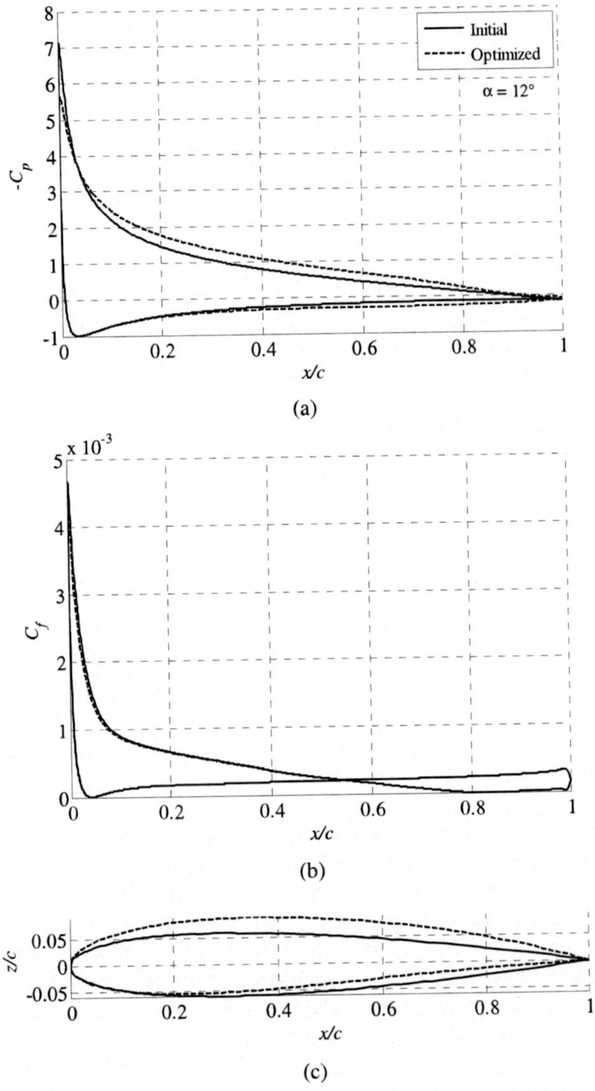

Fig. 4.15 Comparison of initial and optimized designs at $M_\infty = 0.2$, $\alpha = 12$ deg, and $Re = 2.3$ million; (a) Pressure distributions of the initial and optimized designs, (b) skin friction distributions of the initial and optimized designs, (c) initial and optimized airfoil shapes

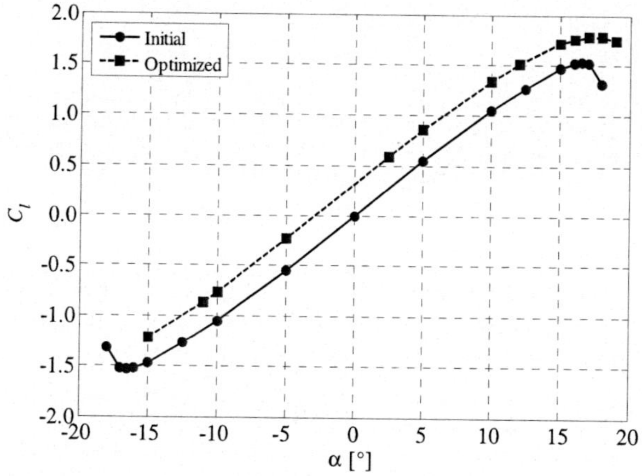

(a)

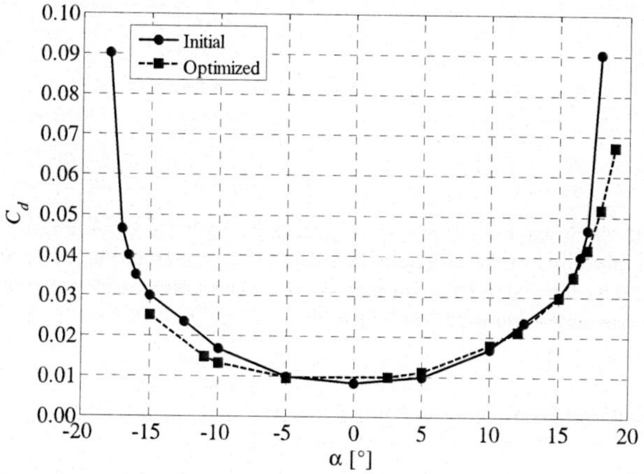

(b)

Fig. 4.16 A comparison of the lift and drag curves of the initial and optimized designs at $M_\infty = 0.2$ and $Re = 2.3$ million

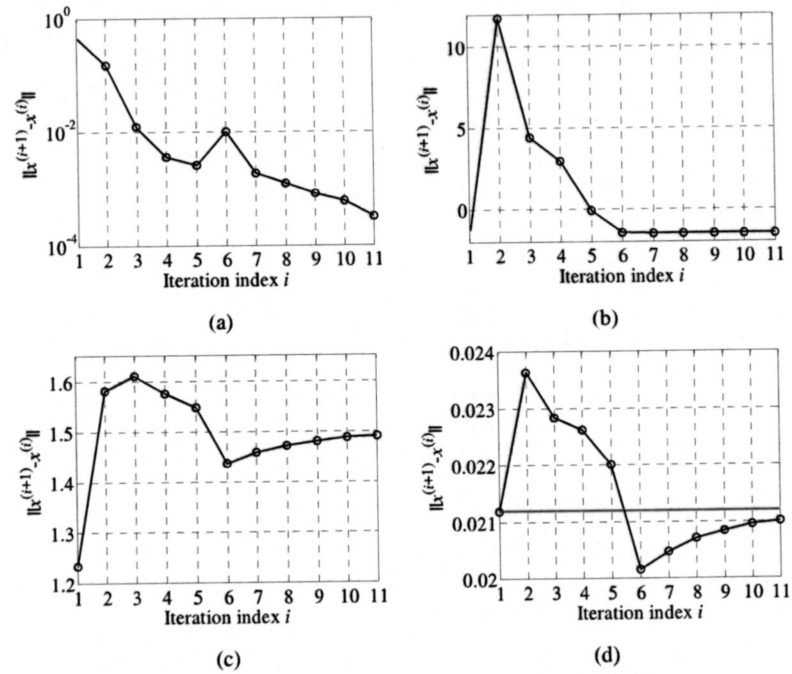

Fig. 4.17 Optimization history; (a) convergence plot; (b) evolution of the objective function; (c) evolution of the lift coefficient; and (d) evolution of the drag coefficient (drag constraint marked using a solid horizontal line). The graphs show all high-fidelity function evaluations performed in the optimization.

4.6 Summary

A variable-fidelity airfoil design optimization algorithm has been presented. The algorithm uses a computationally cheap low-fidelity model to construct a surrogate of an accurate but CPU-intensive high-fidelity model. The low-fidelity model is corrected by aligning the airfoil surface pressure distribution with the corresponding distribution of the high-fidelity model by means of the shape-preserving response correction prediction technique. This ensures a good generalization capability of the surrogate model with respect to both objectives and constraints. The robustness of the algorithm is enhanced by embedding it in the trust region framework. Applications for transonic and high-lift airfoil design are demonstrated with the optimized designs obtained at the computational cost corresponding to of a few high-fidelity model evaluations.

References

1. Hicks, R.M., Henne, P.: A Wing design by numerical optimization. Journal of Aircraft 15(7), 407–412 (1978)
2. Jameson, A.: Aerodynamic design via control theory. Journal of Scientific Computing 3, 233–260 (1988)
3. Reuther, J., Jameson, A.: Control theory based airfoil design for potential flow and a finitie volume discretization. AIAA Paper 94-0499, AIAA 32nd Aerospace Sciences Meeting and Exhibit, Reno, Nevada (1994)
4. Jameson, A., Reuther, J.: Control theory based airfoil design using euler equations. In: Proceedings of AIAA/USAF/NASA/ISSMO Symposium on Multidisciplinary Analsysis and Optimization, Panama City Beach, pp. 206–222 (1994)
5. Kim, S., Alonso, J.J., Jameson, A.: Design optimization of high-lift configurations using a viscous continuous adjoint method. AIAA 40th Aerospace Sciences Meeting & Exhibit, AIAA paper 2002–0844 (2002)
6. Leoviriyakit, K., Kim, S., Jameson, A.: Viscous aerodynamic shape opptimization of wings including planform variables. In: 21st Applied Aerodynamics Conference, Orlando, Florida (2003)
7. Queipo, N.V., Haftka, R.T., Shyy, W., Goel, T., Vaidyanathan, R., Tucker, P.K.: Surrogate-based analysis and optimization. Progress in Aerospace Sciences 41(1), 1–28 (2005)
8. Forrester, A.I.J., Keane, A.J.: Recent advances in surrogate-based optimization. Progress in Aerospace Sciences 45(1-3), 50–79 (2009)
9. Simpson, T.W., Peplinski, J., Koch, P.N., Allen, J.K.: Metamodels for computer-based engineering design: survey and recommendations. Engineering with Computers 17(2), 129–150 (2001)
10. Alexandrov, N.M., Nielsen, E.J., Lewis, R.M., Anderson, W.K.: First-order model management with variable-fidelity physics applied to multi-element airfoil optimization. In: 8th AIAA/USAF/NASA/ISSMO Symposium on Multidisciplinary Design and Optimization, Long Beach, CA, AIAA Paper 2000–4886 (2000)
11. Alexandrov, N.M., Lewis, R.M., Gumbert, C.R., Green, L.L., Newman, P.A.: Optimization with variable-fidelity models applied to wing design. In: 38th Aerospace Sciences Meeting & Exhibit, Reno, NV, AIAA Paper 2000–2841 (2000)
12. Forrester, A.I.J., Bressloff, N.W., Keane, A.J.: Optimization using surrogate models and partially converged computationally fluid dynamics simulations. Proceedings of the Royal Society A: Mathematical, Physical and Engineering Sciences 462(2071), 2177–2204 (2006)
13. Alexandrov, N.M., Lewis, R.M.: An overview of first-order model management for engineering optimization. Optimization and Engineering 2(4), 413–430 (2001)
14. Bandler, J.W., Cheng, Q.S., Dakroury, S.A., Mohamed, A.S., Bakr, M.H., Madsen, K., Søndergaard, J.: Space mapping: the state of the art. IEEE Trans. Microwave Theory Tech. 52(1), 337–361 (2004)
15. Koziel, S., Leifsson, L.: Multi-fidelity high-lift aerodynamic optimization of single-element airfoils. In: Int. Conf. Engineering Optimization, Lisbon (2010)
16. Leifsson, L., Koziel, S.: Multi-fidelity design optimization of transonic airfoils using physics-based surrogate modeling and shape-preserving response prediction. Journal of Computational Science 1(2), 98–106 (2010)

17. Koziel, S., Leifsson, L.: Multi-fidelity high-lift aerodynamic optimization of single-element airfoils. Int. Conf. Engineering Optimization, Lisbon (2010)
18. Koziel, S.: Shape-preserving response prediction for microwave design optimization. IEEE Trans. Microwave Theory and Tech. 58(11), 2829–2837 (2010)
19. ICEM CFD, ver. 12.1, ANSYS Inc., Southpointe, 275 Technology Drive, Canonsburg, PA 15317 (2006)
20. FLUENT.ver. 6.3.26, ANSYS Inc., Southpointe, 275 Technology Drive, Canonsburg, PA 15317 (2006)
21. Murman, E.M., Bailey, F.R., Johnson, M.L.: TSFOIL – A computer code for the two-dimensional transonic calculations, including wind-tunnel wall effects and wave-drag evaluation. In: NASA SP-347, Part II, pp. 769–788 (1975)
22. Koziel, S.: Multi-fidelity multi-grid design optimization of planar microwave structures with Sonnet. In: International Review of Progress in Applied Computational Electromagnetics, Tampere, Finland, pp. 719–724 (2010)
23. Conn, A.R., Gould, N.I.M., Toint, P.L.: Trust Region Methods. MPS-SIAM Series on Optimization (2000)
24. Koziel, S., Bandler, J.W., Madsen, K.: Space-mapping based interpolation for engineering optimization. IEEE Trans. Microwave Theory and Tech. 54(6), 2410–2421 (2006)
25. Spalart, P.R., Allmaras, S.R.: A one-equation turbulence model for aerodynamic flows. In: 30th AIAA Aerospace Sciences Meeting and Exhibit, Reno, Nevada (1992)
26. Koziel, S., Yang, X.-S. (eds.): Computational optimization, methods and algorithms, 1st edn. Springer, Heidelberg (2010)

Chapter 5
Evolutionary Optimisation Techniques to Estimate Input Parameters in Environmental Emergency Modelling

Kerstin Wendt, Mónica Denham, Ana Cortés, and Tomàs Margalef

Abstract. Parameter estimation in environmental modelling is essential for input parameters, which are difficult or impossible to measure. Especially in simulations for disaster propagation prediction, where hard real-time constraints have to be met to avoid tragedy, the additionally introduced computational burden of advanced global optimisation algorithms still hampers their use in many cases and poses an ongoing challenge. In this chapter we demonstrate how modifications of a Genetic Algorithm (GA) are able to decrease time-consuming fitness evaluations and hence to speed up parameter calibration. Knowledge from past observed catastrophe behaviour is used to guide the GA during various phases towards promising solution areas resulting in a fast convergence. Together with parallel computing techniques it becomes a viable estimation approach in environmental emergency modelling. Encouraging results were obtained in predicting forest fire spread.

Keywords: input parameter estimation, environmental modelling, knowledge-guided Genetic Algorithm, forest fire spread prediction.

5.1 Introduction

There exist many environmental models for simulating, explaining, and predicting the behaviour of complex natural phenomena. These comprise models which

Kerstin Wendt · Ana Cortés · Tomàs Margalef
Departament d'Arquitectura de Computadors i Sistemes Operatius,
Escola d'Enginyeria, Universitat Autònoma de Barcelona,
08193 Bellaterra (Barcelona), Spain
email: `kerstin.wendt@caos.uab.es`,
email: `{ana.cortes,tomas.margalef}@uab.es`

Mónica Denham
Facultad de Ingeniería, Universidad Nacional de Río Negro,
Sede Andina, San Carlos de Bariloche, Argentina

X.-S. Yang & S. Koziel (Eds.): Computational Optimization & Applications, SCI 359, pp. 125–143.
springerlink.com

simulate standard events and processes in meteorology, oceanography, groundwater hydrology, and petroleum reservoirs, as well as more specific models which attempt to simulate environmental emergencies such as floods, hurricanes, oil spills, forest fires, or volcanic eruptions. In either case, the precision of simulation output heavily depends on the quality of entered input parameter values. One cannot expect correct results if the entries fed into the simulator were erroneous. Special applications such as propagation predictions of natural catastrophes require the most reliable simulation outcomes to prevent tragedy. Furthermore, these computationally intensive applications have to fulfil stringent real-time constraints to be of use during an ongoing disaster.

The need for input parameter estimation and calibration to improve model output is a long-known and often-tackled problem, particularly in environments where correct and timely input parameters cannot be provided [2, 24, 40]. Therefore, computational parameter estimation and optimisation strategies are required to minimise the deviation between the predicted scenario and the real phenomenon behaviour. Since input parameter calibration adds a significant computational effort to the simulation process, a fast and efficient approach is the more important for time-critical applications.

Many approaches for parameter calibration mainly use standard numerical optimisation techniques, e.g. Kalman filter [22], principal differential analysis [34], which are not fully capable of handling high dimensionality, nonlinearity, and irregularities contained in environmental models [41]. Bayesian calibration methods, including Monte Carlo sampling, as well as stochastic practices, e.g. Simulated Annealing and Genetic Algorithms, are numerically very intensive, but generally deliver good results and tend to find global optima. With the continuous increases in computing power, these calibration methods, especially Genetic Algorithms (GA), have become practicable to solve the parameter problem of environmental models. High performance, parallel, and distributed computing now enable the generation of tractable solutions [7, 17] to expensive and time-consuming optimisation problems.

In this chapter, we summarise how a hybrid GA approach introduces problem-specific knowledge into different phases of the GA and is able to boost its performance. In doing so, online parameter estimation in time-critical applications can be provided. The real case of forest fire spread prediction is chosen to demonstrate how the employment of past observed or simulated disaster behaviour stored in a knowledge base speeds up parameter calibration for environmental emergency models.

The remainder of this work is organised as follows. The next section gives an overview of modelling environmental emergencies and explains details about the prediction of forest fire spread. Afterwards, the input parameter estimation problem is characterised. In section 5.3, we describe the implementation of a parallelised GA for parameter estimation. The benefits of applying domain-specific knowledge, including knowledge representation, retrieval and insertion, are outlined. Experimental results are shown in section 5.4 and section 5.5 comprises main conclusions and briefly discusses future work.

5.2 The Input Parameter Problem in Environmental Emergency Modelling

Environmental emergencies include natural events and also human-induced accidents, that are able to cause severe environmental damage, e.g. loss of ecological resources, air pollution, erosion, water contamination, climate changes, extinction of species, as well as destruction of buildings and infrastructure and, maybe most importantly, loss of human lives. Where environmental emergencies cannot be prevented, their prediction is crucially important. To avoid tragedy, most environmental emergency management systems include effective tools to forecast the propagation of an ongoing event. Based on these forecasts combined with expert user experience, disaster warnings are issued and catastrophe fighting actions are decided. There is no doubt about the importance of quick and most reliable forecasts to minimise the number of victims, the amount of damage caused, and the employed resources for disaster combatting. For prediction purposes mainly computer modelling and simulation applications are used to forecast the state of the event for a given time in the near future.

Basic disaster propagation models for the different emergencies are available since the early seventies, e.g. CLIPER for hurricane track prediction [31], Rothermel's model for forest fire spread behaviour [36], and are subject to continuous enhancements, e.g. NAME III for volcanic ash dispersion [21]. Many environmental models and simulators originate from research activities and are often bound to a specific geographic region or particular vegetational characteristics, e.g. PROMETHEUS, a wildfire growth simulator designed to work in Canadian boreal forest [38], or HFire, a rasterbased model for fire behaviour through Southern California chaparral [30]. To become established in the scientific community, a model should preferably be generally applicable. If the model itself and, moreover, the simulator which implements the model, furthermore fulfil certain end user requirements, e.g. graphical user interface, GIS integration, user support and training, modelling of special disaster occurrences as crown fires or underground peat fires, it is most likely to be applied in daily use in disaster management centres, e.g. FARSITE [13].

5.2.1 Forest Fire Spread Prediction

There exists a large variety of wildfire behaviour and spread simulators (e.g. FARSITE [13], BehavePlus, fireLib [5], NEXUS) and most of them are based on the Rothermel equation model [36]. The two main approaches to propagate fire spread are based on a regular grid system (cellular automata) and on continuous planes (elliptical wave propagation). Fire simulators can be used as stand-alone applications for risk analysis, disaster evolution prediction and fire fighter training. Furthermore, they form a fundamental part of complex decision support systems (DSS), which are typically applied to monitor environmental emergencies such as forest fires. The simulators traditionally work with a set of input parameters describing the environmental conditions of the region where the fire takes place including vegetational, climatological and topographical characteristics. Simulators differ in required model parameters and input and output data formats. The classic way of predicting forest fire behaviour takes the initial state of the fire front (RF = real fire) as input as well as the input parameters given for time t_x. The simulator then returns the prediction (SF = simulated fire) for the state of fire front at a later time t_{x+1} as shown in figure 5.1.

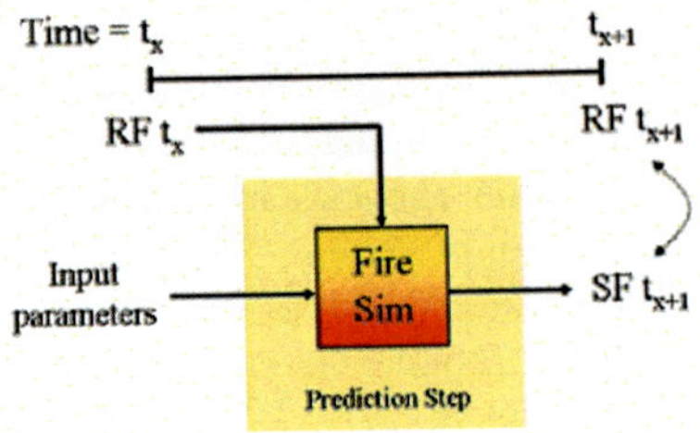

Fig. 5.1 Classical forest fire prediction scheme

Comparing the simulation result SF from time t_{x+1} with the advanced real fire RF at the same instant, the forecasted fire front tends to differ to a greater or lesser extent from the real fire line because the calculation of the simulated fire is based upon a single set of input parameters afflicted with certain insufficiencies. These are explained in section 5.2.2. To enable real-time calibration of model input parameters in each time step during an ongoing prediction, a simulator independent two-stage data-driven prediction scheme was proposed by Abdalhaq et al. in [1]. Introducing a previous calibration step as shown in figure 5.2, the set of input parameters is refined before every prediction step. A similar two-stage data assimilation scheme for ecological modelling has been proposed by Zhu et al. in [44] and delivered promising results.

The objective is to solve an inverse problem: Find a parameter configuration such that, given this configuration as input, the model output matches real disaster

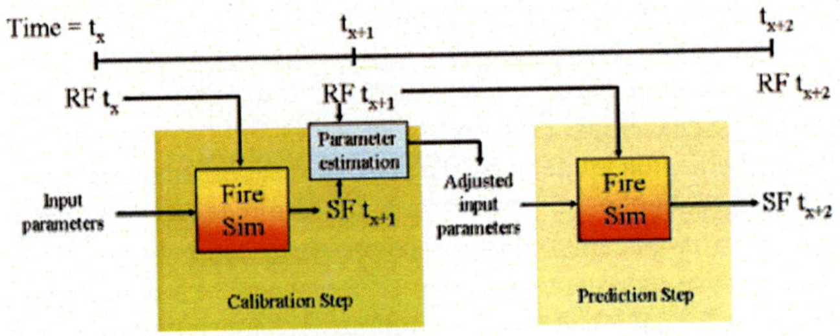

Fig. 5.2 Two-stage data-driven forest fire prediction scheme

behaviour. Having detected the simulator input that describes best current environmental conditions, the same values, it is argued, could also be used to describe best the immediate future assuming stable meteorological conditions during the following prediction interval. Thus, the prediction becomes the result of a series of automatically adjusted input configurations.

Initially, as an optimisation technique during the calibration step, Abdalhaq et al. proposed to employ a Genetic Algorithm (GA) in [1]. Different combinations of input parameter values (scenarios) are generated, evolved, and evaluated. Comparing the simulated fire front at time t_{x+1} to the real fire front at the same time the quality of the corresponding parameter set can be obtained. Lastly, the best fitting scenario is selected to serve as input for the following prediction phase. On the one hand, the data-driven prediction scheme significantly enhances the quality of input parameters and overall prediction results as proven by Denham et al. in [9]. On the other hand, it introduces a significant additional amount of computational effort and consequently increases prediction runtime in a non-negligible way. In real-time disaster modelling applications parameter estimation time needs to be reduced to a practicable minimum. In section 5.3 we will give reason why a modified GA is suited as parameter estimation technique in environmental modelling and explain implementation details.

5.2.2 The Input Parameter Problem

Inaccuracy and uncertainty in the normally large number of input parameters are known and serious problems in environmental modelling leading to unreliable propagation predictions in disaster modelling. Model output is particularly sensitive to those parameters which have a direct impact on model simulation but can not be well determined by direct observations. In disaster propagation models input parameters are often difficult or even impossible to measure in practice and hence are always incomplete and uncertain. Many parameters are highly dynamic and subject to frequent spatiotemporal changes in the microclimate generated by a disaster (e.g. strong wind gusts in forest fires). Recent advances in measurement technologies, remote sensing and power supply for sensors [37] help to remove part of this uncertainty, but installation and maintenance of sufficient sensors remain expensive and therefore a fundamental hindrance in large areas, which are sparsely populated and difficult to access. Instead, input parameter values for an upcoming real-time prediction are initialised with current weather and area forecast data provided by discrete meteorological stations. These might include measurements for uncommon parameters (e.g. fuel moisture) with temporal resolutions too low to be of use during an ongoing hazard prediction.

Assuming observed parameter values with sufficient precision became available as the prediction advances, simulators often work with a static set of input parameters not considering changes in parameter values over time. While some of the available simulators offer integration and support for GIS data, fewest tools present abilities for real-time data assimilation of altered meteorological data during a running simulation. In consequence, the prediction error accumulates gradually and simulation outcomes will deviate from the true state when the model is run for an increased prediction period [44]. A workaround to this problem might be the inclusion of the simulator into external frameworks as proposed by Rodriguez in [35]. Nevertheless, the *correct* or *real* parameter value set might not result in the best overall simulation output as explained in [2]. Simulation errors are not only due to uncertainties in input, but are also the product of model errors (overly simplified description of the natural system) and computation errors (truncation and rounding problems). It is therefore common practice to apply the concept of best fitting input and search for input parameters in the way that they produce the best overall simulation result.

Additionally to imprecision in input parameters, grid-based propagation models have to deal with spatial uncertainty [18]. When real-time constraints have to be met, low resolution representations of the region under consideration are preferable to reduce simulation runtime. Consequently, the contained heterogeneous environments (e.g. different vegetation models) might not be mapped with the necessary degree of exactness. Furthermore, some theoretical model parameters, e.g. arbitrary empirical values, might miss a readily-identifiable counterpart in reality according to [2].

These observations strongly recommend to apply parameter estimation in environmental emergency modelling, but up-to-date only the minority of models and simulators includes estimation techniques by default. Hence, the work with outdated field measures, estimated, extrapolated and missing values remains generating unsatisfactory results and poses an ongoing optimisation challenge. This is why we propose a general and simulator independent approach to enhance the prediction of disaster propagation by using a knowledge-guided GA for input parameter calibration, always following requirements are met: (1) The range of each input parameter is known and defined, (2) The sensitivity index of each model parameter is known or at least the most sensitive input parameters are identified, (3) Information about the true state of the disaster is available in reasonable time intervals, and (4) There exists an initial knowledge base (KB) containing information about past disasters.

5.3 Parameter Estimation with Knowledge-Guided Genetic Algorithm

Annan and Hargreaves state in [2], that the estimation of input parameters in high dimensional models is an inherently intractable problem. Moreover, there is probably no general solution to the problem that will work efficiently in all applications, though there exist efforts in developing universal and model-independent parameter estimation frameworks [12, 33]. Building on that, our focus is on developing methods, which aim to achieve acceptable levels of precision by taking advantage of characteristics that exist in disaster propagation prediction.

Various approximations and heuristic methods have become popular over the years and are now established standards in the scientific community. Especially Genetic Algorithms, together with Monte Carlo sampling and ensemble Kalman filter, are widely applied as an approach to solve global optimisation problems in all areas. GA are a search heuristic and imitate the process of natural evolution [15]. GA are part of the larger class of evolutionary algorithms (EA) and they repetitively apply the methods elitism, selection, crossover, mutation, fitness calculation, and reinsertion. In accordance with [41], GA can be an effective tool for parameter optimisation in environmental modelling and have, like other population-based global-search approaches, desirable properties: They are able to rapidly locate good solutions, even for large search spaces, and especially useful in problem domains that have a complex fitness landscape. GA are therefore applicable and widely-used to solve the parameter problem of environmental models [8, 14, 25, 29, 32].

5.3.1 Parallel Implementation of Hybrid Genetic Algorithm

The major obstacle to utilise parameter estimation practices in time-critical simulations is, first of all, due to the enormously increased computational burden. Most computational time required for calibrating parameters in complex environmental models is spent running the model code and generating the desired output. GA execution requires repeated fitness function evaluations with often very expensive objective functions. In addition, most environmental modelling problems possess a large quantity of input parameters creating a vast search space to be explored by the GA. Relying on fitness approximation methods summarised in [20, 28] can be one possibility to tackle this problem and remains an animated research field. Further criticisms include the complex parameter tuning of GA and their tendency to converge towards local optima or even arbitrary points rather than the global optimum in many problems.

During an ongoing disaster, short simulator response time is a key characteristic to cope with real-time capabilities and that simulation outcomes can be of use. To make evolutionary parameter estimation methods generally applicable to real-time environmental emergency modelling, possibly even granting feasible solutions in case of limited computational resources or high-resolution prediction maps, some modifications and enhancements are needed to speed up GA runtime.

Firstly, instead of using a sequential GA implementation, it is indispensable to make use of high performance computation and apply a parallel GA to substantially reduce global computing time. GA are extremely parallelisable and Denham implemented a parallel GA version based on the *master/worker* paradigm in [10]. MPI [16] libraries manage communication between the configurable number of worker nodes. Data parallelism was chosen and divides the individuals of the GA population into chunks and distributes them to available workers, which carry out the corresponding fitness evaluations. The pseudo code of the parallelised implementation split up into master and worker node operations can be found in [10].

The GA used for parameter calibration in the data-driven forest fire prediction works on a population $\mathbf{x} = [x_1, x_2, ..., x_p]$ consisting of p individuals (chromosomes) $x_k = [x_{k1}, x_{k2}, ..., x_{kd}], \forall k = 1 \leq k \leq p$. Each individual of the population represents a simulator input parameter set (scenario) made up of d parameter values, e.g. fuel model, slope, fuel moistures and wind characteristics. The genes $x_{kj} \in [lb_j, ub_j]$ of each individual are encoded as real values within a previously defined range, where $lb_j, ub_j \in \Re, \forall k = 1 \leq k \leq p, j = 1 \leq j \leq d$ and lb_j being the lower bound of parameter values for gene j and ub_j the upper bound for gene j. Individuals of the initial population are randomly generated within given lower and upper bounds. We apply elitism to keep the most promising individuals throughout the number of generations and guarantee fast and smooth convergence of the GA. Roulette wheel selection and one-point crossover are employed. Details on the mutation operator are given in section 5.3.4.

The goodness of the generated scenarios is evaluated by a problem dependent fitness function. In the fire prediction context, the fittest scenario is the one that generates a simulated fire map the most similar to the real map of fire propagation. To determine the fitness of each scenario, an error function (5.1) based on a cell-by-cell comparison of the affected terrain is applied putting into relation the erroneous burnt cells of the simulated fire with all really burnt cells.

$$error = \frac{(\bigcup - \text{initial\_fire}) - (\bigcap - \text{initial\_fire})}{\text{real\_fire} - \text{initial\_fire}} \tag{5.1}$$

with $\bigcup$ resulting in the number of cells burnt in one or both real and simulated fire map and $\bigcap$ denominating the number of cells burnt in both real and simulated fire map is to be minimised during the optimisation process. For every individual x_k in the population of the GA a simulation has to be run to be able to compute the goodness of this individual, which makes the fitness evaluation process particularly time-consuming and expensive, as explained at the beginning of this section.

Secondly, to further speed up GA execution and reduce the number of fitness evaluations to a ble minimum, decreasing the number of individuals in the population could be considered. This approach normally leads to an unwanted loss of population diversity and is therefore not recommended. On the contrary, population size should be proportional to problem size and therefore increase in high dimensional optimisation problems to cover most parts of the search space.

Thirdly, to provide near-optimal solutions with sufficient precision in early generations, i.e. yield short convergence times of the GA without compromising the good search quality, we embed domain knowledge into various phases of the GA to quickly guide it towards promising solution areas. This technique, together with an adapted mutation probability, not only manages to accelerate GA execution, but also prevents the algorithm from converging towards arbitrary points. The architecture of the resulting hybrid GA is shown in figure 5.3 and our proposals for representation, retrieval, and insertion of problem-specific knowledge are outlined in the following subsections.

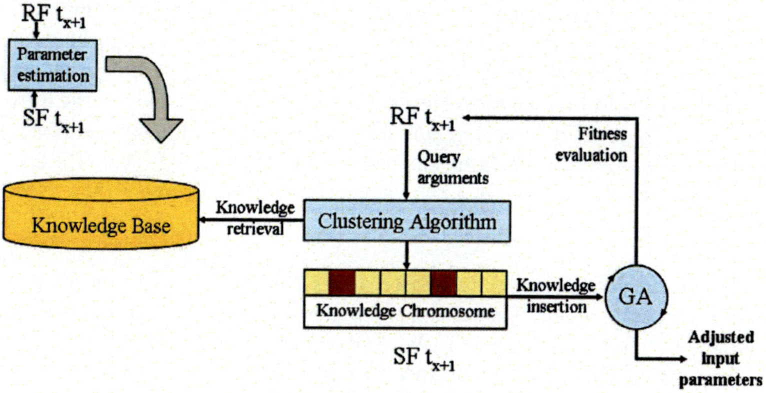

Fig. 5.3 Architecture of the hybridised GA guided with problem-specific knowledge

5.3.2 Knowledge Representation

To make proper use of expert knowledge during GA execution, we maintain a knowledge base (KB), which contains information on the behaviour of past forest fires. It associates model input parameter sets with their model outcome. More precisely, data is stored in form of simulator output parameters determining fire propagation, e.g. spread direction and speed, together with the causing environmental conditions, e.g. slope, fuel model, wind characteristics, as inputs to the simulator. The majority of these parameters are continuous, few are nominal. This approach does not require the complete set of input parameters to be stored, but best results are obtained if the most sensitive and dynamic parameters are available.

Where data of real emergencies was not available, synthetical fires were simulated to ensure that the knowledge base contains a considerable number of input configurations and covers the input parameter space to the highest degree possible. At the moment, there are more synthetical fires present in the KB than information on real catastrophes, as correct and reliable data is difficult to obtain. But due to improved observation technologies and documentation possibilities via satellites and remote sensing, reasonable growth and refinement of knowledge is expected in the near future. The KB was implemented as a standard relational database, which is essential if it reaches a reasonable size. We can thus take advantage of database management techniques enhancing flexibility in general knowledge management and retrieval.

5.3.3 Knowledge Retrieval

It is necessary to find the corresponding knowledge in an efficient manner to avoid a new increase in parameter estimation runtime. For this purpose, model output parameters, which describe the current fire's behaviour (maximum fire spread direction and speed), are obtained from the real fire map RF at time t_{x+1} once in every calibration step before starting the GA. A clustering algorithm takes these values together with available static terrain characteristics as arguments to query the KB. It aims to find the most similar past fire behaviours and returns their causing meteorological conditions, i.e. the query result delivers the input parameter values from forest fires that performed comparably to the ongoing event. The encountered values are then gathered in a knowledge chromosome $kc = [kc_1, kc_2, ..., kc_d]$ which is stored temporarily and thus available in every phase during each iteration of the GA.

The clustering algorithm is based on a k-Nearest Neighbour search and delivers the k input parameter scenarios that are most likely to cause current wildfire propagation. Presently, k is set to 1. We apply a distance function derived from the *Heterogeneous Euclidean-Overlap Metric* (HEOM) [43] to measure similarity between individual events. This distance measure takes into account differences between nominal and linear parameters: For nominal data a value-matching-based metric is defined and linear parameters are compared with the normal Euclidean metric including range normalisation to avoid that parameters with large ranges overpower those with smaller ranges.

The sensitivity analysis conducted by Abdalhaq et al. in [1] shows that fire spread is influenced to varying degrees by the existing input parameters. The parameters which mainly affect fire propagation are wind speed, wind direction and slope characteristics. In order to correctly reflect this sensitivity of inputs and to obtain meaningful results, in [42] Wendt et al. added HEOM a parameter importance ranking by means of a weight factor w_k. The distance measure finally results in

$$WHEOM(\mathbf{x}, \mathbf{y}) = \sqrt{\sum_{k=1}^{d} w_k h_k (x_k - y_k)^2} \quad (5.2)$$

where h_k stands for the two mentioned metrics.

Knowledge retrieval is performed in two steps. First, the SQL mechanism identifies the stored forest fire observations related to the current fire, i.e. fires that present the most similar slope and fuel model characteristics compared to the ongoing disaster. The second step evaluates the similarity of the retrieved observations to the fire under prediction applying WHEOM. Thus, we can avoid that distance calculation is executed on the complete KB dataset, but only performed on the result set of an intermediary range query.

5.3.4 *Knowledge Insertion*

In recent years, it proved of value for certain problems in different areas to hybridise GA by introducing problem-specific knowledge [11, 19, 23, 26, 27]. A special interest in the use of non-random mutation operators could be observed. In general, the quality of the optimisation is remarkably better when domain knowledge is incorporated in the problem solving process as mentioned by [6]. Up-to-date, few general principles, guidelines and best practices on how to incorporate which type of knowledge into GA exist and the efficiency of this method is mainly proven by experiments. This is why current investigation tries to generalise the introduction of domain knowledge in evolutionary algorithms [3, 4, 6].

Using historical incidences can be a good choice to treat uncertainty and missing values. The approach of using information about past experiences furthermore results fully legitimate and reasonable as it tries to imitate behaviour and experience of human system experts. Like domain experts applying their knowledge of observed phenomenon behaviour to rectify decision support from automated prediction systems when forecasting a new disaster, we investigated the injection of domain-specific knowledge into the GA. In doing so, a faster convergence of the GA towards fitter solutions can be reached and the risk of parameter estimation becoming the bottleneck of the overall prediction process is further diminished.

5.3.4.1 Knowledge Insertion during Mutation

In nature, mutation occurs very infrequently and can often result in a weaker individual. Occasionally, the result might be to produce a stronger one. In GA, mutation is an operator that changes the information contained in one or more gene values in an individual according to the defined mutation probability. This probability usually is set fairly low. Directing the mutation process towards promising zones in the search space and transforming randomness into controlled variation, a significantly increased mutation probability up to 0.4 should be considered.

The most common mutation operator for real-valued parameters is the uniform mutation that replaces the value of a chosen gene x_{kj} with a uniform random value selected from the problem-specific parameter range between lower and upper bound $[lb_j, ub_j]$. Our guided mutation approach uses the domain knowledge contained in the knowledge chromosome kc to narrow valid ranges of parameter values. These values then oscillate in their smaller limits, finally, forcing the GA to adopt specific values in certain dimensions from which we know that they will increase an individual's fitness. Thc three steps to follow for each gene during mutation are:

1. Preparation
   Compute the mutation probability for the gene by associating a random number from the interval $[0,1]$ with the gene. The gene is mutated if the associated number is less than the specified mutation rate.

2. Knowledge insertion
   If knowledge is available for this gene in the knowledge chromosome kc, then re-define the gene's range of valid parameter values. Set the new lower bound lb_r_j to

$$lb\_r_j = kc_j - t_j \tag{5.3}$$

   and the re-defined upper bound ub_r_j to

$$ub\_r_j = kc_j + t_j, \tag{5.4}$$

where t_j is a configurable threshold that can be chosen independently for every gene. If necessary, the bounds have to be repaired after knowledge insertion in order to obtain a consistent program: If $lb\_r_j < lb_j$ then reset $lb\_r_j = lb_j$, and if $ub\_r_j > ub_j$ then reset $ub\_r_j = ub_j$.

3. Mutation
If the gene needs to be mutated the gene is modified choosing a random value from the original range, or, if applicable, from the narrowed range of valid values.

5.3.4.2 Knowledge Insertion during Population Initialisation

Further to direction mutation, population initialisation is an obvious phase during GA execution to be supported with available problem-specific knowledge. According to a predefined initialisation probability, part of the individuals could be seeded in areas where optimal solutions are likely to be found. Knowledge injection follows the same three steps as described for mutation.

Again, during guidance, a valid subset of the original range is chosen by adding a threshold t_j to the value of retrieved knowledge, instead of using the raw value without further modification. The knowledge retrieval process only returns parameterisations most similar to the real event and depending on the degree of detailedness of the information in the KB, the configuration of the real fire might be found or not.

At present, we employ knowledge for wind speed and wind direction and cut the ranges for these two parameters because they are highly dynamic and model output proved extremely sensitive to them. The present thresholds used as margin are 5 deg for wind direction and 2 mph for wind speed. There exist other factors influencing fire spread (e.g. fuel moisture), which are, though less determining, also frequently changing and difficult to measure and it should therefore also be considered to guide their values during GA execution.

5.4 Experimental Evaluation

The experimentation's objective is to prove the benefit of introducing domain-specific knowledge. If inserted to guide the GA towards promising solution areas, we can significantly reduce the number of performed fitness evaluations while maintaining the prediction error magnitude. Experiments are divided into three different scenarios, which were all executed using the forest fire spread simulator fireLib [5]. Presented results are based on a series of real fire plots performed as prescribed burns in the frame of the European SPREAD project [39] in Gestosa (Portugal) in 2004. Figure 5.4 shows aerial photographs of three selected plots. Fire front evolution through time of the used fire maps can be seen in figure 5.5. The specific plot characteristics are outlined in table 5.1. In order to get around arbitrary evolution effects and to get more descriptive results, all presented experiments are the averaged outcome of ten different initial populations. The first scenario demonstrates the superiority of population size and diversity over evolutionary methods. In the second

Fig. 5.4 Selected plots from SPREAD project: **(a)** Plot 520 and **(b)** plot 533 and **(c)** plot 751

Table 5.1 Characteristics of selected burns from SPREAD project

Plot	Width *m*	Length *m*	Slope	Ignition type
520	89	91	18°	line ignition
533	95	123	21°	point ignition
751	20	30	6°	line ignition

test scenario we show the effect of knowledge introduction into the GA during different phases and the third experiment scenario illustrates how the same introduction of knowledge is able to reduce error variability.

The first scenario illustrates the importance of population diversity. If numerous individuals are present in a randomly initialised population, these, by default, create a good coverage of the search space and evolutionary methods of the GA tend to have less influence. Figures 5.6 and 5.7 show the calibration step results for plots 520 and 751. Populations of 50 and 500 individuals, respectively, were used and results were obtained applying the described guidance during mutation compared to a non-guided version for both population sizes. Mutation probability was set to 0.2 when guiding, 0.01 otherwise. It can be observed, that best calibration results can be obtained in every case, if many individuals are available. Guidance, depending on the characteristics of each plot, is only able to produce minor enhancements. As excessively increased runtime makes this practice unfeasible in real online applications, an acceptable compromise is to reduce population size to a practicable minimum and guide the GA. Working with a relatively small population generates the highest calibration error in this experiment. But the introduction of knowledge is able to compensate a reduced number of individuals and to gain precision in

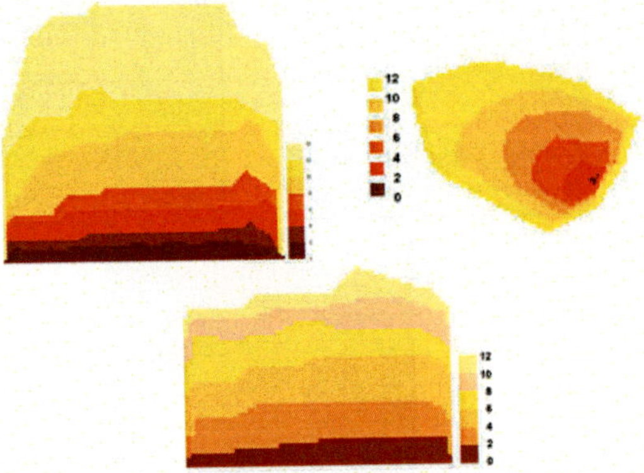

Fig. 5.5 Fire front evolution through time for **(a)** plot 520 and **(b)** plot 533 and **(c)** plot 751

Fig. 5.6 Calibration step results for different GA configurations applied to plot 520 after five iterations

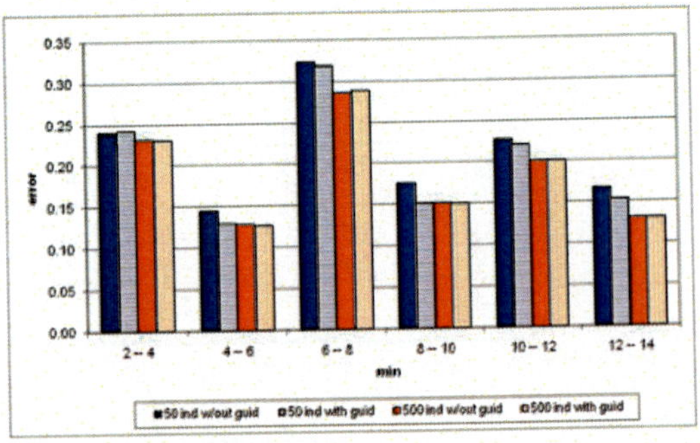

the majority of time steps. Predominant influence of population diversity is further supported by the fact that convergence of the GA is notably slower in bigger populations as exemplarily shown in figure 5.8 for minute 12 to 14 from plot 520. Very fit individuals are already present in the initial population.

The second scenario gives evidence that the introduction of problem-specific knowledge into a GA for input parameter estimation is able to clearly boost its performance in terms of runtime and, though to a minor extent, with regard to prediction quality. Figure 5.9(a) shows the input parameter refinement during calibration stage for plot 520 applying guided (guidance during mutation (0.2), population initialisation (0.25), and mutation and population initialisation) and non-guided GA versions (population size 50). It can be observed that the incorporation of knowledge results in a comparable or smaller calibration error and requires less than half of execution time. Using the optimised input parameter (the best individual after two or five generations, respectively) for the subsequent prediction step, we can note in figure 5.9(b) that the parameter obtained after fewer GA iterations supported by knowledge generates predictions with the same error magnitude as the parameter obtained

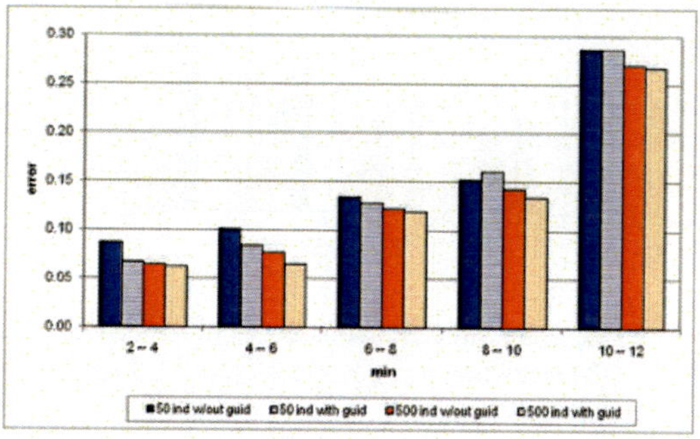

Fig. 5.7 Calibration step results for different GA configurations applied to plot 751 after ten iterations

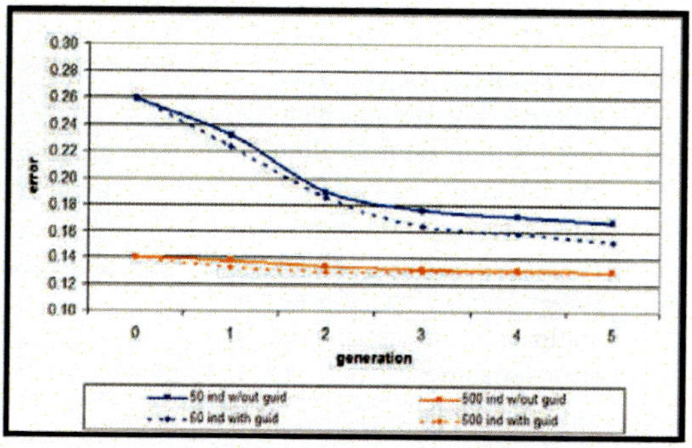

Fig. 5.8 Convergence for different GA configurations during calibration phase for plot 520 in minute 12 to 14

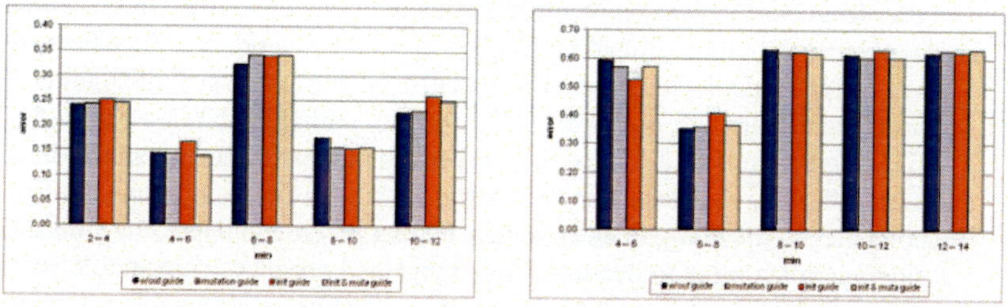

Fig. 5.9 **(a)** Calibration errors (after five (non-guided) and two (guided) GA iterations) and **(b)** prediction phase errors for plot 520 applying different GA configurations

after five generations without supporting knowledge. Figures 5.10(a) and (b) show a similar behaviour for plot 751 comparing the performance of the non-guided GA approach after ten iterations with the guided GA approach after five iterations.

The third scenario analyses error variability of the GA approach in its guided and non-guided version. Figure 5.11 plots the results for map 533 including calibration

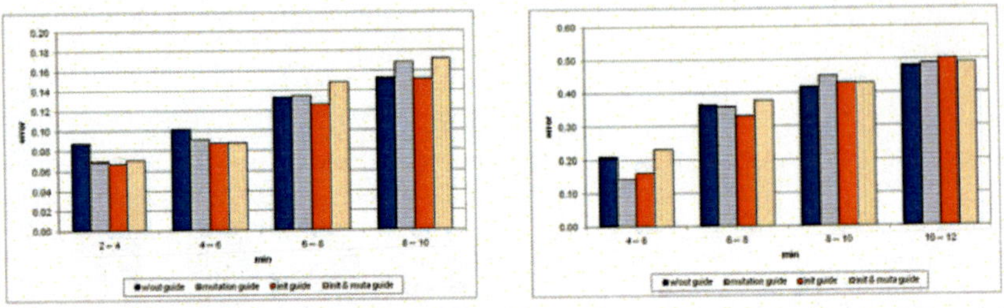

Fig. 5.10 **(a)** Calibration errors (after ten (non-guided) and five (guided) GA iterations) and **(b)** prediction phase errors for plot 5751 applying different GA configurations

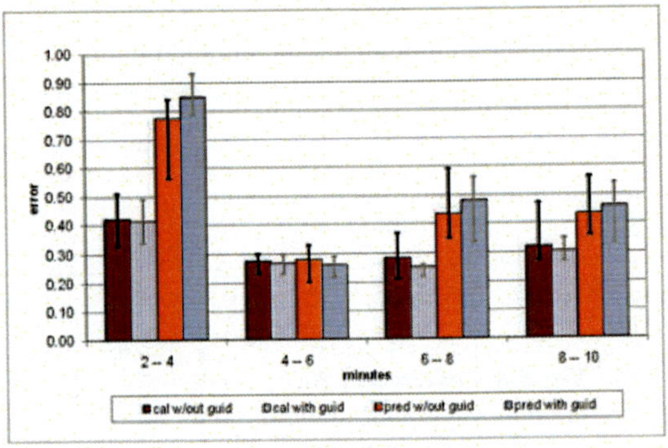

Fig. 5.11 Calibration step and prediction step error for different time steps of plot 533 applying guidance during mutation (observations after 5 generations) compared to the non-guided GA (observations after 10 generations) including observed error variability.

and prediction phase. In each of the two phases the mutation-guided version (results after five GA iterations) is compared to the non-guided version (results after ten GA iterations). Mutation probability was set to 0.4 when guiding, 0.01 otherwise. Average errors are depicted as columns and observed minimum and maximum values are provided as bars. We can clearly observe that, during calibration phase, the guided GA delivers improved results after half of the number of GA iterations. In addition, it significantly reduces error variability. Results oscillate in a smaller error range compared to the non-guided approach. During prediction phase, the guided GA version again manages to decrease error range and produces comparable predictions. The introduction of knowledge thus helps to guarantee acceptable results with low variance after few GA iterations.

5.5 Conclusions and Future Work

In this chapter, we have presented an approach for input parameter estimation for real-time disaster modelling using the example of forest fire spread prediction. The calibration process takes advantage of domain-specific knowledge guiding a parallel GA during mutation and population initialisation towards promising solution areas. This notably speeds up the calibration of highly dynamic or unavailable model input.

The method is able to ensure that resulting parameter value sets oscillate in a certain error range. The gain in estimation speed helps the simulation application to cope with real-time requirements and to deliver predictions results for online decision support systems closer to reality.

Future work aims to determine the benefit of including additional domain-dependent information (e.g. fire shape, geographical and seasonal information) into the KB and to prove the potential of the approach with a different fire simulator or another disaster application. Experiments with larger real plots are ongoing. The combination of GA with an intelligent paradigm (IP, e.g. fuzzy inference system, neural network) to form an Evolutionary Intelligent System delivered first promising results and the application of further IP is currently investigated.

Acknowledgements. This research has been supported by the MICINN-Spain under contract TIN 2007-64974.

References

1. Abdalhaq, B., Cortés, A., Margalef, T., Luque, E.: Accelerating optimization of input parameters in wildland fire simulation. In: Wyrzykowski, R., Dongarra, J., Paprzycki, M., Waśniewski, J. (eds.) PPAM 2004. LNCS, vol. 3019, pp. 1067–1074. Springer, Heidelberg (2004)
2. Annan, J.D., Hargreaves, J.C.: Efficient estimation and ensemble generation in climate modelling. Philosophical Transactions of the Royal Society A: Mathematical, Physical and Engineering Sciences 365(1857), 2077–2088 (2007)
3. Berlik, S.: A directed mutation framework for evolutionary algorithms. In: Proceedings of the International Conference on Soft Computing, MENDEL (2004)
4. Berry, A., Vamplew, P.: PoD can mutate: A simple dynamic directed mutation approach for genetic algorithms. In: Proceedings AISAT 2004 The 2nd International Conference on Artificial Intelligence in Science and Technology, pp. 200–205 (2004)
5. Bevins, C.D.: firelib user manual and technical reference (October 1996), `http://www.fire.org/downloads/fireLib/1.0.4/doc.html` (Cited October 28, 2010)
6. Bonissone, P.P., Subbu, R., Eklund, N., Kiehl, T.R.: Evolutionary algorithms + domain knowledge = real-world evolutionary computation. IEEE Transactions on Evolutionary Computation 10(3), 256–280 (2006)
7. Bulatewicz, T., Andresen, D., Welcha, S., Jina, W., Dasb, S., Miller, M.: A software system for scalable parameter estimation on clusters. In: Proceedings of the 8th LCI International Conference on High-Performance Clustered Computing (2007)
8. Crowe, A.M., McClean, C.J., Cresser, M.S.: An application of genetic algorithms to the robust estimation of soil organic and mineral fraction densities. Environmental Modelling & Software, Evolutionary Optimisation in Environmental Emergency Modelling 21(10), 1503–1507 (2006)
9. Denham, M., Cortés, A., Margalef, T., Luque, E.: Applying a dynamic data driven genetic algorithm to improve forest fire spread prediction. In: Computational Science-ICCS 2008. LNCS, pp. 36–45. Springer, Heidelberg (2008)

10. Mónica, M.D.: Predicción de la evolución de los incendios forestales guiada dinámicamente por los datos. PhD thesis, Universitat Autónoma de Barcelona, Spain (July 2009)
11. Divina, F., Marchiori, E.: Knowledge-based evolutionary search for inductive concept learning. In: Jin, Y. (ed.) Knowledge Incorporation in Evolutionary Computation, pp. 237–254. Springer, Heidelberg (2004)
12. Doherty, J.: PEST: Model-Independent Parameter Estimation. Watermark Numerical Computing, Brisbane, Australia (2009)
13. Finney, M.A.: Farsite: Fire area simulator - model development and evaluation. In: Forest Service Research Paper RMRS-RP-4, Department of Agriculture, Ogden, UT (2004)
14. Giacobbo, F., Marseguerra, M., Zio, E.: Solving the inverse problem of parameter estimation by genetic algorithms: the case of a groundwater contaminant transport model. Annals of Nuclear Energy 29(8), 967–981 (2002)
15. Goldberg, D.E.: Genetic Algorithms in Search, Optimization and Machine Learning. Kluwer Academic Publishers, Dordrecht (1989)
16. Gropp, W., Huss-Lederman, S., Lumsdaine, A., Lusk, E., Nitzberg, B., Saphir, W., Snir, M.: MPI: The Complete Reference, vol. 2. MIT Press, Cambridge (1998)
17. He, K., Dong, S., Zheng, L.: Service-oriented grid computation for large-scale parameter estimation in complex environmental modeling. In: Proceedings of the 2006 ACM symposium on applied computing SAC 2006, pp. 741–745. ACM, New York (2006)
18. Jager, H.I., King, A.W.: Spatial uncertainty and ecological models. Ecosystems 7(8), 841–847 (2004)
19. Jin, Y. (ed.): Knowledge Incorporation in Evolutionary Computation. Studies in Fuzziness and Soft Computing, vol. 167. Springer, Heidelberg (2004)
20. Jin, Y.: A comprehensive survey of fitness approximation in evolutionary computation. Soft Computing - A Fusion of Foundations, Methodologies and Applications 9, 3–12 (2005)
21. Jones, A.R., Thomson, D.J., Hort, M., Devenish, B.: The U.K. Met Office's next-generation atmospheric dispersion model, NAME III. In: Air Pollution Modeling and its Application XVII (Proc. of the 27th NATO/CCMS Int. Technical Meeting on Air Pollution Modelling and its Application), pp. 580–589. Springer, Heidelberg (2007)
22. Kalman, R.E.: A new approach to linear filtering and prediction problems. Transactions of the ASME - Journal of Basic Engineering 82(Series D), 35–45 (1960)
23. Kanoh, H., Sakamoto, Y.: Knowledge-based genetic algorithm for university course timetabling problems. International Journal of Knowledge-based and Intelligent Engineering Systems 12(4), 283–294 (2007)
24. Kavetski, D., Franks, S.W., Kuczera, G.: Confronting input uncertainty in environmental modelling. In: Calibration of Watershed Models. Water Science and Applications Series, American Geophysical Union (2003)
25. Lee, Y.H., Park, S.K., Chang, D.-E.: Parameter estimation using the genetic algorithm and its impact on quantitative precipitation forecast. Annales Geophysicae 24(12), 3185–3189 (2006)
26. Li, F., Lindquist, T.M.: Knowledge guided genetic algorithm for optimal contracting strategy in a typical standing reserve market. In: Power Engineering Society General Meeting, vol. 2, pp. 859–863 (2003)
27. Li, Y., Yao, D., Zheng, J., Yao, J.: A modified genetic algorithm for the beam angle optimization problem in intensity-modulated radiotherapy planning. In: Talbi, E.-G., Liardet, P., Collet, P., Lutton, E., Schoenauer, M. (eds.) EA 2005. LNCS, vol. 3871, pp. 97–106. Springer, Heidelberg (2006)

28. Liu, Y., Khu, S.-T.: Automatic calibration of numerical models using fast optimization by fitness approximation. In: International Joint Conference on Neural Networks IJCNN 2007, pp. 1073–1078 (2007)
29. Montero, G., Rodriguez, E., Montenegro, R., Escobar, J.M., Gonzalez-Yuste, J.M.: Genetic algorithms for an improved parameter estimation with local refinement of tetrahedral meshes in a wind model. Advances in Engineering Software 36(1), 3–10 (2005)
30. Morais, M.E.: Comparing spatially explicit models of fire spread through chaparral fuels: A new model based upon the rothermel fire spread equation. M.a. thesis, University of California, Santa Barbara, USA (June 2001)
31. Neumann, C.J.: An alternate to the HURRAN tropical cyclone forecast system. Tech. Memo. NWS SR-62, National Oceanic and Atmospheric Administration (1972)
32. Pelletier, G.J., Chapra, S.C., Tao, H.: Qual2kw - a framework for modeling water quality in streams and rivers using a genetic algorithm for calibration. Environmental Modelling & Software 21(3), 419–425 (2006)
33. Poeter, E.P., Hill, M.C., Banta, E.R., Mehl, S., Christensen, S.: Ucode 2005 and six other computer codes for universal sensitivity analysis, calibration, and uncertainty evaluation. In: Techniques and Methods 6-A11, U.S. Geological Survey, Reston, Virginia (2005)
34. Poyton, A.A., Varziri, M.S., McAuley, K.B., McLellan, P.J., Ramsay, J.O.: Parameter estimation in continuous-time dynamic models using principal differential analysis. Computers& Chemical Engineering 30(4), 698–708 (2006)
35. Rodríguez, R., Cortés, A., Margalef, T.: Injecting dynamic real-time data into a DDDAS for forest fire behavior prediction. In: Allen, G., Nabrzyski, J., Seidel, E., van Albada, G.D., Dongarra, J., Sloot, P.M.A. (eds.) ICCS 2009. LNCS, vol. 5545, pp. 489–499. Springer, Heidelberg (2009)
36. Rothermel, R.C.: A mathematical model for predicting fire spread in wildland fuels. Forest Service Research Paper INT-115, U.S. Dept. of Agriculture, Ogden (1972)
37. ScienceDaily and MIT. Preventing forest fires with tree power: Sensor system runs on electricity generated by trees (September 2008), `http://www.sciencedaily.com/2008/09/080922095435.htm` (Cited October 28, 2010)
38. Tymstra, C., Bryce, R.W., Wotton, B.M., Taylor, S.W., Armitage, O.B.: Development and structure of prometheus: the canadian wildland fire growth simulation model. Information Report NOR-X-417, Natural Resources Canada, Canadian Forest Service, Edmonton, Alberta (2010)
39. Viegas, D.X.: Spread project (October 2003), `http://www.algosystems.gr/spread` (Cited October 28, 2010)
40. Vrugt, J.A., Nuallin, B., Robinson, B.A., Bouten, W., Dekker, S.C., Sloot, P.M.A.: Application of parallel computing to stochastic parameter estimation in environmental models. Computers & Geosciences 32(8), 1139–1155 (2006)
41. Wang, Q.J.: Using genetic algorithms to optimise model parameters. Environmental Modelling & Software 12(1), 27–34 (1997)
42. Wendt, K., Cortés, A., Margalef, T.: Knowledge-guided genetic algorithm for input parameter optimisation in environmental modelling. In: Procedia Computer Science ICCS 2010, vol. 1(1), pp. 1367–1375 (2010)
43. Randall Wilson, D., Martinez, T.R.: Improved heterogeneous distance functions. Journal of Artificial Intelligence Research 6, 1–34 (1997)
44. Zhu, L., Chen, J.M., Qin, Q., Li, J., Wang, L.: Optimization of ecosystem model parameters using spatio-temporal soil moisture information. Ecological Modelling 220(18), 2121–2136 (2009)

Chapter 6
Harmony Search Algorithms in Structural Engineering

M.P. Saka, I. Aydogdu, O. Hasancebi, and Z.W. Geem

Abstract. Harmony search method is widely applied in structural design optimization since its emergence. These applications have shown that harmony search algorithm is robust, effective and reliable optimization method. Within recent years several enhancements are suggested to improve the performance of the algorithm. Among these Mahdavi has presented two versions of harmony search methods. He named these as improved harmony search method and global best harmony search method. Saka and Hasancebi (2009) have suggested adaptive harmony search where the harmony search parameters are adjusted automatically during design iterations. Coelho has proposed improved harmony search method. He suggested an expression for one of the parameters of standard harmony search method. In this chapter, the optimum design problem of steel space frames is formulated according to the provisions of LRFD-AISC (Load and Resistance Factor Design-American Institute of Steel Corporation). The weight of the steel frame is taken as the objective function to be minimized. Seven different structural optimization algorithms arc developed each of which are based on one of the above mentioned versions of harmony search method. Three real size steel frames are

M.P. Saka · I. Aydogdu
Department of Engineering Sciences,
Middle East Technical University, Ankara, Turkey
e-mail: mpsaka@metu.edu.tr, aydogdu@metu.edu.tr

O. Hasancebi
Department of Civil Engineering,
Middle East Technical University, Ankara, Turkey
e-mail: oguzhan@metu.edu.tr

Z.W. Geem
Information Technology Program,
iGlobal University, Annandale, Virginia, USA
e-mail: zwgeem@gmail.com

X.-S. Yang & S. Koziel (Eds.): Computational Optimization & Applications, SCI 359, pp. 145–182.
springerlink.com

designed using each of these algorithms. The optimum designs obtained by these techniques are compared and performance of each version is evaluated.

Keywords: Structural Optimization, Metaheuristic Techniques, Harmony Search Algorithm.

6.1 Introduction

Building construction causes use of large amount of land, consumption of energy and water. Furthermore, production of the materials used in the building construction add large amount of pollution into the atmosphere. It is important to reduce the amount of natural resources utilized in building construction if we would like to reduce the environmental impact and have a sustainable development. Green construction is becoming a common practice all over the world which intends to construct buildings that are environmentally friendly and resource-efficient throughout its life cycle. This practice covers all the stages from setting to design, construction and operation. Overdesigns and use of excessive materials are not desired because they consume more of natural sources and add more pollution to the atmosphere. Hence it is clear that in order to have a sustainable development structures are required to be designed and built using sufficient amount of material but not more. Structural design optimization tools exactly try to achieve this goal. They aim to design the steel structures such that the steel frame has the minimum weight and in the mean time the response of the frame under the external loads that the frame may be subjected to during its life time is within the design code limitations. Design of steel structures has its own features and not similar to the design of other structures. Designer cannot use any section she/he may desire but to select among the set of steel profiles available in practice for beams and columns of the frame under consideration. This selection is required to be carried out such that the frame with the selected steel profiles should have the displacements less than those prescribed in the design code and its members have sufficient strength to satisfy the strength limitations under the external loads. In the mean time its cost is the minimum.

In this chapter firstly the design optimization problem of steel space frames according to the provisions of LRFD-AISC (Load and Resistance factor Design-American Institution of Steel Corporation) [1] is presented. The weight of the steel frame is taken as the objective function to be minimized. Such formulation of the optimum design problem yields a discrete programming problem. The solution of this programming problem is obtained by harmony search algorithm [2-6]. This method is one of the recent combinatorial optimization techniques that belong to general class of what is called metaheuristic algorithms. Metaheuristic algorithms [7-11] finds the solution of optimization problems by utilizing certain tactics that are generally inspired from the nature, though not limited to, instead of classical procedures that move along the descending direction of gradient of objective function. Harmony search method mimics music improvisation. Harmony search method is widely applied in structural design optimization since its emergence [12-19]. These applications have shown that harmony search algorithm is robust, effective and reliable optimization method. Within recent years several enhancements are suggested to improve the performance of the harmony search method. Among these Mahdavi [20, 21] has presented two versions of harmony search methods. He named them as improved harmony search method and global best harmony search method. Hasancebi et. al. (2010) [22, 23] has suggested adaptive harmony search where the harmony search parameters are adjusted automatically during design iterations. Coelho [24] has proposed improved harmony search method. He suggested an expression for one the parameters of standard harmony search method. In this chapter seven different structural optimization algorithms are developed each of which is based on one of the above mentioned versions of harmony search method. Three steel space frames are designed using each of these algorithms. The optimum designs obtained by each of these techniques are compared and performance of each version is evaluated.

6.2 Discrete Optimum Design of Steel Space Frames to LRFD-AISC

The design of steel space frames necessitates the selection of steel sections for its columns and beams from a standard steel section tables such that the frame satisfies the serviceability and strength requirements specified by the code of practice while the economy is observed in the overall or material cost of the frame. When the design constraints are implemented from LRFD-AISC the following discrete programming problem is obtained.

6.2.1 The Objective Function

The objective function is taken as the minimum weight of the frame which is expressed as in the following.

$$\text{Minimize} \quad W = \sum_{r=1}^{ng} m_r \sum_{s=1}^{t_r} \ell_s \tag{6.1}$$

where; W defines the weight of the frame, m_r is the unit weight of the steel section selected from the standard steel sections table that is to be adopted for group r. t_r is the total number of members in group r and ng is the total number of groups in the frame. l_s is the length of member s which belongs to group r.

6.2.2 Strength Constraints

For the case where the effect of warping is not included in the computation of the strength capacity of W-sections that are selected for beam-column members of the frame the following inequalities given in Chapter H of LRFD-AISC are required to be satisfied.

$$\text{for}\ \frac{P_u}{\phi P_n} \geq 0.2\,;\ g_{s,i} = \frac{P_u}{\phi P_n} + \frac{8}{9}\left(\frac{M_u}{\phi_b M_{nx}} + \frac{M_u}{\phi_b M_{ny}}\right) \leq 1{,}0 \tag{6.2}$$

$$\text{for}\ \frac{P_u}{\phi P_n} \geq 0.2\,;\ g_{s,i} = \frac{P_u}{2\phi P_n} + \left(\frac{M_u}{\phi_b M_{nx}} + \frac{M_u}{\phi_b M_{ny}}\right) \leq 1{,}0 \tag{6.3}$$

where, M_{nx} is the nominal flexural strength at strong axis (x axis), M_{ny} is the nominal flexural strength at weak axis (y axis), M_{ux} is the required flexural strength at strong axis (x axis), M_{uy} is the required flexural strength at weak axis (y axis), P_n is the nominal axial strength (tension or compression) and P_u is the required axial strength (tension or compression) for member i. l represents the loading case. The values of M_{ux} and M_{uy} are required to be obtained by carrying out $P-\Delta$ analysis of the steel frame. This is an iterative process which quite time consuming. In Chapter C of LRFD-AISC an alternative procedure is suggested for the computations of M_{ux} and M_{uy} values. In this procedure, two first order elastic analyses are carried out. In the first, frame is analyzed under the gravity loads only where the sway of the frame is prevented to obtain M_{nt} values. In the second, the frame is analyzed only under the lateral loads to find M_{lt} values. These moment values are then combined using the following equation as given in the design code.

$$M_u = B_1 M_{nt} + B_2 M_{lt} \tag{6.4}$$

where B_1 is the moment magnifier coefficient and B_2 is the sway moment magnifier coefficient. The details of how these coefficients are calculated are given in Chapter C of LRFD-AISC [1].

Eqns. (6.2) and (6.3) represents strength constraints for doubly and singly symmetric steel members subjected to axial force and bending. If the axial force in member k is tensile force the terms in these equations are given as: P_{uk} is the required axial tensile strength, P_{nk} is the nominal tensile strength, ϕ becomes ϕ_t in the case of tension and called strength reduction factor which is given as 0.90 for yielding in the gross section and 0.75 for fracture in the net section, ϕ_b is the strength reduction factor for flexure given as 0.90, M_{uxk} and M_{uyk} are the required flexural strength M_{nxk} and M_{uyk} are the nominal flexural strength about major and minor axis of member k respectively. It should be pointed out that required flexural bending moment should include second-order effects. LRFD suggests an approximate procedure for computation of such effects which is explained in C1 of LRFD. In the case the axial force in member k is compressive force the terms in Eqns. (6.2) and (6.3) are defined as: P_{uk} is the required compressive strength, P_{nk} is the nominal compressive strength, and ϕ becomes ϕ_c which is the resistance factor for compression given as 0.85. The remaining notations in Eqns. (6.16) and (6.17) are the same as the definition given above.

The nominal tensile strength of member k for yielding in the gross section is computed as $P_{nk} = F_y A_{gk}$ where F_y is the specified yield stress and A_{gk} is the gross area of member k. The nominal compressive strength of member k is computed as

$P_{nk} = A_{gk}F_{cr}$ where $F_{cr} = \left(0.658^{\lambda_c^2}\right)F_y$ for $\lambda_c \le 1.5$ and $F_{cr} = \left(0.877/\lambda_c^2\right)F_y$ for $\lambda_c > 1.5$ and $\lambda_c = \frac{Kl}{r\pi}\sqrt{F_y/E}$. In these expressions E is the modulus of elasticity, K and l are the effective length factor and the laterally unbraced length of member k respectively.

6.2.3 Displacement Constraints

The lateral displacements and deflection of beams in steel frames are limited by the steel design codes due to serviceability requirements. According to the ASCE Ad Hoc Committee report [25], the accepted range of drift limits in the first-order analysis is 1/750 to 1/250 times the building height H with a recommended value of H/400. The typical limits on the inter-story drift are 1/500 to 1/200 times the story height. Based on this report the deflection limits recommended are proposed in [26, 27, 28] for general use which is repeated in Table 6.1.

Table 6.1 Displacement limitations for steel frames

	Item	Deflection Limit
1	Floor girder deflection for service live load	L/360
2	Roof girder deflection	L/240
3	Lateral drift for service wind load	H/400
4	Inter-story drift for service wind load	H/300

6.2.3.1 Deflection Constraints

It is necessary to limit the mid-span deflections of beams in a steel space frame not to cause cracks in brittle finishes that they may support due to excessive displacements. Deflection constraints can be expressed as an inequality limitation as shown in the following.

$$g_{dj} = \frac{\delta_{jl}}{\delta_j^u} - 1 \le 0 \quad j = 1, \ldots, n_{sm}, \quad l = 1, \ldots, n_{lc} \tag{6.5}$$

where, δ_{jl} is the maximum deflection of jth member under the lth load case, δ_j^u is the upper bound on this deflection which is defined in the code as span/360 for beams carrying brittle finishers, n_{sm} is the total number of members where deflections limitations are to be imposed and n_{lc} is the number of load cases.

6.2.3.2 Drift Constraints

These constraints are of two types. One is the restriction applied to the top story sway and the other is the limitation applied on the inter-story drift.

Top Story Drift Constraint

Top story drift limitation can be expressed as an inequality constraint as shown in the following.

$$g_{td\,j} = \frac{\left(\Delta_{top}\right)_{jl}}{H\,/\,Ratio} - 1 \le 0 \quad j = 1,\ldots,n_{jtop}\,, \quad l = 1,\ldots,n_{lc} \tag{6.6}$$

where H is the height of the frame, n_{jtop} is the number of joints on the top story, n_{lc} is the number of load cases, $(\Delta_{top})_{jl}$ is the top story drift of the jth joint under l^{th} load case. Ratio is a constant value given in ASCE Ad Hoc Committee report [25].

Inter-story Drift

In multi-story steel frames the relative lateral displacements of each floor is required to be limited. This limit is defined as the maximum inter-story drift which is specified as h_{sx}/Ratio where h_{sx} is the story height and Ratio is a constant value given in ASCE Ad Hoc Committee report [25].

$$g_{id\,j} = \frac{\left(\Delta_{oh}\right)_{jl}}{h_{sx}\,/\,Ratio} - 1 \le 0 \quad j = 1,\ldots.,n_{st}\,, \quad l = 1,\ldots\ldots n_{lc} \tag{6.7}$$

where n_{st} is the number of story, n_{lc} is the number of load cases and $(\Delta_{oh})_{jl}$ is the story drift of the jth story under l^{th} load case.

6.2.4 Geometric Constraints

In steel frames it is not desired that column section for upper floor should not have a larger section than the lower story column for practical reasons. Because having a larger section for upper floor requires a special joint arrangement which is neither preferred nor economical. The same applies to the beam-to-column connections. The W-section selected for any beam should have a flange width smaller than or equal to the flange width of the W-section selected for the column to which the beam is to be connected. These are shown in Fig. 6.1 and named as geometric constraints. These limitations are included in the design optimization model to satisfy practical requirements. Two types of geometric constraints are considered in the mathematical model. These are column-to-column geometric constraints and beam-to-column geometric limitations.

6.2.4.1 Column-to-Column Geometric Constraints

The depth and the unit weight of W sections selected for the columns of two consecutive stores should be either equal to each other or the one in the upper story should be smaller than the one in the lower story. These limitations are included in the design problem as inequality constraints as shown in the following.

$$g_{cdi} = \frac{D_i}{D_{i-1}} - 1 \leq 0 \quad i = 2......n_j \tag{6.8}$$

$$g_{cmi} = \frac{m_i}{m_{i-1}} - 1 \leq 0 \quad i = 2......n_j \tag{6.9}$$

where n_j is the number of stories, m_i is the unit weight of W section selected for column story i, m_{i-1} is the unit weight of W section selected for of column story (i–1), D_i is the depth of W section selected for of column story i and D_{i-1} is the depth of W section selected for of column story (i-1).

6.2.4.2 Beam-to-Column Geometric Constraints

When a beam is connected to a flange of a column, the flange width of the beam should be less than or equal to the flange width of the column so that the connection can be made without difficulty. In order to achieve this, the flange width of the beam should be less than or equal to $(D - 2t_b)$ of the column web dimensions in the connection where D and t_b are the depth and the flange thickness of W section respectively as shown in Fig. 6.1.

$$g_{bci} = \frac{(B_f)_{bi}}{D_{ci} - 2(t_{bc})_i} - 1 \leq 0 \quad i = 1......n_{j1} \tag{6.10}$$

or

$$g_{bbi} = \frac{(B_f)_{bi}}{(B_f)_{ci}} - 1 \leq 0 \quad i = 1......n_{j2} \tag{6.11}$$

where n_{j1} is the number of joints where beams are connected to the web of a column, n_{j2} is the number of joints where beams connected to the flange of a column, D_{ci} is the depth of W section selected for the column at joint i, $(t_{bc})_i$ is the flange thickness of W section selected for the column at joint i, $(B_f)_{ci}$ is the flange width of W section selected for the column at joint i and $(B_f)_{bi}$ is the flange width of W section selected for the beam at joint i.

The above optimum design of steel space frames problem where the objective function is given in Eqn. (6.1) and the constraints are described from Eqn. (6.2) to Eqn. (6.11) is a combinatorial optimization problem of discrete optimization. This is because the solution of the problem necessitates the selection of appropriate steel sections for the beams and columns of the frame from W-sections list such that the objective function described in Eqn. (6.1) has the minimum value while the design constraints given in inequalities from Eqn. (6.2) to Eqn. (6.11) are satisfied. Designer has to find out the suitable combination of W-sections that makes the frame weight minimum in the same time the design code provisions are all satisfied. Here the selection of a W-section from an available steel profile list is carried out by choosing an integer number from a set which consist of integer numbers starting 1 to the total number of sections in the list. This integer number is the sequence number of that particular W-section. Hence the design solution is a set of integer numbers each of which represents the sequence number of W-section in the design pool. This is a combinatorial optimization problem [7,8].

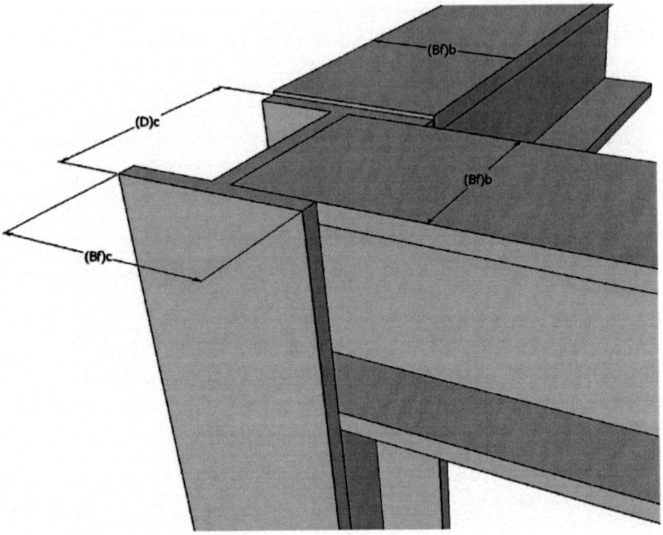

Fig. 6.1 Beam column geometric constraints

In last two decades a new kind of algorithms have emerged which make use of certain heuristics in order to explore a search space efficiently. These methods are called metaheuristic algorithms [7-11]. A metaheuristic is an iterative process with set of concepts that are used for exploring and exploiting the search space to determine the best solution among the alternative solutions. Metaheuristic algorithms are not problem specific, approximate and usually non-deterministic. It is important that there should be a dynamic balance between diversification and intensification in metaheuristic procedure. Diversification generally refers to the exploration of the search space and intensification refers to the exploitation of the accumulated search experience [7]. The balance between these two concepts is important so as not waste too much time in regions of the search space which does not possess high quality solutions while the algorithm can quickly find out the regions of high quality solutions. Some of the metaheuristic algorithms employ strategies that are inspired from nature. They simulate natural phenomena such as survival of the fittest, immune system, swarm intelligence and the cooling process of molten metals through annealing into a numerical algorithm. They are named according to the natural phenomena their search strategy is based such as evolutionary algorithms, immune system algorithm, particle swarm optimization and simulated annealing. Metaheuristic methods are non-traditional stochastic search and optimization methods and they are very suitable and efficient in finding the solution of combinatorial optimization problems.

It is shown in the literature that harmony search method which is one of the recently-developed metaheuristic techniques is quite effective and robust in solving structural optimization problems [12-19]. Performance evaluation of seven metaheuristic technique used in optimum design of pin jointed and rigidly jointed real size steel frames is carried in [16, 19]. In these studies it is shown that harmony search algorithm is quite successful stochastic search method and its performance in some problems is better than some other metaheuristic methods. Since its emergence, numbers of enhancements are suggested in order to improve the performance of the standard harmony search method. In this chapter these improvements are employed in solving the optimum design problem of steel space frames described above and their performances are compared.

6.3 Harmony Search Algorithms

The harmony search algorithm (HS) is originated by Geem et al. [2]. The algorithm was inspired by using the musical performance processes that occur when a musician searches for a perfect state of harmony, such as during jazz improvisation [2-6]. The analogy between finding a pleasing harmony in music and the optimum solution in an optimization problem is illustrated in Figure 6.2. A musician always intends to procedure a piece of music with perfect harmony. On the other hand, the optimal solution of an optimization problem should be the best solution available to the problem under given objective and limited by constraints. Both processes aim at reaching the best solution that is the optimum.

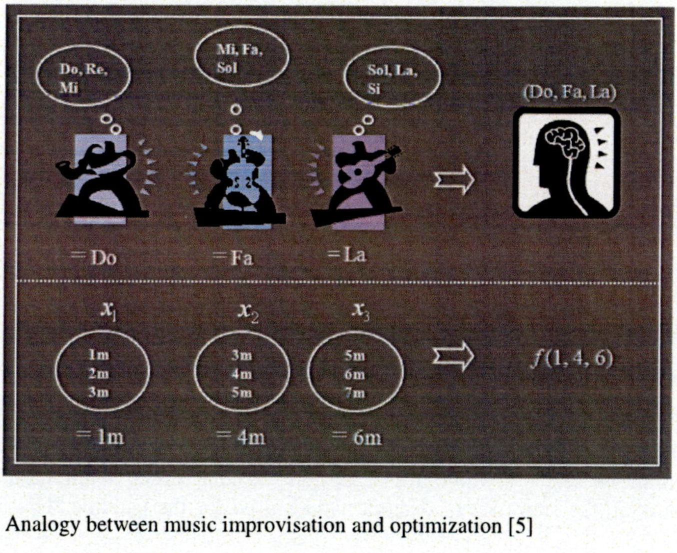

Fig. 6.2 Analogy between music improvisation and optimization [5]

6.3.1 Standard Harmony Search Algorithm

Harmony search method imitates the improvisation process of a skilled musician. When a musician is improvising, he or she has three possible choices: (a) can play any tune from his or her memory; (b) can play something similar to aforementioned tune by just adjusting pitch slightly; (c) can play a tune completely new. These three options are simulated in three components in harmony search method. These are usage of harmony memory matrix (**H**), pitch adjusting (*par*) and randomization. Before initiating the design process, a set of steel sections selected from an available profile list are collected in a design pool. Each steel section is assigned a sequence number I that varies between 1 to total number of sections (N_{sec}) in the list. It is important to note that during optimization process selection of sections for design variables is carried out using these numbers. The steps of the algorithm are outlined in the following as given in [2]:

6.3.1.1 Initialization of Harmony Memory Matrix

A harmony memory matrix **H** given in Eqn. (6.12) is randomly generated. The harmony memory matrix simply represents a design population for the solution of a problem under consideration, and incorporates a predefined number of solution vectors referred to as harmony memory size (hms). Each solution vector (harmony vector, $\mathbf{I}^i$) consists of ng design variables, and is represented in a separate row of the matrix; consequently the size of **H** is $(hms \times ng)$.

$$\mathbf{H} = \begin{bmatrix} I_1^1 & I_2^1 & \dots & I_{ng}^1 \\ I_1^2 & I_2^2 & \dots & I_{ng}^2 \\ \dots & \dots & \dots & \dots \\ I_1^{hms} & I_2^{hms} & \dots & I_{ng}^{hms} \end{bmatrix} \begin{matrix} \phi(\mathbf{I}^1) \\ \phi(\mathbf{I}^2) \\ \dots \\ \phi(\mathbf{I}^{hms}) \end{matrix} \tag{6.12}$$

6.3.1.2 Evaluation of Harmony Memory Matrix

(hms) solutions are then analyzed, and their objective function values are calculated. The solutions evaluated are sorted in the matrix in the increasing order of objective function values, that is $\phi(\mathbf{I}^1) \leq \phi(\mathbf{I}^2) \leq \ldots \leq \phi(\mathbf{I}^{hms})$.

6.3.1.3 Improvising a New Harmony

In harmony search algorithm the generation of a new solution (harmony) vector is controlled by two parameters ($hmcr$ and par) of the technique. The harmony memory considering rate ($hmcr$) refers to a probability value that biases the algorithm to select a value for a design variable either from harmony memory or from the entire set of discrete values used for the variable. That is to say, this parameter decides in what extent previously visited favorable solutions should be considered in comparison to exploration of new design regions while generating new solutions. At times when the variable is selected from harmony memory, it is checked whether this value should be substituted with its very lower or upper neighboring one in the discrete set. Here the goal is to encourage a more explorative search by allowing transitions to designs in the vicinity of the current solutions. This phenomenon is known as pitch-adjustment in harmony search method, and is controlled by pitch adjusting rate parameter (par). In the standard algorithm both of these parameters are set to suitable constant values for all harmony vectors generated regardless of whether an exploitative or explorative search is indeed required at a time during the search process. Accordingly a new harmony $\mathbf{I}' = \left[I_1', I_2', .., I_{ng}'\right]$ is improvised (generated) by selecting each design variable from either harmony memory or the entire discrete set. The probability that a design variable is selected from the harmony memory is controlled by harmony memory considering rate $(hmcr)$. To execute this probability, a random number r_i is generated between 0 and 1 for each variable I_i. If r_i is smaller than or equal to $hmcr$, the variable is chosen from harmony memory. Otherwise, a random value is assigned to the variable from the entire discrete set as shown in Eqn. (6.13).

$$I_i' = \begin{cases} I_i' \in \left\{I_i^1, I_i^2, ..., I_i^{hms}\right\} & if\ r_i \le hmcr \\ I_i' \in \{1, ..., ng\} & if\ r_i > hmcr \end{cases} \tag{6.13}$$

If a design variable attains its value from harmony memory, it is checked whether this value should be pitch-adjusted or not. In pitch adjustment, the value of a design variable is altered to its very upper or lower neighboring value obtained by adding ± 1 to its current value. Similar to $(hmcr)$ parameter, it is operated with a probability known as pitch adjustment rate (par). If not activated by (par), the value of the variable does not change as given in Eqn. (6.14).

$$I_i'' = \begin{cases} I_i' \pm bw & if\ r_i \le par \\ I_i' & if\ r_i > par \end{cases} \tag{6.14}$$

where, bw is arbitrary distance bandwidth which is taken as 1 in the standard harmony search method.

6.3.1.4 Update of Harmony Matrix

After generating the new harmony vector, its objective function value is calculated. If this value is better (lower) than that of the worst harmony vector in the harmony memory, it is then included in the matrix while the worst one is discarded out of the matrix. The updated harmony memory matrix is then sorted in ascending order of the objective function value.

6.3.1.5 Termination

The steps 3.1.2 and 3.1.3 are repeated until a pre-assigned maximum number of cycles are reached.

6.4 Various Harmony Search Algorithms

Within the recent years, number of enhancements is suggested to standard harmony search method in order to improve its performance. In this study, seven variations of harmony search algorithms are considered to determine the solution of the optimum design problem of steel space frames. These techniques are summarized in the following.

6.4.1 Standard Harmony Search with Adaptive Error Strategy (SHSAES)

This version is same as the standard harmony search method. It follows the same steps explained above. The only difference is that in addition to feasible solution vectors slightly infeasible solution vectors are also included in the harmony memory matrix. The candidate solution vectors that violate one or more design constraints slightly are also accepted as solutions due to the fact that they may possess some appropriate values for some of the design variables which can be used in pitch adjusting in the next iteration. It should be noticed that such candidate design vectors are allowed in the beginning phase of the design process but they are required to be taken out from the harmony memory matrix towards final phases of design cycles. This achieved by using larger error value initially and then this value is adjusted during the design cycles according to the expression given below.

$$Tol(i) = Tol_{\max} - \frac{(Tol_{\max} - Tol_{\min}) \cdot i^{0.5}}{(iter_{\max})^{0.5}} \tag{6.15}$$

where, $Tol(i)$ is the error value in iteration i, $Tol_{\max}$ and $Tol_{\min}$ are the maximum and the minimum error values defined in the algorithm respectively, $iter_{\max}$ is the maximum iteration number until which tolerance minimization procedure continues. Equation (6.15) provides larger error values in the beginning of the design cycles and quite small error values towards the final design cycles. Hence when the maximum design cycles are reached the acceptable design vectors remain in the harmony memory matrix and the ones which do not satisfy one or more design constraints smaller than the error tolerance would be pushed out during the design iterations.

6.4.2 Standard Harmony Search with Penalty Function (SHSPF)

In this application of the harmony search method also follows the steps of the standard harmony search technique. It only differs from the standard one in the acceptance of the candidate solution vectors. All the design vectors selected randomly are included in the harmony memory matrix regardless of whether they satisfy the design constraints in design problem or not. However, a penalty function is constructed as shown in the following.

$$W_p = W(1+C)^{\varepsilon} \tag{6.16}$$

where, W_p is the objective function that contain the penalty and W is the original objective function which is taken as the minimum weight of the steel space frames as given in Eqn. (6.1). C is the total constraint violation value calculated from the sum of the values of constraints function violations as given in equation (6.17). ε is the penalty coefficient taken as 2.

$$C = \sum C_s + \sum C_d + \sum C_{id} + \sum C_{td} + \sum C_{cd} + \sum C_{cc} + \sum C_{bc} \tag{6.17}$$

where, C_s, C_d, C_{id}, C_{td}, C_{cd}, C_{cc} and C_{bb} are the constraint functions violations for strength, deflection, inter-story drift, top story drift, column-to-column depth and unit weight and beam-to-column geometric constraint functions given in inequalities (6.2), (6.3), (6.5), (6.6), (6.7), (6.8), (6.9), (6.10) and (6.11) respectively. In general form, constraint function violation is calculated as:

$$C_i = \begin{cases} 0 & if\ g_i(x_j) \le 0 & i = 1, \dots nc \\ g_i(x_j) & if\ g_i(x_j) > 0 & j = 1, \dots ng \end{cases} \tag{6.18}$$

where, $g_i(x)$ is ith constraint function, x is the vector of design variables, nc is the total number of constraint functions and ng is the total number of member groups (the total number of design variables) in the optimum design problem. It is apparent from Eqn. (6.18) that feasible solutions will not be subjected to any penalty and their objective function value will be equal to the original objective function value given in Eqn. (6.1). The harmony search method seeks solution vectors in the design space that have smaller objective function values and stores these in the harmony memory matrix during the design cycles. As a result those solution vectors that have larger objective function values are eliminated from the harmony memory matrix within the harmony search iterations. Towards the end of design cycles only those solution vectors that do not have any penalty remains in the harmony memory matrix and among these that have the least weight represents the optimum solution.

6.4.3 Adaptive Harmony Search with Penalty Function (AHSPF)

In standard harmony search method there are two parameters known as harmony memory considering rate ($hmcr$) and pitch adjusting rate (par) that play an important role in obtaining the optimum solution. These parameters are assigned to constant values that are arbitrarily chosen within their recommended ranges by Geem [2-6] based on the observed efficiency of the technique in different problem fields. It is observed through the application of the standard harmony search method that the selection of these values is problem dependent. While a certain set of values yields a good performance of the technique in one type of design problem, the same set may not present the same performance in another type of design problem. Hence it is not possible to come up with a set of values that can be used in every optimum design problem. In each problem a sensitivity analysis is required to be carried out determine what set of values results a good performance. Adaptive harmony search method eliminates the necessity of finding the best set of parameter values [22, 23]. It adjusts the values of these parameters automatically during the optimization process. The basic components of the adaptive harmony search algorithm are outlined as follows.

6.4.3.1 Initialization of a Parameter Set

Harmony search method uses four parameters values of which are required to be selected by the user. This parameter set consists of a harmony memory size (hms), a harmony memory considering rate ($hmcr$), a pitch adjusting rate (par) and a maximum search number (N_{max}). Out of these four parameters, ($hmcr$) and (par) are made dynamic parameters in adaptive harmony search method that vary from one solution vector to another. They are set to initial values of $hmcr^{(0)}$ and $par^{(0)}$ for all the solution vectors in the initial harmony memory matrix. In the standard harmony search algorithm these parameters are treated as static quantities, and they are assigned to suitable values chosen within their recommended ranges of $hmcr \in [0.70, 0.95]$ and $par \in [0.20, 0.50]$ [2-6].

6.4.3.2 Initialization and Evaluation of Harmony Memory Matrix

The harmony memory matrix is established randomly as explained in section 3.1.1 which contains candidate design vectors for the optimum design problem under consideration. The structural analysis of each solution is then performed with the set of steel sections selected for design variables, and responses of each candidate solution are obtained under the applied loads. The objective function values of the feasible solutions that satisfy all problem constraints are directly calculated from Eqn. (6.1). However, infeasible solutions that violate some of the problem constraints are penalized using external penalty function approach, and their objective function values are calculated according to Eqn. (6.19).

$$\phi = W\left[1+\alpha\left(\sum_i g_i\right)\right] \tag{6.19}$$

In Eqn. (6.19), ϕ is the constrained objective function value, g_i is the i-th problem constraint value and α is the penalty coefficient used to tune the intensity of penalization as a whole. This parameter is set to an appropriate static value of $\alpha = 1$ in the numerical examples. Finally, the solutions evaluated are sorted in the matrix in the descending order of objective function values, that is, $\phi(\mathbf{I}^1) \leq \phi(\mathbf{I}^2) \leq \ldots \leq \phi(\mathbf{I}^{hms})$.

6.4.3.3 Generating a New Harmony Vector

In the adaptive algorithm a new set of values is sampled for *hmcr* and *par* parameters each time prior to improvisation (generation) of a new harmony vector, which in fact forms the basis for the algorithm to gain adaptation to varying features of the design space. Accordingly, to generate a new harmony vector in the algorithm proposed, a two-step procedure is followed consisting of (i) sampling of control parameters, and (ii) improvisation of the design vector.

6.4.3.3.1 Sampling of Control Parameters

For each harmony vector to be generated during the search process, first a new set of values are sampled for *hmcr* and *par* control parameters by applying a logistic normal distribution based variation to the average values of these parameters within the harmony memory matrix, as formulated in Eqns. (6.20 and 6.21).

$$(hmcr)^k = \left(1+\frac{1-(hmcr)'}{(hmcr)'}.e^{-\gamma.N(0,1)}\right)^{-1} \tag{6.20}$$

$$(par)^k = \left(1+\frac{1-(par)'}{(par)'}.e^{-\gamma.N(0,1)}\right)^{-1} \tag{6.21}$$

In Eqns. (6.20) and (6.21), $(hmcr)^k$ and $(par)^k$ represent the sampled values of the control parameters for a new harmony vector. The notation $N(0,1)$ designates a normally distributed random number having expectation 0 and standard deviation 1. The symbols $(hmcr)'$ and $(par)'$ denote the average values of control parameters within the harmony memory matrix, obtained by averaging the corresponding values of all the solution vectors within the $\mathbf{H}$ matrix, that is,

$$(hmcr)' = \frac{\sum_{i=1}^{hms} (hmcr)^i}{(hms)} \quad , \quad (par) = \frac{\sum_{i=1}^{hms} (par)^i}{(hms)} \tag{6.22}$$

Finally, the factor γ in Eqns. (6.20) and (6.21) refers to the learning rate of control parameters, which is recommended to be selected within a range of [0.25, 0.50]. In the numerical examples this parameter is set to 0.35.

In this implementation, for each new vector a probabilistic sampling of control parameters is motivated around average values of these parameters $(hmcr)'$ and $(par)'$ observed in the $\mathbf{H}$ matrix. Considering the fact that the harmony memory matrix at an instant incorporates the best (hms) solutions sampled thus far during the search, to encourage forthcoming vectors to be sampled with values that the search process has taken the most advantage in the past. The use of a logistic normal distribution provides an ideal platform in this sense because not only it guarantees the sampled values of control parameters to lie within their possible range of variation, i.e., [0, 1], but also it permits occurrence of small variations around $(hmcr)'$ and $(par)'$ more frequently than large ones. Accordingly, sampled values of control parameters mostly fall within close vicinity of the average values, yet remote values are occasionally promoted to check alternating demands of the search process.

6.4.3.3.2 Improvisation of the Design Vector

Upon sampling of a new set of values for control parameters, the new harmony vector $\mathbf{I}^k = \left[I_1^k, I_2^k, .., I_{ng}^k\right]$ is improvised in such a way that each design variable is selected at random from either harmony memory matrix or the entire discrete set. Which one of these two sets is used for a variable is determined probabilistically in conjunction with harmony memory considering rate $(hmcr)^k$ parameter of the solution. To implement the process a uniform random number r_i is generated between 0 and 1 for each variable I_i^k. If r_i is smaller than or equal to $(hmcr)^k$, the variable is chosen from harmony memory in which case it is assigned any value from the i-th column of the $\mathbf{H}$ matrix, representing the value set of the variable in (hms) solutions of the matrix (Eqn. 6.12). Otherwise (if $r_i > (hmcr)^k$), an arbitrary value is assigned to the variable from the entire design set.

$$I_i^k = \begin{cases} I_i^k \in \left\{I_i^1, I_i^2, ..., I_i^{hms}\right\} & \text{if } r_i \le (hmcr)^k \\ I_i^k \in \left\{1, ..., N_{sec}\right\} & \text{if } r_i > (hmcr)^k \end{cases} \tag{6.23}$$

If a design variable attains its value from harmony memory, it is checked whether this value should be pitch-adjusted or not. In pitch adjustment the value of a design variable ($I_i^{k'}$) is altered to its very upper or lower neighboring value obtained by adding ± 1 to its current value. This process is also operated probabilistically in conjunction with pitch adjusting rate $(par)^k$ parameter of the solution, Eqn. (6.21). If not activated by $(par)^k$, the value of the variable does not change. Pitch adjustment prevents stagnation and improves the harmony memory for diversity with a greater chance of reaching the global optimum.

$$I_i^{k'} = \begin{cases} I_i^k \pm 1 & if\ r_i \leq (par)^k \\ I_i^k & if\ r_i > (par)^k \end{cases} \tag{6.24}$$

6.4.3.4 Update of the Harmony Memory and Adaptivity

After generating the new harmony vector, its objective function value is calculated as per Eqn. (6.19). If this value is better (lower) than that of the worst solution in the harmony memory matrix, it is included in the matrix while the worst one is discarded out of the matrix. It follows that the solutions in the harmony memory matrix represent the best (hms) design points located thus far during the search. The harmony memory matrix is then sorted in ascending order of objective function value. Whenever a new solution is added into the harmony memory matrix, the $(hmcr)'$ and $(par)'$ parameters are recalculated using Eqn. (6.22). This way the harmony memory matrix is updated with the most recent information required for an efficient search and the forthcoming solution vectors are guided to make their own selection of control parameters mostly around these updated values. It should be underlined that there are no single values of control parameters that lead to the most efficient search of the algorithm throughout the process unless the design domain is completely uniform. On the contrary, the optimum values of control parameters have a tendency to change over time depending on various regions of the design space in which the search is carried out. The update of the control parameters within the harmony memory matrix enables the algorithm to catch up with the varying needs of the search process as well. Hence the most advantages values of control parameters are adapted in the course of time automatically (i.e., by the algorithm itself), which plays the major role in the success of *adaptive* harmony search method discussed in the paper.

6.4.3.5 Termination

The steps 4.3.4 and 4.3.5 are iterated in the same manner for each solution sampled in the process, and the algorithm terminates when a predefined number of solutions ($N_{\max}$) is sampled.

6.4.4 Improved Harmony Search (IHS, Mahdavi)

Standard harmony search method uses fixed values for both pitch adjustment rate *(par)* and arbitrary distance width *(bw)*. Prior to the application of the algorithm some appropriate values are selected for these parameters and they are kept the same until the end of the iterations. For example the value of the arbitrary distance bandwidth *(bw)* is taken as ±1 in the standard harmony search method, although some other value can also be used if preferred. It is also stated in the work of Mahdavi et. al. [20] that the use of for example small fixed values for pitch adjustment rate *(par)* with large values of arbitrary distance bandwidth *(bw)* can cause considerable increase in the total number of iterations required to reach the optimum solution, resulting in an undesirable poor performance of the algorithm. In order to avoid such a poor performance of the algorithm they have suggested adaptive expressions for both of these parameters instead of fixed values. The values of these parameters are adjusted dynamically by using Equations (6.25) and (6.26) during the harmony search iterations. However, the fixed value is used for the harmony memory consideration rate *(hmcr)* which is kept the same until the end of the iterations as in the standard harmony search method.

$$par(i) = par_{\min} + \frac{(par_{\max} - par_{\min})}{Iter_{\max}} \times i \qquad i = 1,2,\ldots,Iter_{\max} \tag{6.25}$$

where; $par(i)$ is pitch adjusting rate at iteration i, $par_{\max}$ and $par_{\min}$ are the maximum and the minimum values of pitch adjusting rates, $Iter_{\max}$ is the maximum iteration number.

$$bw(i) = bw_{\max} \exp\left(\frac{\ell n\left(\frac{bw_{\min}}{bw_{\max}}\right)}{Iter_{\max}} \times i \right) \tag{6.26}$$

where; $bw(i)$ is bandwidth in iteration i, $bw_{\max}$ and $bw_{\min}$ are the maximum and the minimum distance bandwidth. $par_{\max}$, $par_{\min}$, $bw_{\max}$ and $bw_{\min}$ are specified prior to the application of the algorithm. They are taken as 0.5, 0.05, 1 and 5 respectively. In this study this technique is used with adaptive error strategy explained in section 4.1 not with penalty function concept.

6.4.5 Global Best Harmony Search (GBHS, Mahdavi)

Mahdavi [5] also suggested another enhancement which makes use of the concept of particle swarm optimizer [29, 30] to the improved harmony search method [20]. In particle swarm optimizer system, a swarm of particles fly through the search space. Each particle represents a candidate solution to the optimization problem. The position of a particle is influenced by its best position and also the position of the best particle in the swarm. The global best harmony search modifies the pitch adjustment step of the harmony search method similar to particle swarm optimizer. It replaces the arbitrary distance width *(bw)* altogether and adds a social dimension to the harmony search by selecting the best harmony for the new harmony vector. The algorithm computes *(par)* from the equation (6.25) and but does not use *(bw)* at all. Instead, it employs the following equation to construct the new harmony vector, not the one given in Eq. (6.14).

$$I_i^{new} \rightarrow I_k^{best} \tag{6.27}$$

where; *best* is the index of the best harmony in the harmony memory matrix and k is the variable number randomly selected between 1 to ng which is the total number of design variables in the optimum design problem. In this study this technique is used with adaptive error strategy explained in section 4.1 not with penalty function concept.

6.4.6 Improved Harmony Search (IHSC, Coelho)

In another enhancement to standard harmony search algorithm Coelho et al. [24] has suggested another adaptive expression given in Eqn. (6.28) for the pitch adjustment rate *(par)*. This version of harmony search method is also called improved harmony search method and it has the same steps of the standard harmony search algorithm with the exception that it changes the value of the pitch adjustment rate *(par)* each iteration with the value computed from the following equation.

$$par\,(i) = par_{\min} + \left(par_{\max} - par_{\min}\right) \times \frac{\left(HMVal_{\max}\,(i) - Mean\,(HMVal\,)\right)}{\left(HMVal_{\max}\,(i) - HMVal_{\min}\,(i)\right)} \tag{6.28}$$

where; $par(i)$ is the value of the pitch adjusting rate in iteration i, $par_{\max}$ and $par_{\min}$ are the maximum and the minimum values of pitch adjusting rates respectively, $HMVal_{\max}$, $HMVal_{\min}$ and $Mean\,(HMVal)$ are the minimum, the maximum and the mean values of objective function in the harmony memory matrix respectively. The values of par_{max} and par_{min} are taken as 0.99 and 0.01 respectively in this study. In this study this technique is used with adaptive error strategy explained in section 4.1 not with penalty function concept.

6.4.7 Dynamic Harmony Search (DHS)

Dynamic harmony search is suggested in this study. This version of the harmony search method has the same steps of the standard harmony search algorithm with the exception that instead of using fixed values for both parameters of harmony memory considering rate *(hmcr)* and pitch adjusting rate *(par)* their values are calculated by means of adaptive expressions. The value of *(hmcr)* is computed from Eqn. (6.20) and *(par)* is calculated from Eqn. (6.28). In other words dynamic harmony search method is a mixture of adaptive harmony and Coelho's improved harmony search algorithms. The adaptive error strategy explained in section 4.1 but not the penalty function concept is employed in this technique as well.

6.5 Design Examples

Seven different structural optimization programs are coded each of which is based on one of the above explained versions of the harmony search algorithms. Three steel space frames are designed using these seven different versions of harmony search algorithms and the optimum solutions determined are compared with each other in order to evaluate the performance of each version.

6.5.1 Five-Story, Two-Bay Regular Steel Space Frame

The plan and 3D views of the five-story, two-bay steel frame shown in the Figures 6.3 and 6.4 is a regular steel frame with 54 joints and 105 members that are grouped into 11 independent design variables. The frame is subjected to gravity loads as well as lateral loads that are computed as per ASCE 7-05 [28]. The design dead and live loads are taken as 2.88kN/m^2 and 2.39kN/m^2 respectively. The ground snow load is considered to be 0.755kN/m^2 and a basic wind speed is 105mph (65 m/s). The un-factored distributed gravity loads on the beams of the roof and floors are tabulated in Table 6.2. The following load combinations are considered in the design of the frame according to the code specification. 1.2D+1.6L+0.5S, 1.2D+0.5L+1.6S, 1.2D+1.6W+0.5L+0.5S where D is the dead load, L represents the live load, S is the snow load and W is the wind load. The drift ratio limits of this frame are defined as 1.33 cm for inter story drift and 6.67 cm for top story drift. Maximum deflection of beam members is restricted as 1.67 cm.

Table 6.2 Beam gravity loading of the five-story, two bay steel frame

Beam Type	Uniformly distributed load (kN/m)		
	Dead Load	Live Load	Snow Load
Roof Beams	4.78	-	1.508
Floor Beams	4.78	5.76	-

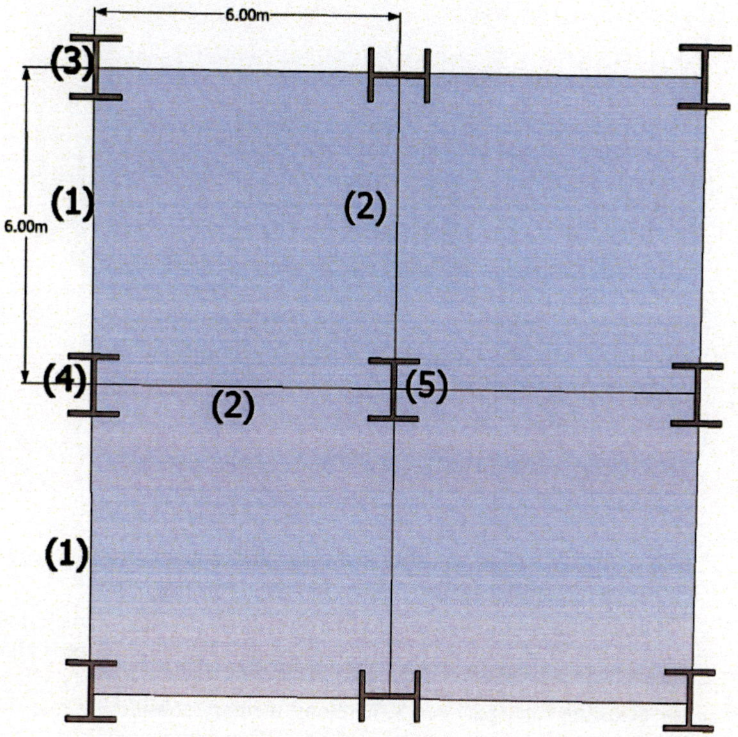

Fig. 6.3 Plan view of five-story, two bay steel frame

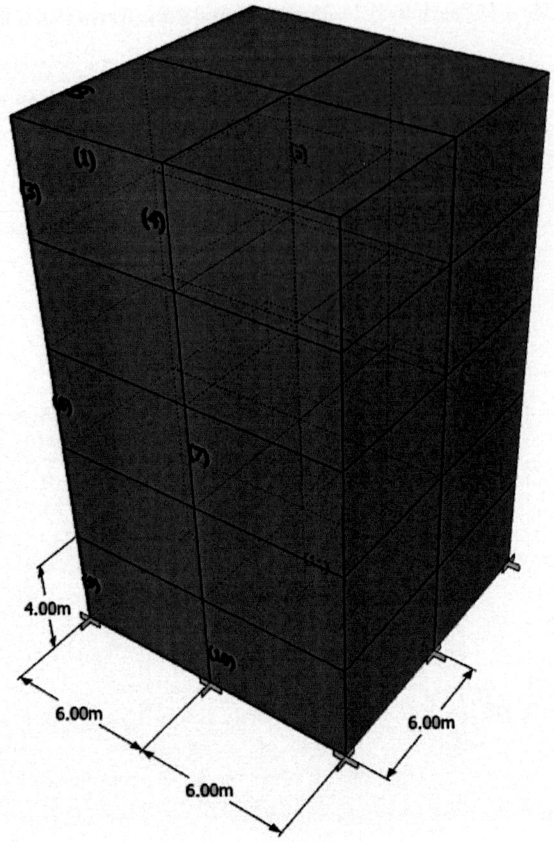

Fig. 6.4 3D View of the five-story, two bay steel frame

Optimum design problem of the five-story, two-bay steel frame is solved by using seven different versions of harmony search algorithms. In these algorithms the following harmony search parameters are used: harmony memory size (hms) = 20, pitch adjusting rate (par) = 0.3, harmony memory considering rate (hmcr) = 0.9 and maximum iteration number = 50000. The optimum designs obtained from each of these algorithms are shown in Table 6.3. It is apparent from the table that the lightest weight is 261.128 kN which is obtained by the adaptive harmony search algorithm and the second lightest design is 261.360kN attained by the dynamic harmony search method suggested in this study. The design histories of these algorithms for the best solutions are plotted in Fig. 6.5. It is apparent from the figure that the dynamic harmony search and adaptive harmony search algorithms show better performance than others. It is noticed that the minimum weight determined by the dynamic harmony search and adaptive harmony search algorithms are 12.3% less than the heaviest frame.

Table 6.3 Design results of the five-story, two bay steel frame

Member Group	Type	SHSAES	SHSPF	AHSPF
1	Beam	W530X66	W360X39	W410X46.1
2	Beam	W310X38.7	W310X38.7	W310X38.7
3	Column	W200X35.9	W360X39	W250X32.7
4	Column	W200X35.9	W200X46.1	W200X46.1
5	Column	W360X44	W610X101	W460X52
6	Column	W310X38.7	W530X66	W410X53
7	Column	W360X72	W410X60	W360X64
8	Column	W610X92	W1000X222	W920X201
9	Column	W410X53	W610X92	W410X53
10	Column	W360X72	W410X60	W460X74
11	Column	W760X147	W1100X390	W1000X258
Max. Strength Ratio		0.979	0.986	0.936
Top Drift(cm)		4.837	5.264	4.81
Inter Story Drift(cm)		1.333	1.329	1.331
Maximum Iteration		50000	50000	50000
Weight (kN)		278.196	268.172	261.128

Member Group	Type	IHS (Mahdavi)	GBHS (Mahdavi)	IHSC (Coelho)	DHS Present Study
1	Beam	W530X66	W530X74	W530X74	W460X52
2	Beam	W310X38.7	W360X44	W360X44	W250X38.5
3	Column	W200X35.9	W200X41.7	W200X41.7	W310X38.7
4	Column	W200X35.9	W200X41.7	W200X41.7	W200X35.9
5	Column	W360X44	W360X44	W360X44	W460X52
6	Column	W310X38.7	W360X44	W360X44	W360X51
7	Column	W360X64	W360X44	W310X44.5	W250X73
8	Column	W610X82	W610X92	W610X113	W610X101
9	Column	W410X53	W360X51	W360X51	W360X51
10	Column	W360X64	W410X60	W410X60	W460X74
11	Column	W840X193	W840X193	W760X134	W840X193
Max. Strength Ratio		0.989	0.986	0.979	0.975
Top Drift(cm)		4.763	4.579	4.73	5.074
Inter Story Drift(cm)		1.331	1.325	1.333	1.33
Maximum Iteration		50000	50000	50000	50000
Weight (kN)		275.46	297.928	297.424	261.36

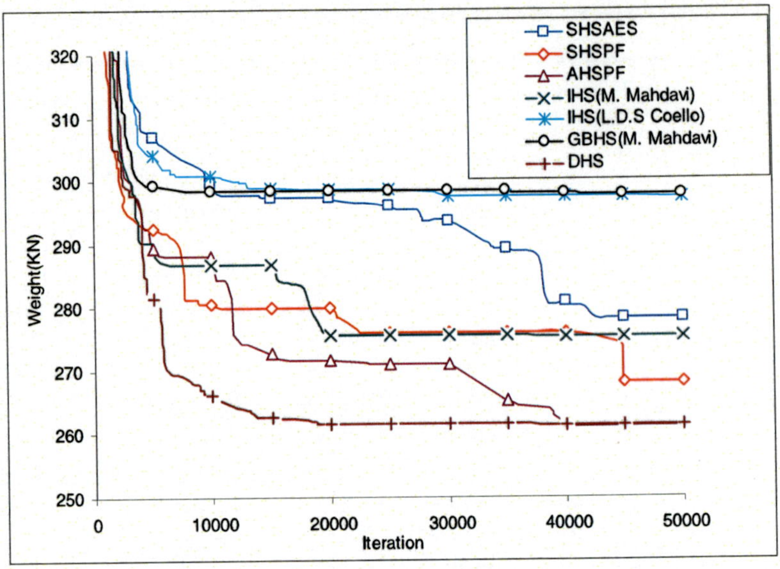

Fig. 6.5 Design histories of the five-story, two bay steel frame

6.5.2 *Ten-Story, Four-Bay Steel Space Frame*

The three dimensional view, side view and the plan of ten-story four-bay steel frame shown in Figures 6.6 and 6.7 is taken from [19, 22]. This frame has 220 joints and 568 members which are collected in 25 member groups which are the independent design variables as shown in Figure 6.6. Inner roof beams, outer roof beams, inner floor beams and outer floor beams are subjected to 14.72kN/m, 7.36kN/m, 21.43kN/m and 10.72kN/m uniformly distributed gravity loads respectively. Lateral forces acting at the level of each story of the steel space frame are tabulated in Table 6.3. Drift ratio limits are defined as $h/400$ where h is the story height for inter story drift and $H/400$ for top story drift where H is the total height of the structure.

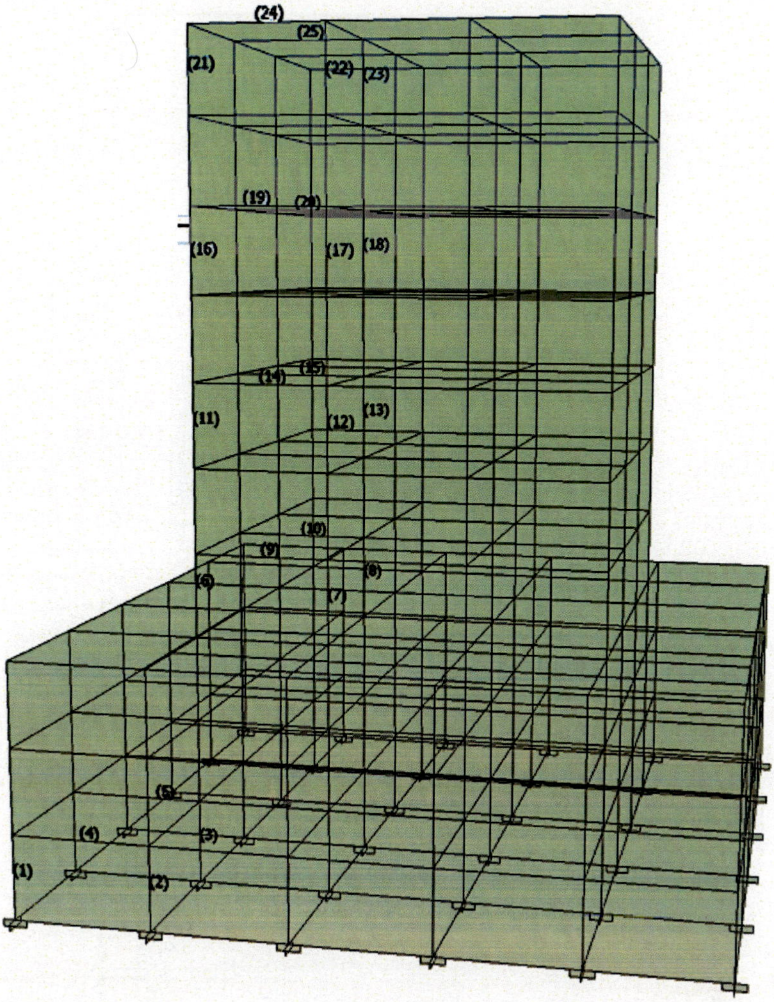

Fig. 6.6 3-D view of ten-story, four-bay steel space frame

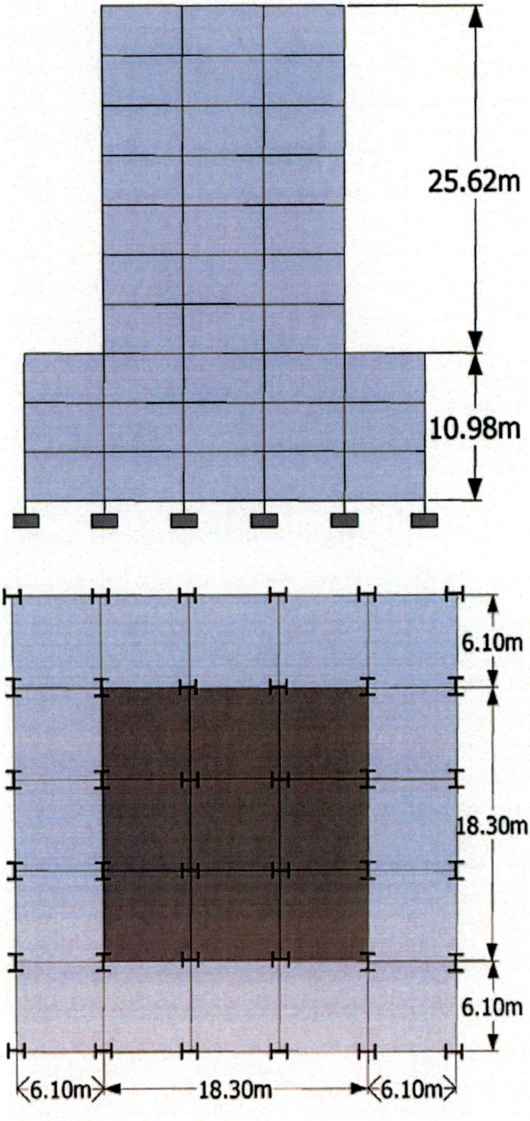

Fig. 6.7 Plan and side view of ten-story, four-bay steel space frame

Table 6.4 Lateral loads acting at the level of each story of ten-story, four-bay steel space frame

Story Number	Windward		Leeward	
	(lb/ft)	(kN/m)	(lb/ft)	(kN/m)
1	12.51	0.1825	127.38	1.8585
2	28.68	0.4184	127.38	1.8585
3	44.68	0.6519	127.38	1.8585
4	156.86	2.2886	127.38	1.8585
5	167.19	2.4393	127.38	1.8585
6	176.13	2.5698	127.38	1.8585
7	184.06	2.6854	127.38	1.8585
8	191.21	2.7897	127.38	1.8585
9	197.76	2.8853	127.38	1.8585
10	101.9	1.5743	127.38	1.8585

Optimum design problem of this frame is solved under the design constraints described in section 2 by using seven different versions of harmony search algorithms described. In these algorithms the following harmony search parameters are used: harmony memory size (hms) = 30, pitch adjusting rate (par) = 0.3, harmony memory considering rate (hmcr) = 0.9, and maximum iteration number = 80000. The optimum designs obtained by each of these algorithms are shown in Table 6.5. It is clear from the table that the lightest weight is 1699.88kN which is obtained by the dynamic harmony search method and the second lightest weight of the frame is 1714.46kN attained by the adaptive harmony search algorithm. The design histories of these algorithms for the best solutions are plotted in Fig. 6.8. It is apparent from the figure that the dynamic harmony search algorithms shows steady convergence and outperforms others. It is noticed that the minimum weight determined by the dynamic harmony search is 7.3% less than the heaviest frame.

Table 6.5 Design results for ten-story, four-bay steel space frame

Member Group	Type	SHSAES	SHSPF	AHSPF
1	Column	W310X28.3	W150X22.5	W310X28.3
2	Column	W310X28.3	W200X86	W310X28.3
3	Column	W360X32.9	W760X173	W360X39
4	Beam	W410X46.1	W250X25.3	W410X46.1
5	Beam	W410X46.1	W410X38.8	W460X52
6	Column	W410X46.1	W410X114	W410X38.8
7	Column	W460X52	W760X196	W410X38.8
8	Column	W530X66	W840X176	W610X82
9	Beam	W310X23.8	W360X110	W310X23.8
10	Beam	W460X60	W690X152	W460X52
11	Column	W250X67	W410X100	W200X35.9
12	Column	W250X73	W460X128	W250X80
13	Column	W310X44.5	W690X170	W360X44
14	Beam	W310X97	W310X60	W460X113
15	Beam	W460X128	W530X85	W460X113
16	Column	W530X85	W310X97	W530X85
17	Column	W310X107	W310X117	W460X128
18	Column	W530X150	W530X85	W610X217
19	Beam	W690X170	W250X32.7	W760X173
20	Beam	W310X117	W410X46.1	W530X150
21	Column	W760X196	W310X97	W690X217
22	Column	W840X176	W200X59	W760X173
23	Column	W150X29.8	W410X60	W150X24
24	Beam	W250X73	W250X32.7	W250X49.1
25	Beam	W410X132	W310X38.7	W360X134
Max. Strength Ratio		0.99	1	0.995
Top Drift(cm)		8.158	7.639	7.695
Inter Story Drift(cm)		0.914	0.914	0.914
Maximum Iteration		80000	80000	80000
Weight (kN)		1756.56	1800.28	1714.46

Table 6.5 (*continued*)

Member Group	Type	IHS (Mahdavi)	GBHS (Mahdavi)	IHSC (Coelho)	DHS Present Study
1	Column	W310X28.3	W310X23.8	W200X19.3	W310X23.8
2	Column	W310X28.3	W360X32.9	W530X85	W310X28.3
3	Column	W360X39	W460X52	W410X132	W410X46.1
4	Beam	W410X38.8	W310X38.7	W310X23.8	W410X46.1
5	Beam	W530X66	W360X64	W460X60	W410X46.1
6	Column	W530X66	W530X150	W310X107	W410X46.1
7	Column	W410X46.1	W310X32.7	W760X196	W530X66
8	Column	W410X38.8	W690X125	W760X257	W530X66
9	Beam	W310X23.8	W310X23.8	W530X109	W310X23.8
10	Beam	W460X60	W360X32.9	W360X134	W460X52
11	Column	W250X67	W250X73	W310X97	W250X73
12	Column	W250X73	W250X58	W250X101	W250X67
13	Column	W360X44	W360X51	W690X170	W360X44
14	Beam	W310X97	W250X80	W410X60	W310X97
15	Beam	W410X100	W610X113	W530X85	W460X113
16	Column	W530X85	W530X85	W250X73	W610X92
17	Column	W310X107	W610X101	W250X101	W360X134
18	Column	W530X150	W690X140	W530X92	W460X128
19	Beam	W690X170	W610X174	W410X38.8	W840X176
20	Beam	W360X162	W610X155	W410X46.1	W360X134
21	Column	W760X196	W920X201	W250X73	W760X196
22	Column	W690X170	W1000X296	W200X59	W840X176
23	Column	W150X29.8	W150X24	W410X60	W150X24
24	Beam	W410X53	W250X49.1	W310X28.26	W250X49.1
25	Beam	W310X129	W840X193	W310X28.3	W530X150
Max. Strength Ratio		0.988	0.961	0.965	0.976
Top Drift(cm)		7.774	8.077	7.589	8.069
Inter Story Drift(cm)		0.915	0.914	0.913	0.912
Maximum Iteration		80000	80000	80000	80000
Weight (kN)		1739.47	1842.95	1773.51	1699.88

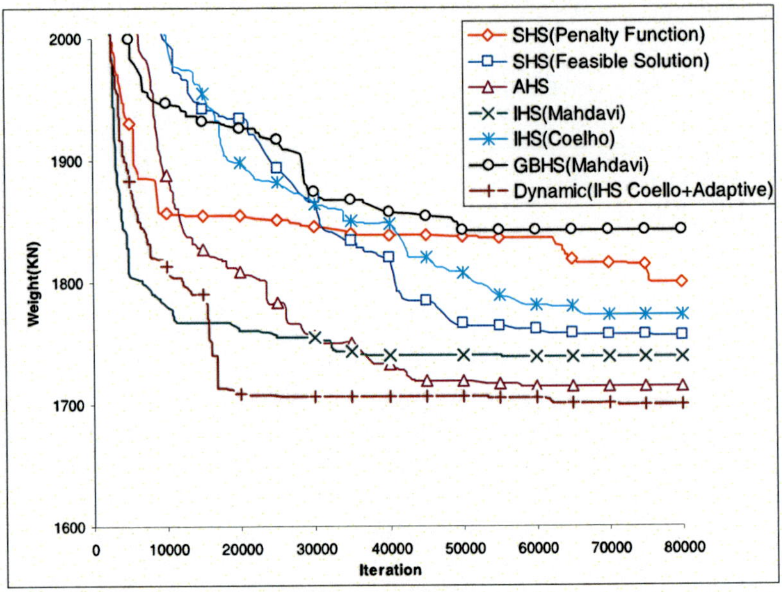

Fig. 6.8 Design histories of ten-story, four-bay steel space frame

6.5.3 Twenty-Story, 1860–Member, Steel Space Frame

The three dimensional view and plan of twenty-story, 1860-member steel space frame are illustrated in Figures 6.9 and 6.10. The frame has 820 joints and 1860 members which are collected in 86 independent design variables. The member grouping is given in Figure 6.9. The frame is subjected to gravity loads as well as lateral loads that are computed according to ASCE 7-05 [28]. The design dead and live loads are taken as 2.88kN/m^2 and 2.39kN/m^2 respectively. Basic wind speed is considered as 85mph (38 m/s). The following load combinations are considered in the design of the frame according to code specification [25]. 1.2D+1.3WZ+0.5L+0.5S and 1.2D+1.3WX+0.5L+0.5S where D is the dead load, L represents the live load, S is the snow load and WX, WZ are the wind loads in the global X and Z axis respectively. Drift ratio limits are defined as $h/400$ where h is the story height for inter story drift and $H/400$ for top story drift where H is the height of structure.

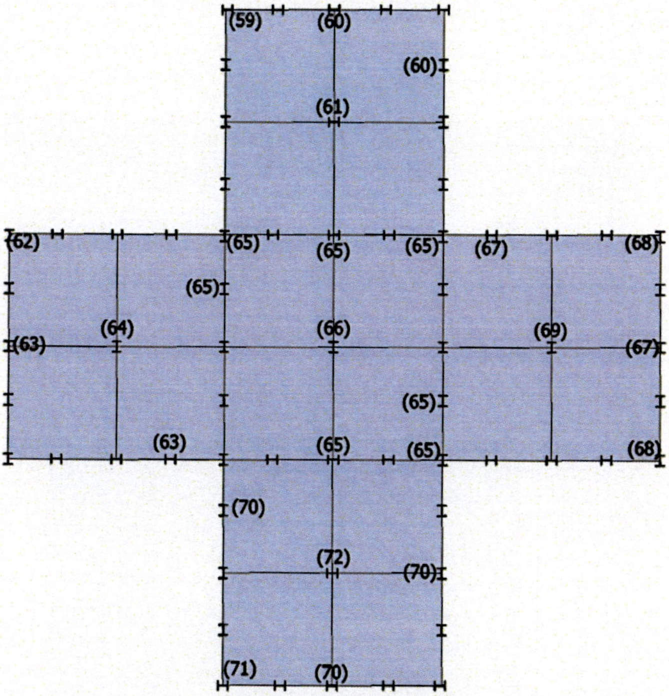

(a) Plan view of 1-4th story

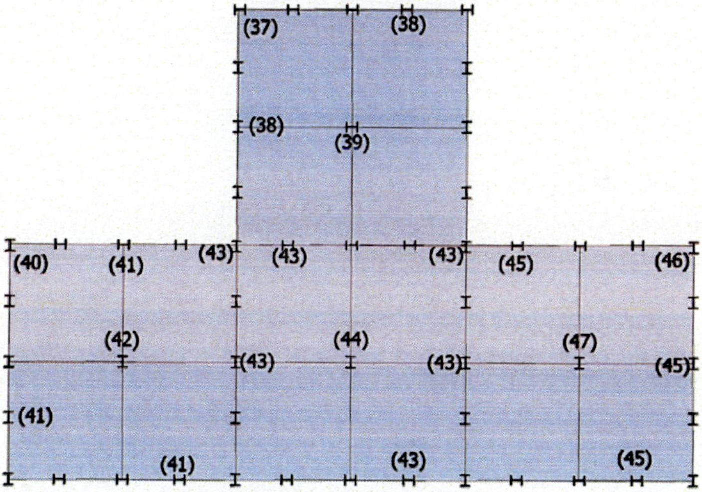

(b) Plan view of 5-8th story

Fig. 6.9 Plan view of twenty-story, 1860 member steel space frame

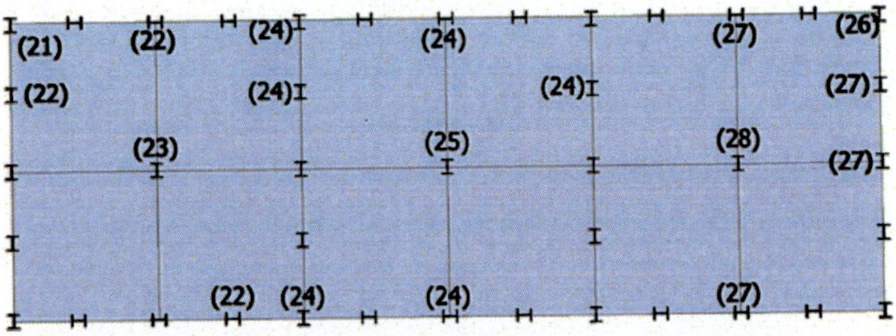

(c) Plan view of 5-8th story

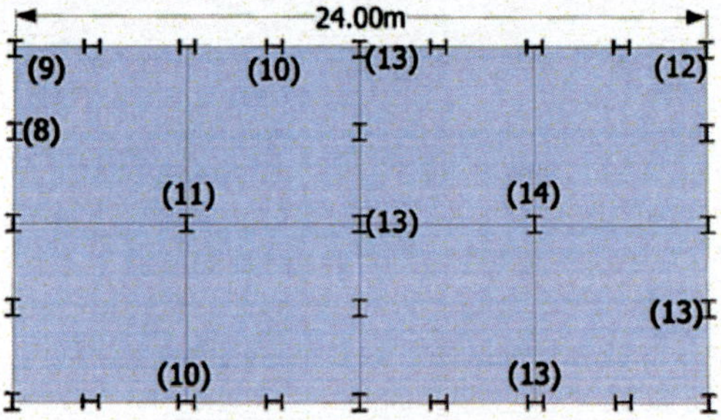

(d) Plan view of 13-16th story

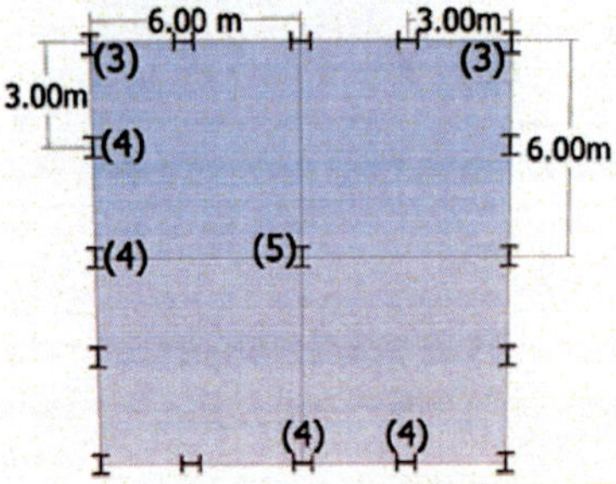

(e) Plan view of 17-20th story

Fig. 6.9 (*continued*)

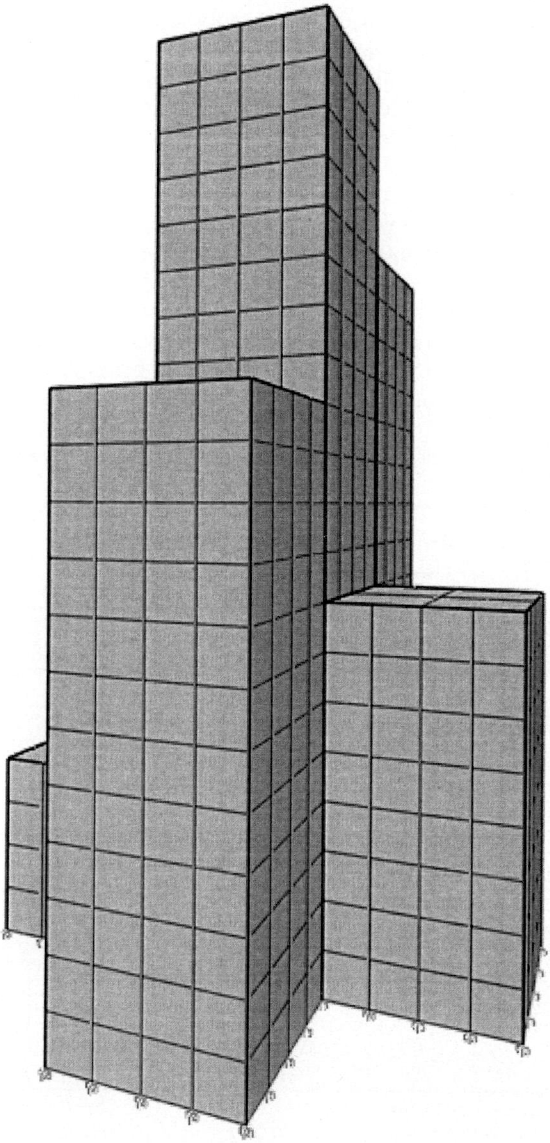

Fig. 6.10 3-D view of twenty-story, 1860 member steel space frame

This frame which has 1860 members is also designed seven times using different versions of harmony search algorithms. In these runs the harmony search parameters are selected as: harmony memory size (hms) = 50, the pitch adjusting rate (par) = 0.3, the harmony memory considering rate (hmcr) = 0.9, the maximum iteration number = 80000. The optimum designs obtained by each of these algorithms are given in Table 6.6. It is apparent from the table that the best design is obtained by dynamic harmony search method which has the minimum weight of 4716.576kN. The second best design is obtained by the adaptive harmony search algorithm (AHS) as 4932.012kN. Difference between these results is only 4%. However the minimum weights of best designs obtained by other harmony search algorithms are around 6000kN. Therefore, it can be stated that the dynamic and adaptive harmony search methods demonstrated better performance than the other versions of harmony search methods. The design histories of each harmony search method are shown in Fig. 6.11. The figure clearly reveals the fact that the dynamic and adaptive harmony search methods perform better than the other versions of the harmony search algorithms from the beginning of the design cycles.

Table 6.6 Design results for twenty-story, 1860-member steel space frame

Beam Type	Member Group	SHSAES	SHSPF	AHSPF
Outer	1	W410X67	W410X53	W460X60
Interior	2	W460X52	W460X68	W460X60
Columns Story	Member Group	SHSAES	SHSPF	AHSPF
20,19	3	W410X85	W250X73	W310X38.7
19,18	6	W410X132	W690X125	W410X38.8
16,15	9	W410X60	W200X41.7	W200X22.5
14,13	15	W920X223	W460X68	W200X26.6
12,11	21	W920X271	W610X174	W360X39
10,9	29	W1000X314	W840X176	W460X60
8,7	37	W150X29.8	W200X22.5	W250X25.3
6,5	48	W410X46.1	W460X128	W310X28.3
4,3	59	W1000X272	W840X176	W360X51
4,3	72	W310X44.5	W200X86	W250X32.7
2,1	73	W1000X272	W840X251	W690X125
Top Story Drift (cm)		9.013	8.777	9.809
Inter-Story Drift (cm)		0.742	0.738	0.75
Max. Strength Ratio		0.84	0.837	1
Weight (kN)		6319.554	6204.204	4932.012

Table 6.6 (*continued*)

Beam Type	Member Group	HIS (Mahdavi)	GBHS (Mahdavi)	HIS (Coelho)	DHS Present Study
Outer	1	W410X60	W410X60	W360X51	W410X53
Interior	2	W460X60	W410X60	W460X60	W460X60
Columns Story	**Member Group**	**HIS (Mahdavi)**	**GBHS (Mahdavi)**	**HIS (Coelho)**	**DHS (Present Study)**
20,19	3	W200X46.1	W250X80	W360X51	W410X53
19,18	6	W310X52	W460X82	W460X128	W410X53
16,15	9	W200X41.7	W200X22.5	W150X37.1	W200X22.5
14,13	15	W920X201	W760X161	W410X60	W310X28.3
12,11	21	W920X201	W920X345	W690X265	W310X32.7
10,9	29	W1000X258	W1100X499	W1000X412	W360X39
8,7	37	W250X73	W610X82	W360X39	W310X28.3
6,5	48	W840X193	W760X147	W760X196	W310X38.7
4,3	59	W1000X222	W1000X249	W760X257	W610X101
2,1	73	W1100X343	W1000X249	W1100X433	W610X101
Top Story Drift (cm)		8.954	8.76	9.576	10.02
Inter-Story Drift (cm)		0.75	0.744	0.733	0.748
Max. Strength Ratio		0.795	0.69	0.892	1
Weight (kN)		6259.736	6431.886	6337.728	4716.756

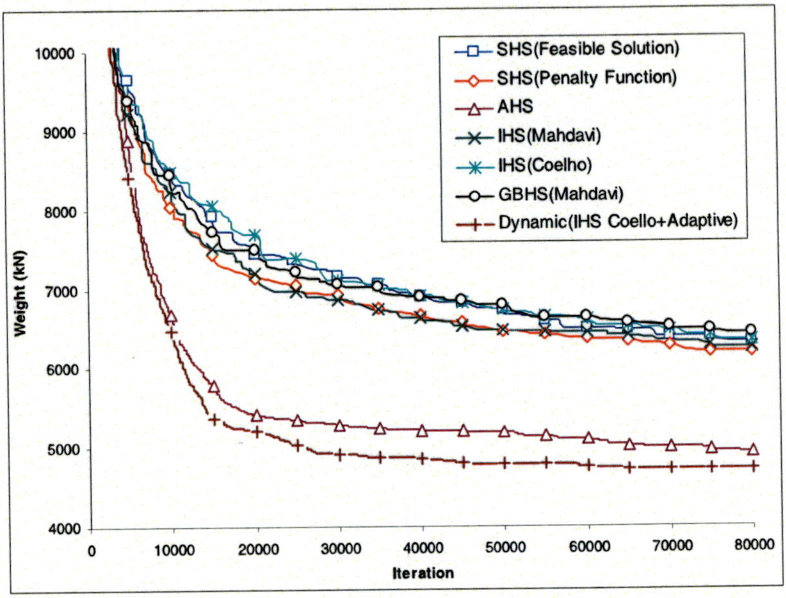

Fig. 6.11 Design histories of for twenty-story, 1860-member steel space frame

Table 6.7 Performance evaluation of seven different versions of the harmony search algorithms in the design examples

Design Examples	SHSAES	SHSPF	AHSPF	IHS (Mahdavi)	GBHS (Mahdavi)	IHSC (Coelho)	DHS Present Study
Five-story frame	5	3	**1**	4	7	6	2
Ten-story frame	4	6	2	3	7	5	**1**
Twenty-story frame	5	3	2	4	7	5	**1**

6.6 Conclusions

Seven different structural optimization algorithms are developed that are based on seven different versions of the harmony search algorithms that are recently developed. Three steel space frames are designed by these algorithms to evaluate their performance in finding the optimum solutions. All of these alternative harmony search algorithms are shown to be reliable, robust and effective algorithms. However, two versions among the all; adaptive harmony search method and dynamic harmony search method show better performance than the other versions. Particularly in the third design example where there are relatively large number of design variables and bigger design domain, the dynamic harmony search method has succeeded to find the optimum weight which is 25.36% less than the one determined by the standard harmony search algorithm. The performance evaluation of all these techniques in the design of three steel space frames considered is summarized in Table 6.7.

References

1. Load and Resistance Factor Design, Structural Members Specifications Codes, 3 edn.,vol.1 American Institute of Steel Construction (2001)
2. Lee, K.S., Geem, Z.W.: A new structural optimization method based on the harmony search algorithm. Computers and Structures 82, 781–798 (2004)
3. Lee, K.S., Geem, Z.W.: A new meta-heuristic algorithm for continuous engineering optimization: Harmony search theory and practice. Computer Methods in Applied Mechanics and Engineering 194, 3902–3933 (2005)
4. Geem, Z.W., Lee, K.S., Tseng, C.-L.: Harmony Search for Structural Design. In: Proceedings of 2005 Genetic and Evolutionary Computation Conference (GECCO-2005), Washington, DC, USA, June 25–29 pp. 651–652 (2005)
5. Geem, Z.W. (ed.): Music-Inspired Harmony Search Algorithm; Theory and Applications, Studies in Computational Intelligence. SCI, vol. 191. Springer, Heidelberg (2009)
6. Geem, Z.W. (ed.): Harmony Search Algorithm for Structural Design Optimization. SCI, vol. 219. Springer, Heidelberg (2009)
7. Kochenberger, G.A., Glover, F.: Handbook of Meta-Heuristics. Kluwer Academic Publishers, Dordrecht (2003)
8. Blum, C., Roli, A.: Metaheuristics in combinatorial optimization: Overview and conceptual comparison. ACM Computing Surveys 35(30), 268–308 (2003)
9. Gonzales, T.F.: Handbook of Approximation Algorithms and Metaheuristics. CRC Press, Chapman & Hall (2007)
10. Yang, X.-S.: Nature Inspired Metaheuristic Algorithms. Luniver Press (2008)
11. Luke, S.: Essentials of Metaheuristic. Lulu Press (2011)
12. Erdal, F., Saka, M.P.: Optimum design of grillage systems using harmony search algorithm. In: Proceedings of 8th International Conference on Computational Structures Technology (CST 2006), Las Palmas de Gran Canaria, Spain September 12–15 (2006)
13. Saka, M.P.: Optimum design of steel sway frames to BS5950 using harmony search algorithm. In: Proceedings of The Eleventh International Conference on Civil, Structural and Environmental Engineering Computing (CC-2007) Civil-Comp Press, St. Julians (2007)
14. Saka, M.P.: Optimum geometry design of geodesic domes using harmony search algorithm. An International Journal, Advances in Structural Engineering 10(6), 595–606 (2007)
15. Carbas, S., Saka, M.P.: Optimum design of single layer network domes using harmony search method. Asian Journal of Civil Engineering, 10(1), 97–112 (2009)
16. Hasançebi, O., Çarbaş, S., Doğan, E., Erdal, F., Saka, M.P.: Performance evaluation of metaheuristic techniques in the optimum design of real size pin jointed structures. Computers and Structures 87, 284–302 (2009)
17. Saka, M.P.: Optimum design of steel sway frames to BS5950 using harmony search algorithm. Journal of Constructional Steel Research 65, 36–43 (2009)

18. Saka, M.P., Erdal, F.: Harmony search based algorithm for the optimum design of grillage systems to LRFD-AISC. Structural and Multidisciplinary Optimization 38, 25–41 (2009)
19. Hasançebi, O., Çarbaş, S., Doğan, E., Erdal, F., Saka, M.P.: Comparison of non-deterministic search techniques in the optimum design of real size steel frames. An International Journal, Computers and Structures 88(17-18), 1033–1048 (2010)
20. Mahdavi, M., Fesanghary, M., Damangir, E.: An improved harmony search algorithm for solving optimization problems. Applied Mathematics and Computation 188, 1567–1579 (2007)
21. Omran, M.G.H., Mahdavi, M.: Global-Best harmony search. Applied Mathematics and Computation 198, 643–656 (2008)
22. Saka, M.P., Hasancebi, O.: Adaptive Harmony Search Algorithm for Design Code Optimization of Steel structures. In: Geem, Z.W. (ed.) Harmony Search Algorithms for Structural Design Optimization. SCI, vol. 239, pp. 79–120. Springer, Heidelberg (2009)
23. Hasançebi, O., Erdal, F., Saka, M.P.: An Adaptive Harmony Search Method for Structural Optimization. Journal of Structural Engineering, ASCE 136(4), 419–431 (2010)
24. Coelho, L.D.S., Bernett, D.L.D.A.: An improvement harmony search algorithm for synchronization of discrete-time chaotic systems. Chaos, Solutions and Fractals 41, 2526–2532 (2009)
25. Ad Hoc Committee on Serviceability: Structural Serviceability: A critical Appraisal and Research needs. Journal of Structural Engineering, ASCE 112(12), 2646–2664 (1986)
26. Ellingwood, B.: Serviceability guidelines for steel structures. Engineering Journal, AISC 26,1st Quarter 1–8 (1989)
27. Chen, W.F., Kim, S.-E.: LRFD Steel Design Using Advanced Analysis. CRC Press, Boca Raton (1997)
28. ASCE 7-05. Minimum Design Loads for Building and Other Structures (2005)
29. Bonabeau, E., Dorigo, M., Theraulaz, G.: Swarm Intelligence: FromNatural to Artificial System. Oxford University Press, U.K (1999)
30. Kennedy, J., Eberhart, R., Shi, Y.: Swarm Intelligence. Morgan KaufmanPublishers, San Francisco (2001)

Chapter 7
Waveform Optimization for Integrated Radar and Communication Systems Using Meta-Heuristic Algorithms

Momin Jamil and Hans-Jürgen Zepernick

Abstract. Integration of multiple functions such as navigation and radar tasks with communication applications has attracted substantial interest in recent years. In this chapter, we therefore focus on the waveform optimization for such integrated systems based on Oppermann sequences. These sequences are defined by a number of parameters that can be chosen to design sequence sets for a wide range of performance characteristics. It will be shown that meta-heuristic algorithms are well-suited to find the optimal parameters for these sequences. The motivation behind the use of biologically inspired heuristic and/or meta-heuristic algorithms is due to their ability to solve large, complex, and dynamic problems.

7.1 Introduction

In recent years, integration of multiple functions such as navigation and radar tasks with communication applications has sparked a number of research initiatives. This includes research on future signals for hybrid receivers for Global Navigation Satellite Systems (GNSS)/communication and others tasks. The many benefits of multi-functionality range from reducing costs and probability of intercept to offering tolerable co-site interference. While navigation and radar applications require waveform designs that offer excellent autocorrelation characteristics, the target for communication applications is on sets of waveforms with minimum crosscorrelation among the sequences in the set. In the former case, typically only a single sequence

Momin Jamil
Harman International, Automotive Division,
Becker-Göring Str. 16, D-76307 Karlsbad, Germany
e-mail: momin.jamil@harman.com

Hans-Jürgen Zepernick
Blekinge Institute of Technology,
SE-37179 Karlskrona, Sweden
e-mail: hans-jurgen.zepernick@bth.se

X.-S. Yang & S. Koziel (Eds.): Computational Optimization & Applications, SCI 359, pp. 183–203.
springerlink.com

is needed while in the latter case many sequences are required to support access of multiple users to the common transmission medium. As excellent autocorrelation properties come at the expense of crosscorrelation characteristics and vice versa, a related waveform optimization problem has to be posed and solved taking into account these conflicting requirements.

As far as the integration of radar and communication functionalities are concerned, the Office of Naval Research in 1996 launched the Advanced Multifunction Radio Frequency Concept (AMRFC) program [18,53]. This major program was motivated by the lack of integration of radar, communications, and electronic warfare functions which resulted in a significant increase of the number of topside antennas. Furthermore, it was realized that the lack of integration may also cause severe problems with antenna blockage and difficulties with own-ship electromagnetic interference. Also, a large number of antennas puts stress on maintenance resources. The concepts developed within the AMRFC program are centered around suitable broadband RF apertures that can cope with simultaneous operation of multiple functions and as such focuses on the rather expensive radio frequency (RF) front-end. A different approach on the basis of linear frequency modulated (LFM) waveforms, also known as chirps, has been proposed in [49]. In order to enhance the orthogonality among the signals and to support distinct separation of the different functions, it uses up-chirps for the communications component and down-chirps for the radar functionality of the integrated system. In this way, the suggested chirp signals allow for the radar and communication data to be simultaneously transmitted and received using some standard antenna array. Noting the inherent connection of the chirp-based integration concept to spread spectrum techniques, the work of [59, 60] investigated integrated radar and communication systems with the help of bipolar pseudo noise (PN) sequences, namely m-sequences [14, 63]. However, one of the severe drawbacks of m-sequences with respect to radar applications is their poor Doppler tolerance [32] and related problems of detecting multiple targets. These and related designs such as polyphase Barker sequences are optimized only with respect to the zero Doppler cut of the ambiguity function but produce much higher interference levels in the presence of Doppler shifted waveforms. As for the application to communications, large sets of m-sequences that would be needed to support multiple-access of many users have typically rather poor crosscorrelation properties [63]. As a consequence, they are generally only used as components of more complex designs such as Gold sequences. On the other hand, the large advances in modern integrated circuit technologies would facilitate an efficient implementation of more advanced sequence designs such as complex-valued sequences. Clearly, efficient optimization methods are needed to find suitable waveform and sequence designs for different applications.

Over the last few decades, researchers around the world have developed a vast number of algorithms to solve different optimization problems. Many of these algorithms are based on numerical linear and non-linear programming methods. As a result, the related algorithms require substantial gradient information and try to improve the solution in the proximity of an initial starting point. As a consequence, these methods provide useful strategies to find the global optimum for rather ideal and simple models. However, if the objective function and constraints have multiple or sharp peaks, these methods tend to become unstable. Most of the real world problems turn out to be too complex and difficult to solve using numerical based optimization methods as these tend to fail or are even unable to solve them. There exist also several direct search approaches which use no gradient information such as the Hooke and Jeeves method [17], Nelder-Mead simplex method [44], the Rosenbrock method [50], and the Powell method [47]. Common to these methods is that they take some basic approach of heading downhill from an arbitrary starting point but differ in deciding in which direction to and how far to move. Accordingly, the final outcome depends somewhat on the initial guess of the starting point. This would not be a major shortcoming if the parameter space is well behaved, i.e. if it contains a single, well-defined minimum. However, if the parameter space contains many local minima, as may be the case in waveform optimization, it can be more difficult for such traditional approaches to find the global minimum. In contrast to population based algorithms, these direct searches cannot explore the search space effectively in different directions simultaneously. Successive improvements can be made to speed up the downhill movement of the algorithms but this does not improve the algorithms ability to find the global minimum instead of converging to a local minimum.

The drawbacks of numerical methods motivated researchers to adopt ideas from nature and to translate them to solve problems in engineering sciences. This has led to the inception of many biologically inspired heuristic or meta-heuristic algorithms to solve challenging optimization problems. The word "meta" means beyond or higher and "heuristic" means to find or to discover by trial and error. These methods have proven to be efficient in handling computationally complex problems. They aim at defining effective general purpose methods to explore the solution space and avoid tailoring them to a specific problem. Due to their general purpose nature, they can be applied to a wide range of problems. Meta-heuristic algorithms are also referred to as black-box algorithms as they exploit limited knowledge about the problem to be solved. As no gradient or Hessian matrix information is required for their operation, they are also referred to as derivative-free or zero-order algorithms [5]. The term zero-order implies that only the function values are used to establish the search vector. Moreover, the function to be optimized does not necessarily have to be continuous or differentiable and may also be accompanied by a set of constraints. The choice of method for solving a particular problem depends primarily on the type and characteristics of the problem at hand. It must be stressed that the goal of a particular method used is to find the "best" solution of some sort to a problem compared to finding the optimal solution. In this context, the term "best" refers to an acceptable or satisfactory solution to the problem. This could be the absolute best solution from a set of candidate solutions or may be any of the candidate solutions. The requirements and characteristics of the problem determine if the overall best solution can be found [10, 54].

Nature has an evolution span of millions or even billions of years. In all these years, it has mastered the art of finding a perfect solution to almost all the problems it has been confronted with. As mentioned above, the development of nature inspired optimization algorithms has been an area of active research during recent years and resulted in many approaches such as genetic algorithms (GA), ant colony optimization (ACO), bee algorithms (BA), artificial bee algorithms (ABC), particle swarm optimization (PSO), simulated annealing (SA), harmony search (HS), firefly algorithms (FA), and artificial immune systems (AIS). The interested reader may be referred to [4, 10, 15, 31, 51, 56, 62] and the reference therein for more details and discussions on these topics.

Given the vast amount of available optimization methods, their application in waveform design also stretches from simple searches over more sophisticated and computational demanding realizations to the use of meta-heuristic algorithms. A simple computer search has been used in [45] to obtained sets of sequences with various combinations of sequence parameters. In [58], the optimization of orthogonal polyphase spreading sequences for wireless data applications is reported. It uses a built-in standard 'fmin' function provided in the numerical computing environment MATLAB. In particular, the related functions support multidimensional unconstrained nonlinear minimization including the Nelder-Mead direct search method. As the utilized cost functions in terms of average mean-square autocorrelation and crosscorrelation are very irregular and may have several local minima, the authors report the dependency of the optimization outcome on the starting point and corresponding convergence to different local minima. A similar optimization problem for complex-valued spreading sequences has been investigated in [9] using a global optimization method based on a modified bridging method. In order to solve the related complex optimization problem having a non-linear cost function and a non-linear constraint, a bridged function is used in the search for the global minimum such that the algorithm does not get stuck in a local minimum. Given that cost functions in waveform optimization are often highly irregular with many local minima or are even discontinuous, evolutionary algorithms have gained increased attention in the design of waveforms with respect to communication and radar applications. An evolutionary approach for designing complex spreading codes for direct sequence code-division multiple-access (DS-CDMA) systems has been proposed in [42, 43]. In particular, a multi-objective evolutionary approach is used to search for solutions that satisfy simultaneous objectives posed on autocorrelation and crosscorrelation properties. This approach turned out to be beneficial in the communications field for designing large number of spreading sequence sets with a wide range of correlation properties. In [7], genetic algorithms have been used to design PN sequence families with bounded correlation properties. It is claimed that this approach can produce sequences of any length and superior performance compared to the well-known Gold sequences. A number of recent works has also been reported for the use of evolutional algorithms in the field of radar applications. In [2], an evolutionary algorithm is applied to determine a suite of optimal waveforms to simultaneously perform different surveillance missions such as ground moving target indication, airborne moving target indication, and synthetic aperture radar. The authors have shown that evolutionary algorithms are well suited to design optimal waveforms for multi-mission objectives such as peak sidelobe levels, integrated sidelobe levels, pulse integration, and revisit time. The work reported in [38] used meta-heuristic algorithms to optimize waveforms with sparse spectrum for radar applications in the high frequency band. In particular, a genetic algorithm and particle swarm optimization are used to produce optimal waveforms with acceptable autocorrelation sidelobes. It is concluded that the particle swarm optimization is simpler and faster than the genetic algorithm. They are of the opinion that computational efficiency of particle swarm optimization is comparable or would be even better than the adaptive method of [40].

In view of the above, this chapter considers integrated radar and communication systems based on waveforms known as polyphase sequences. In order to account for the waveform design challenges associated with such integrated systems, we have compared performance and potential application scenarios of different classes of polyphase pulse compression sequences in our earlier studies reported in [25, 26]. Specifically, Oppermann sequences have been revealed in these studies to potentially better support the considered integration as these allow for the design not only of families with a wide range of correlations but also support a variety of characteristics with respect to the ambiguity function, i.e. delay-Doppler tolerance. These sequences provide a number of parameters that can be chosen to design sequences for a wide range of performance characteristics. It will be shown that meta-heuristic algorithms are well-suited to find the optimal parameters for these sequences. Numerical results will be provided for optimal Oppermann sequences obtained with meta-heuristic algorithms.

The rest of this chapter is organized as follows. In Section 7.2, an overview of meta-heuristic algorithms is presented. A brief discussion of polyphase sequences and the definition of Oppermann sequences is provided in Section 7.3. In Section 7.4, performance measures are introduced. Numerical examples are given in Section 7.5. In Section 7.6, conclusions are drawn.

7.2 Meta-Heuristic Algorithms

Meta-heuristic algorithms, also referred to as meta-heuristics for brevity, belong to a branch of stochastic optimization. They are utilized by both engineers and scientists wishing to optimize solutions to problems that are intractable by conventional methods. Meta-heuristic methods consist of two major components known as randomization and selection of the best solutions. The first component avoids that an algorithm gets trapped in a local optimum but also increases the diversity of the potential solutions while the latter component ensures convergence towards the optimal value [10, 61, 62]. A good combination of these two components usually ensures that the global optimum is achievable. The popularity of these algorithms stems from their ability to solve large, complex and dynamic problems. The efficiency of these algorithms or solutions they provide is a measure of their ability to reach an acceptable solution within a reasonable time frame.

The applications of meta-heuristics are broad, versatile and diverse. Application areas include controller design, applied mathematics, power systems, physics, data mining, fuzzy systems and many others. In this chapter, we will apply some of these algorithms to pseudo random signal processing with focus on waveform design for integrated radar and communication systems. For this purpose, meta-heuristic algorithms may be classified as being either population-based or flight/trajectory-based. Genetic algorithms, for example, can be classified as a population-based method while particle swarm optimization utilizes multiple particles to reach the optimal solution. On the other hand, simulated annealing uses a single solution that moves through the search space or design space in a piecewise manner. The essence of the algorithm is always to accept a better solution, whereas a not-so-good solution is accepted with certain probability. In the sequel, selected state-of-the-art zero order and meta-heuristic algorithms are presented.

7.2.1 Particle Swarm Optimization

The PSO is a population-based stochastic optimization technique which has been inspired by social behavior of a flock of birds, school of fishes and swarm of bees as proposed by Eberhart and Kennedy [30]. Since its inception, there have now as many as about 20 different variants of PSO been proposed while remaining still an active area of research. It shares many similarities with genetic and virtual ant algorithms including concepts such as population initialization with random solutions and search for a global optimum solution in successive generations. However, the evolution operators like mutation and crossover as well as encoding or decoding of the parameters into binary strings are not used with PSO algorithms. Instead, it uses a real-number randomness and global communication among the swarm population. Accordingly, each member in the swarm adapts its search patterns by learning from its own experiences of the other members. A member in the swarm is referred to as a particle and represents a potential solution which is a point in the search space. The global optimum is regarded as the location of food [37]. Each particle has a fitness value and a velocity to adjust its flying direction by learning from the best experiences of the swarm to search for the global optimum in the D-dimensional solution space. In our case, the dimension D of the problem is given by the number of parameters that are available for optimization for a given class of sequences. In order to avoid haphazard movements of the particles in the search space, upper and lower bounds are usually specified on the velocity. If the velocity v falls below the specified lower bound, it is set to v_{min} as a measure to prevent in-sufficient exploration of the search space. On the other hand, if the velocity exceeds the specified upper bound, it is set to v_{max} in order to avoid particles moving away from or past a good solution. Similarly, the actual search range for a D-dimensional problem is usually also constrained to a given interval $[c_{min}, c_{max}]^D$, in order to restrain the particles moving on the search boundary.

The standard PSO uses both the personal best, *pbest*, with respect to the location achieved by an individual particle and the global best, *gbest*, referring to the best solution/location among all particles in the swarm [10, 30]. The concept of personal best is primarily used to increase the diversity in finding a solution and to avoid pulling all the particles to the global best. This may cause the algorithm to converge prematurely without finding the overall best solution. However, such diversity can also be simulated by using some kind of randomness [61, 62]. Based on this observation, [62] argues that there is no need to use the personal best, unless the optimization problem is highly nonlinear and multi-modal. This version of the PSO is known as accelerated PSO (APSO) [61, 62].

7.2.2 *Harmony Search*

A new type of heuristic optimization algorithms known as harmony search (HS) was developed by Lee and Geem [31]. It formalizes the musician improvisation process, i.e. inventing music while performing, into a quantitative optimization process. It comprises of the following parts: (1) Usage of harmony; (2) pitch adjustment; and (3) randomization. In an HS algorithm, each musician (decision variable) plays (generates) a note (value) for finding a best harmony (global optimum). In other words, a harmony translates to an optimization solution vector and the musician's improvisation corresponds to local and global search schemes in terms of optimization. Solutions of the optimization process correspond to a musician while the harmony of the notes generated by a musician corresponds to the fitness of the solution. The pitch adjustment rate $r_{pa} \in [0.1, 0.5]$ and so-called harmony memory $r_{accept} \in [0.7, 0.95]$ ensure that the best harmonies established at some point will be carried over to a new harmony memory. For a detailed discussion on harmony search, the interested reader is referred to [31, 61, 62] and the references therein.

7.2.3 *Adaptive Simulated Annealing*

The classical SA algorithm [10, 54, 61, 62] relies on the Boltzmann sampling distribution. It comprises of components such as the probability density function of the state space $g(\gamma)$ with γ being the current solution, an acceptance probability function $h(\Delta E)$ with respect to the difference in system energy ΔE between two design vectors, and an annealing schedule for temperature $T(k)$ with annealing time k using Boltzmann annealing. An enhanced version of the classical SA known as adaptive SA (ASA) has been proposed in [20, 21, 22, 23] including comparisons, test case studies and applications. In contrast to SA, the annealing schedule for temperature $T(k)$ decreases exponentially in annealing time k. In addition, re-annealing and quenching is introduced with ASA that allows for adaptation to changing sensitivities in multidimensional parameter spaces.

7.2.4 *Artificial Bee Colony Algorithm*

The ABC algorithm was proposed by Karaboga [27] in 2005. It simulates the foraging behavior associated with bee colonies. A colony of honey bees can extend itself over long distances, sometimes more than 10 kilometers and in multiple directions simultaneously to exploit a large number of food sources. In a bee colony, tasks are divided among the specialized individuals or bees, namely employed, onlooker and scout bees. The population in a bee colony is divided into two halves. The first half of the population is comprised of employed bees while the second half includes the onlooker bees. The foraging process begins in a colony by scout bees being sent to search for promising food sources. Scout bees move from one food source to another in a random fashion. Employed bees perform duties of exploiting the possible food sources and passing on the information about the quality of the food source to the onlookers bee. The decision taken by onlooker bees to exploit a potential food source depends on the information provided by the employed bees. ABC algorithms have been used to solve both unconstrained and constrained optimization problems [3,27,28,29]. It requires only a few control parameters such as the colony size and maximum number of cycles [29].

7.2.5 Preliminaries for Waveform Design

From this point onwards, we will consider two-dimensional optimization problems unless otherwise specified. In the context of waveform design using Oppermann sequences, the term swarm in APSO, harmonies in HS, bees in ABC and candidate points in ASA relate to the parameters m and n which define a specific sequence family. In all these algorithms, the control parameters are defined in the initialization phase. Initially, all the algorithms start with a population randomly distributed except for ASA, which starts with the initial guess in the search space. In each step of the algorithms, there is always a solution or a set of solutions, representing the current state of the algorithm. These solutions are used to generate phases of the Oppermann sequences (see Section 7.3). In order to distinguish good waveform designs from inferior designs, waveform characteristics such as aperiod correlations, figure of merit, and integrated sidelobe measures are computed. The interested reader can find pseudo code of HS in [62], ASA in [52], and ABC in [27] while details of the APSO can be found in [61,62].

7.3 Polyphase Sequences and Their Applications

The history of complex-valued sequences ranges back as far as the 1950s when polyphase sequences where considered in many research laboratories. As the related research outcomes were reported mainly in classified documents with limited access, a broader audience was first reached with the work in [16] on phase shift pulse sequences. In the following decades, many complex-valued sequences have been proposed and analyzed with their applications ranging from radar systems to spread-spectrum communication systems. In particular, polyphase sequences have gained increased attention due to their ability to match regular phase shift keying modulation schemes. In addition, the advances in integrated circuit technologies have paved the way for moving from simple binary sequences to implementations of complex-valued sequences and related more involved pseudo random signal processing. In the sequel, we consider polyphase sequences and will shed some light on their potential to serve in integrated radar and communication systems. In particular, the family of Oppermann sequences [45] are considered in more detail as they offer the system designer large sets of sequences with a wide range of correlation properties compared to other classes of polyphase sequences.

7.3.1 Polyphase Sequences for Radar Systems

Pseudo random sequences and the related signal processing have emerged from space and military applications. In this context, the concept of pulse compression, i.e. expanded pulses with large time-bandwidth products, has been utilized in radar systems. This type of signals offer high range resolution as they can obtain high pulse energy and large pulse width. As an alternative to frequency-modulated signals, pulse compression sequences have been subject of many studies [14, 32]. Polyphase sequences are known to have better Doppler tolerance for a broader range-Doppler coverage than binary sequences [8, 32, 41, 46]. These sequences can be derived from the phase history of chirp or step chirp analog signals and can be processed digitally [36]. In radar applications, the performance of different polyphase sequences can be compared in terms of delay or range tolerance using measures such as the autocorrelation function, mainlobe-to-total-sidelobe ratio and peak-to-sidelobe ratio. The sensitivity of a particular waveform design towards Doppler shifts in case of moving targets can be characterized by using the ambiguity function. As there exist no analytical method that would allow for synthesizing the desired waveform given its desired ambiguity function, more practical optimization approaches are needed to facilitate such designs. For example, the design of a particular radar waveform may be first aiming for optimization of autocorrelation properties with respect to range characteristics followed by evaluating the ambiguity function to identify the Doppler tolerance of the deduced sequence.

As far as radar applications are concerned, Frank sequences [12] were the first polyphase sequences used in pulse compression radar [46]. They can only be designed for perfect square lengths, therefore, they have limited family size. Later in [34] modified versions of Frank sequences were obtained by permuting their phase history. The modified versions are referred to as P1 and P2 sequences. Rapajic and Kennedy in [48] proposed a new class of sequences, known as Px sequences. These sequences have superior performance in terms of integrated sidelobe levels compared to Frank, P1, and P2 sequences. However, for even square root sequence lengths, their performance is the same as for P2 sequences. In [35], the families of P3 and P4 sequences were proposed that can be constructed for any length. The authors of [6, 13] generalized the ideas behind Frank sequences resulting in Frank-Zadoff-Chu (FZC) sequences which can also be designed for any length. Several performance aspects of the aforementioned classes of polyphase sequences with respect to radar applications have been discussed in literature [34, 36, 48].

7.3.2 Polyphase Sequences for Communication Systems

A major boost for the application of pseudo random sequences in the field of communication systems was given by the development of cellular mobile communication systems and spread-spectrum based radios for indoor communication. In particular, the CDMA system for digital cellular phone applications by Qualcomm Incorporated and the family of IEEE802.11 standards for wireless local area networks (WLANs) has taken the theoretical concepts into practical systems. The main classes of sequences used with these systems are Walsh-Hadamard sequences [11, 55], m-sequences [11, 63], Barker codes [11, 63], and complementary code keying based modulation [19]. Subsequently, with the advent of the third generation of mobile communication systems, more advanced spread-spectrum techniques such as orthogonal variable spreading factor sequences [1] and complex-valued short scrambling sequences have been utilized. In contrast to radar applications where it is usual sufficient to have a single sequence with good autocorrelation characteristics, communication systems require a set of sequences to facilitate simultaneous channel access to a number of users. Clearly, minimum crosscorrelation among the sequences is a major design consideration in this case. Given the large advances in modern integrated circuit technologies, it has become feasible to implement complex-valued sequence designs including polyphase sequences such as Frank sequences, FZC sequences, and Oppermann sequences.

7.3.3 Application of Oppermann Sequences for Integrated Radar and Communication Systems

Given the insights from the brief overview on polyphase sequences from the viewpoint of radar and communication applications, it can be concluded that more flexible waveform designs are needed to address the conflicting objectives of these two applications. Our earlier research [25, 26] on this topic has revealed that Oppermann sequences may serve favorable in such integrated radar and communication systems compared to conventional waveform designs. This is mainly due to the fact that families of Oppermann sequences can be designed for a wide range of correlation properties. For any given sequence length, Oppermann sequences are defined by three parameters. These parameters can be used in an optimization process to control the progression of the autocorrelation function, crosscorrelation function, the power spectral density and characteristics of the ambiguity function. Due to space limitations, however, we will concentrate here on range (autocorrelation) and multiple access (crosscorrelation) characteristics. On the other hand, inclusion of moving targets and the related Doppler shifts into the framework of meta-heuristic algorithms may be addressed in our future research considering ambiguity and cross-ambiguity functions.

In this chapter, we consider weighted pulse trains that can be described by a complex envelope as

$$U_x(t) = \frac{1}{\sqrt{T}} \sum_{i=0}^{N-1} u_x(i)\,\mathrm{rect}\left(\frac{t - iT_c}{T_w}\right) \tag{7.1}$$

where $T = NT_c$ is the duration of the xth pulse train while T_c and $T_w \le T_c$, respectively, denote the repetition period and the width of each rectangular pulse

$$\mathrm{rect}\left(\frac{t}{T_w}\right) = \begin{cases} 1 \text{ for } -\frac{T_w}{2} \le t \le \frac{T_w}{2} \\ 0 \text{ otherwise} \end{cases} \tag{7.2}$$

The elements $u_x(i)$, $i = 0, 1, \ldots, N-1$, of the xth complex-valued sequence $\mathbf{u}_x$ of length N represent the weights of the pulse train in (7.1). In general, these elements are given for a polyphase sequence as

$$u_x(i) = \exp\left[j\varphi_x(i)\right], \quad j = \sqrt{-1} \tag{7.3}$$

where the set of N phases $\{\varphi_x(0), \varphi_x(1), \ldots, \varphi_x(N-1)\}$ are referred to as phase sequence. In particular, the phase $\varphi_x(i)$ of the ith element $u_x(i)$ of the xth Oppermann sequence $\mathbf{u}_x = [u_x(0), u_x(1), \ldots, u_x(N-1)]$ of length N taken from a family or set $\mathcal{U}$ of sequences is given as

$$\varphi_x(i) = \frac{\pi}{N}\left[x^m (i+1)^p + (i+1)^n + x(i+1)N\right] \tag{7.4}$$

where $1 \le x \le N-1$, $0 \le i \le N-1$ and integers i are relatively prime to the length N. The maximum size of a family $\mathcal{U}$ of Oppermann sequences is obtained as $N-1$ when the length N of the sequences is a prime number. A particular family of Oppermann sequences is defined by the real-valued parameters m, n, and p. All the sequences in a family have the same magnitude of the autocorrelation function for a fixed combination of these three parameters. In [45], it has been shown that the magnitude of the autocorrelation function depends only on the parameter n if the parameter $p = 1$. For this case, the autocorrelation magnitude follows the expression

$$|C_x(l)| = \left| \frac{1}{N} \sum_{i=0}^{N-1-l} \exp\left\{ \frac{j\pi}{N}\left[(i+1)^n - (i+l+1)^n\right] \right\} \right| \tag{7.5}$$

In the sequel, we therefore focus on the case of $p = 1$ which leaves us with m and n as free parameters for use in an optimized waveform design.

Due to the general definition of Oppermann sequences, they include some more specific sequences. For example, for the parameters $m = 2$, $n = -\infty$, $p = 1$, FZC sequences can be generated. As such, application of the considered meta-heuristic algorithms to these more specific sequences is straightforward.

7.4 Performance Measures

In the following sections, the definitions of the measures used in the performance comparison of the considered Oppermann sequences will be given. Specifically, let an Oppermann sequence of length N be denoted as $\mathbf{u}_x = [u_x(0), u_x(1), \ldots, u_x(N-1)]$ where subscript $1 \leq x \leq U$ relates to the xth sequence $\mathbf{u}_x$ taken from a given set $\mathcal{U}$ of size U.

7.4.1 *Aperiodic Correlation Measures*

In order to quantify the degree of similarity between different sequences from a given set or between a given sequence and a shifted version of it, respectively, autocorrelation and crosscorrelation measures are usual considered. In many fields, aperiodic signals need to be processed which occur only once within a considerable time span and appear to the application as more or less singular events. Accordingly, the aperiodic crosscorrelation (ACC) between two complex-valued sequences $\mathbf{u}_x = [u_x(0), u_x(1), \ldots, u_x(N-1)]$ and $\mathbf{u}_y = [u_y(0), u_y(1), \ldots, u_y(N-1)]$ of length N at discrete shift l is given as [11, 63]

$$C_{xy}(l) = \begin{cases} \frac{1}{N} \sum_{i=0}^{N-1-l} u_x(i) u_y^*(i+l), & 0 \leq l \leq N-1 \\ \frac{1}{N} \sum_{i=0}^{N-1+l} u_x(i-l) u_y^*(i), & 1-N \leq l < 0 \\ 0, & |l| \geq N \end{cases} \tag{7.6}$$

where $(\cdot)^*$ denotes the complex conjugate of the argument $(\cdot)$. In case of $\mathbf{u}_x = \mathbf{u}_y$, (7.6) is referred to as aperiodic autocorrelation (AAC) and is denoted as $C_x(l) = C_{xx}(l)$.

In addition to ACC and AAC, it is often more realistic to incorporate the whole range of possible correlation values into the performance evaluation of a given set of sequences rather than considering only peak values of aperiodic correlations. In this context, mean-square values from AAC and ACC may be used in favor of worst case scenarios. For this purpose, let us introduce the mean-square out-of-phase autocorrelation (MSAC), R_{ac}, and mean-square crosscorrelation (MSCC), R_{cc}, respectively, of a given set $\mathcal{U}$ of size U as

$$R_{ac} = \frac{1}{U} \sum_{x=1}^{U} \sum_{\substack{l=1-N \\ l \neq 0}}^{N-1} |C_x(l)|^2 \tag{7.7}$$

$$R_{cc} = \frac{1}{U(U-1)} \sum_{x=1}^{U} \sum_{\substack{y=1 \\ y \neq x}}^{U} \sum_{l=1-N}^{N-1} |C_{xy}(l)|^2 \tag{7.8}$$

7.4.2 Sidelobe Measures

The figure of merit (FOM) of a sequence $\mathbf{u}_x \in \mathscr{U}$, $1 \leq x \leq U$ of length N with aperiodic autocorrelation function $C_x(l)$ measures the ratio of energy in the mainlobe to the energy in the sidelobe of the autocorrelation function. It is defined as

$$FOM_x = \frac{C_x(0)}{2\sum\limits_{l=1}^{N-1} |C_x(l)|^2}, \quad \forall x \tag{7.9}$$

Alternatively, the integrated sidelobe level (ISL) is often used for radar applications in the context of distributed target environments. The ISL of a sequence $\mathbf{u}_x \in \mathscr{U}$, $1 \leq x \leq U$ of length N is defined as

$$ISL_x = \frac{1}{FOM_x}, \quad \forall x \tag{7.10}$$

Another important measure in relation to radar applications is the peak-to-sidelobe ratio (PSLR) which relates to the ability of detecting targets without masking interfering targets. For example, if an AAC has large sidelobes, it will mask nearby targets and leave them undetected. Specifically, the PSLR of a sequence $\mathbf{u}_x$ measures the ratio of the in-phase value $C_x(0)$ to the maximum sidelobe magnitude $|C_x(l)|$ of the periodic autocorrelation function $C_x(l)$. It is defined as

$$PSLR_x = \frac{C_x(0)}{\max\limits_{1 \leq l < N} |C_x(l)|}, \quad \forall x \tag{7.11}$$

7.5 Numerical Examples

In the sequel, some numerical examples are provided to illustrate the application of meta-heuristic algorithms for waveform optimization for integrated radar and communication systems. For this purpose, we consider the class of Oppermann sequences as defined in (7.4) of length $N = 31$. It is noted that the maximum number of $N - 1 = 30$ sequences in the designed set is obtained as N is chosen as a prime number. Furthermore, the considered sequence family offers parameters m and n for optimization given the case of parameter $p = 1$. Accordingly, the following optimization problems may be posed:

$$P1: \quad \min_{n \in [n_1, n_2]} ISL(\mathscr{U}) \tag{7.12}$$

$$P2: \quad \max_{n \in [n_1, n_2]} PSLR(\mathscr{U}) \tag{7.13}$$

$$P3: \quad \min_{m \in [m_1, m_2], n \in [n_1, n_2]} [R_{ac}(\mathscr{U}) + \alpha R_{cc}(\mathscr{U})] \tag{7.14}$$

where $m \in [m_1, m_2]$ and $n \in [n_1, n_2]$ are the search regions for m and n, respectively, and α is a weighting factor. While problems $P1$ and $P2$ given in (7.12) and (7.13), respectively, relate strongly to radar applications, problem $P3$ formulated in (7.14) can be used to find a trade-off between conflicting objectives of radar and communication applications. Especially, the weighting factor α may be chosen with respect to desirable system specifications. In contrast to [25], where we have used a two-step approach to first optimize autocorrelation properties by a simple brute-force search over parameter n followed by tuning m towards favorable delay-Doppler properties, we consider here two-dimensional optimization to simultaneously find the optimal values of n and m for problem $P3$. On the other hand, in view of the independence of the autocorrelation of Oppermann sequences on parameter m as shown in (7.5), problems $P1$ and $P2$ remain one-dimensional as PSLR and ISL only involve the aperiodic autocorrelation.

In order to solve the problems formulated in (7.12)-(7.14), we use APSO, HS ASA and ABC. The two-dimensional search space was constrained to the interval $m \in [0,4]$ and $n \in [0,4]$. The algorithms were executed on a laptop computer with Intel Pentium M 740 Processor running at 1.73 GHz and 2048 Megabytes of RAM. With the exception of ASA, where we used a C-routine called from MATLAB, all the other algorithms have been implemented in MATLAB. As for the translation of the notions from meta-heuristics to the optimization problem at hand, the following interpretation can be given.

- **APSO:** Initially, particles in a swarm are randomly distributed in a D-dimensional search space. In APSO, the parameter D refers to the dimension of the problem, swarm refers to a population, and particle is similar to an individual. Alternatively, each solution (or particle) flies through the search space and looks for an optimal position to land. In terms of Oppermann sequences, particles are represented by the values of m and n in a two-dimensional search space and are used to generate the phases of Oppermann sequences as defined in (7.4). The search for the optimal landing position, i.e. finding optimal values of m and n will continue until the criteria selected from (7.7) to (7.11) are met.
- **HS:** Initially, harmonies are randomly generated in a D-dimensional space and are stored in a harmony memory (HM). The use of HM ensures that the best harmonies will be carried over to the HM. As for the optimization of Oppermann sequences, the parameters m and n are represented by the obtained harmonies to generate phases as defined in (7.4). Then, pitch adjustment is used to control the convergence of the algorithm. Randomization introduced in the algorithm drives the algorithm to search previously unexplored areas in the search space until the criteria selected from (7.7) to (7.11) are met.
- **ASA:** This algorithm starts with the initial guess of the parameters in the D-dimensional search space. In terms of Oppermann sequences, the initial guess represents values of the parameters m and n. Each step of the ASA algorithm replaces the current solution by a random nearby solution. The obtained solutions are used to generate Oppermann sequences. The process of finding optimal values of m and n continues by generating feasible points in the search space and

acceptance probability including annealing and re-annealing temperatures until criteria selected from (7.7) to (7.11) are met.

- **ABC:** It is recalled that food sources are randomly distributed in the D-dimensional search space at the start of the search. Here, bees refer to a population of bees (employed, onlookers and scout) which are in the search of the best food position. Employed bees search for new food sources within their neighborhood that have more nectar compared to the food sources they have previously visited. These food sources represent the values of the parameters m and n of Oppermann sequences to generate the phases defined in (7.4). If during the optimization process the criteria set for (7.7) to (7.11) are not met, it will represent abandoned food source or bad sequence designs.. The search for the final food position represent optimal values of m and n that satisfy the criteria set for (7.7) to (7.11).

Figure 7.1 compares the performance of Oppermann sequences obtained through meta-heuristics in terms of PSLR with the brute-force search method with fixed step size reported in [25]. Clearly, the random search strategy employed in meta-heuristics widens the search area allowing the particles to explore the search space more effectively compared to an optimization using fixed step size. As can be seen from the figure, PSLR values can be improved for those prime length that would have inferior performance using brute-force search with fixed increment on n. In this case, meta-heuristic algorithms improve the performance of the designed set of Oppermann sequences to be comparable to other families such as the FZC sequences (see also [25]).

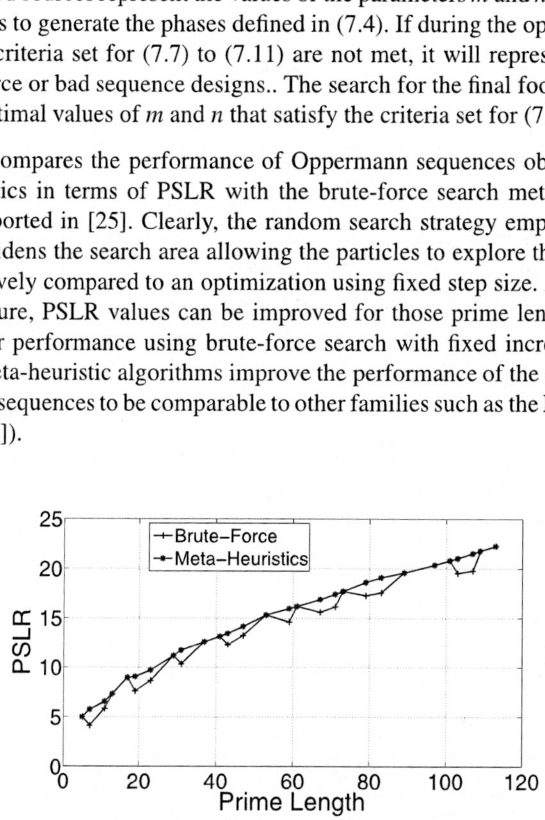

Fig. 7.1 Performance comparisons between brute-force search with fixed increment and meta-heuristic algorithms in terms of PSLR

The convergence behavior of the considered algorithms for the example of optimizing PSLR is illustrated in Fig. 7.2. It can be seen from the progressions in terms of iterations shown in the figure that ASA achieves the fastest convergence to the optimal values followed by APSO, ABC and HS. The fast convergence of ASA may be attributed to the fact that exponential annealing permits the algorithm to adaptively re-anneal and pacing the convergence in the search space in all dimensions. It

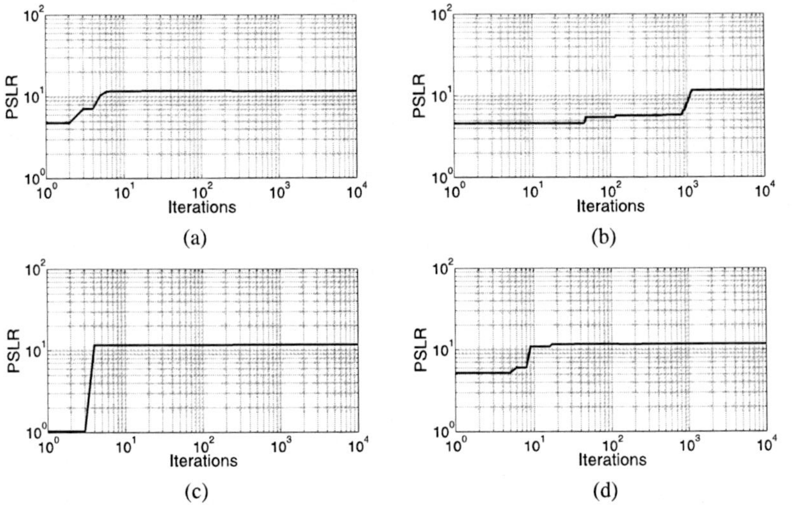

Fig. 7.2 Convergence of different meta-heuristic algorithms towards optimal PSLR: (a) APSO, (b) HS, (c) ASA, (d) ABC

should be mentioned that the similar convergence behavior and ranking among the algorithms can be observed when applied to optimize FOM, ISL, and mean-square aperiodic correlation measures.

Tables 7.1(a)-(e) show numerical results of optimal designs for Oppermann sequences of length $N = 31$ with respect to the optimization problems posed in (7.12), (7.13), and (7.14) using APSO, HS, ASA, ABC. As for the optimal designs presented in Table 7.1(a) and Table 7.1(b) for PSLR and ISL, respectively, it is sufficient to consider only the parameter n as these metrics involve only the AAC (see also (7.10) and (7.11)). It is recalled that according to (7.5), the AAC is independent of the parameter m for the considered case of parameter $p = 1$. Also, all $N-1=30$ Oppermann sequences in an optimized set achieve the same PSLR and ISL. Clearly, all considered meta-heuristic algorithms converge towards very similar results for these two classical design objectives of radar systems.

In order to illustrate the trade-off in waveform optimization for integrated radar and communication systems, let us focus now on the results presented in Tables 7.1(c)-(e) with respect to the optimization problem posed in (7.14). In particular, we have chosen $\alpha = 0$ relating to radar systems, $\alpha = 60$ emphasizing on communication systems, and $\alpha = 1$ as an example of an integrated radar and communication scenario. Clearly, the autocorrelation properties indicated by the small R_{ac} values in Table 7.1(c) are beneficial for radar systems and are independent of parameter m. On the other hand, good crosscorrelation characteristics are shown Table 7.1(d) for use with communication systems but these come at the expense of poor autocorrelation properties quantified by high values of R_{ac}. The results of

Table 7.1 Optimal designs for Oppermann sequences of length $N = 31$

(a) Peak-to-sidelobe ratio

Algorithm	n	PSLR
APSO	2.000	11.735
HS	2.000	11.734
ASA	2.000	11.735
ABC	2.000	11.735

(b) Integrated sidelobe level

Algorithm	n	ISL
APSO	2.007	0.110
HS	2.007	0.110
ASA	2.000	0.116
ABC	2.007	0.110

(c) MSAC; $\alpha = 0$

Algorithm	m	n	R_{ac}	R_{cc}
APSO	2.597	2.007	0.110	1.000
HS	2.744	2.007	0.110	1.001
ASA	2.000	2.000	0.116	1.000
ABC	0.614	2.007	0.110	1.005

(d) MSCC; $\alpha = 60$

Algorithm	m	n	R_{ac}	R_{cc}
APSO	1.003	1.002	19.676	0.341
HS	1.003	1.000	19.677	0.341
ASA	1.000	1.000	19.677	0.344
ABC	1.003	1.000	19.677	0.341

(e) MSAC+MSCC; $\alpha = 1$

Algorithm	m	n	R_{ac}	R_{cc}
APSO	0.930	2.007	0.110	0.997
HS	1.000	2.007	0.110	0.996
ASA	1.000	2.000	0.116	0.996
ABC	0.999	2.007	0.110	0.996

the trade-off example shown in Table 7.1(e) may perform favorable with integrated radar and communication systems keeping autocorrelation values low and driving crosscorrelation values smaller. An additional increase of α would result in an increase of autocorrlelation values and further reduce crosscorrelation values. Also, all four considered meta-heuristic algorithms provide very similar outcomes to the different optimization problems.

7.6 Conclusions

In this chapter, we have focused on the waveform optimization for integrated radar and communication systems. Given the conflicting requirements on autocorrelation and crosscorrelation characteristics, meta-heuristic algorithms are considered to basically perform a multidimensional optimization. Specifically, the selected class of

Oppermann sequences allows for designing families with a wide range of correlations with respect to a two-dimensional search space. The numerical results illustrate the potential of meta-heuristic algorithms for designing sequences for radar, communications, as well as integrated systems. By way of example with respect to PSLR, it is shown that meta-heuristics can improve performance compared to search methods with fixed increment.

References

1. Adachi, F., Sawahashi, M., Okawa, K.: Tree-structured Generation of Orthogonal Spreading Codes with Different Lengths for Forward Link DS-CDMA Mobile. IEE Electronics Letters 33(1), 27–28 (1997)
2. Amuso, V.J., Enslin, J.: An Evolutionary Algorithm Approach to SimultaneousMulti-Mission RadarWaveform Design. In: Wicks, M., Mokole, E., Blunt, S., Schneible, R., Amuso, V. (eds.) Principles ofWaveform Diversity and Design, pp. 110–125. SciTech Publishing, Rayleigh (2011)
3. Basturk, B., Karaboga, D.: An artificial bee colony (ABC) algorithm for numeric function optimization. In: IEEE Swarm Intelligence Symposium, Indianapolis, USA (2006)
4. Bergh, F.: An Analysis of Particle Swarm Optimizers. Ph.D. thesis, University of Pretoria, Pretoria, South Africa (2001)
5. Brent, R.P.: Algorithms forMinimizationwithout Derivatives. Prentice Hall, Englewood Cliffs (1973)
6. Chu, D.C.: Polyphase codes with good periodic correlation properties. IEEE Trans. on Inf. Theory 18(4), 531–532 (1972)
7. Cinteza, M., Marghescu, I., Radulescu, T.: Design of PN Sequence Families with Bounded Correlation Properties Using Genetic Algorithms. In: IEEE EUROCON, Belgrade, Serbia and Montenegro, pp. 1362–1365 (2005)
8. Cook, C.E., Bernfeld, M.: Radar Signals: An Introduction to Theory and Applications. Academic Press, New York (1967)
9. Dam, H.H., Zepernick, H.-J., Nordholm, S.: Spreading Code Design Using a Global Optimization Method. In: Annals of Operations Research, vol. 123, pp. 249–264. Springer, New York (2005)
10. Engelbrecht, A.P.: Fundamentals of Computational Swarm Intelligence. John Wiley & Sons, Chichester (2005)
11. Fan, P., Darnell, M.: Sequence Design for Communications Applications. Research Studies Press, Taunton (1996)
12. Frank, R.L.: Polyphasae codes with good nonperiodic correlation properties. IEEE Trans. On Inf. Theory 9(1), 43–45 (1963)
13. Frank, R.L., Zadoff, S.A.: Phase shift pulse codes with good periodic correlation properties. IEEE Trans. on Inf. Theory 19(1), 115–120 (1975)
14. Golomb, S.W., Gong, G.: Signal Design for Good Correlation for Wireless Communications, Cryptography and Radar. Cambridge University Press, Cambridge (2005)
15. Haupt, R.L., Haupt, S.E.: Practical Genetic Algorithms. John Wiley & Sons, Chichester (2004)
16. Heimiller, R.C.: Phase shift pulse codes with good periodic correlation properties. IRE Trans. on Inf. Theory 7(10), 254–257 (1961)
17. Hooke, R., Jeeves, T.A.: Direct Search Solution of Numerical and Statistical Problems. Journal of the ACM 8, 212–229 (1961)

18. Hughes, P.K., Choe, J.Y.: Overview of Advanced Multifunction RF System (AMRFS). In: IEEE International Conference on Phased Array Systems and Technology, Dana Point, USA, pp. 21–24 (2000)
19. IEEE Std 802.11. Wireless LAN Medium Access Control (MAC) and Physical Layer (PHY) Speccification - Higher-Speed Physical Layer Extension in the 2.4 GHz Band (2000)
20. Ingber, L.: Adaptive Simulated Annealing (ASA): Global Optimization C-Code. Technical Report, Caltech Alumni Association (1993)
21. Ingber, L.: Adaptive simulated annealing (ASA): Lessons learned. Control and Cybernetics 25, 33–54 (1996)
22. Ingber, L.: Adaptive Simulated Annealing (ASA) and Path-Integral Algorithms: Generic Tools for Complex Systems. Technical Report, Chicago, USA (2001)
23. Ingber, L., Rosen, B.: Genetic algorithms and very fast simulated reannealing: A comparison. Mathematical Computer Modeling 16(11), 87–100 (1992)
24. Ipatov, V.P.: Spread Spectrum and CDMA: Principles and Applications. John Wiley & Sons, Chichester (2005)
25. Jamil, M., Zepernick, H.-J., Pettersson, M.I.: Performance assessment of polyphase pulse compression codes. In: IEEE International Symposium on Spread Spectrum Techniques and Applications, Bologna, Italy, pp. 166–172 (2008)
26. Jamil, M., Zepernick, H.-J., Pettersson, M.I.: On integrated radar and communication systems. In: IEEE Military Communications Conference, San Diego, USA, pp. 1–6 (2008)
27. Karaboga, D.: An Idea Based on Honey Bee Swarm for Numerical Optimization. Technical Report TR06, Erciyes University, Turkey (2005)
28. Karaboga, D., Basturk, B.: A powerful and efficient algorithm for numerical function optimization: Artificial bee colony (ABC) algorithm. Journal of Global Optimization 39(3), 459–471 (2007)
29. Karaboga, D., Basturk, B.: On the performance of artificial bee colony (ABC) algorithm. Applied Soft Computing 8(1), 687–697 (2008)
30. Kennedy, J., Eberhart, R.C.: Particle swarm optimization. In: IEEE International Conference on Neural Networks, Piscataway, USA, pp. 1942–1948 (1995)
31. Lee, K.S., Geem, Z.W.: A new meta-heuristic algorithm for continuous engineering optimization: Haromny search theory and practice. Computer Methods Appl. Mech. Eng. 194, 3902–3933 (2005)
32. Levanon, N., Mozeson, E.: Radar Signals. John Wiley & Sons, Chichester (2004)
33. Levanon, N., Mozeson, E.: Radar Signals. John Wiley & Sons, Chichester (2004)
34. Lewis, B.L., Kretschmer, F.F.: A new class of polyphase pulse compression codes and techniques. IEEE Trans. on Aerospace and Electronic Systems 17(3), 364–372 (1981)
35. Lewis, B.L., Kretschmer, F.F.: Linear frequency modulation derived polyphase compression codes. IEEE Trans. on Aerospace and Electronic Systems 18(4), 637–641 (1982)
36. Lewis, B.L., Kretschmer, F.F., Shelton, W.W.: Aspects of Radar Signal Processing. Artech House, London (1986)
37. Liang, J.J., Qin, A.K., Suganthan, P.N., Baskar, S.: Comphrehensive learning particle swarm optimizer for global optimization of multimodal functions. IEEE Trans. on Evolutionary Computation 10(3), 281–295 (2006)
38. Liu, W., Lu, Y.L., Lesturgie, M.: Evolutionary Algorithms Based Sparse Spectrum Waveform Optimization. In: Wicks, M., Mokole, E., Blunt, S., Schneible, R., Amuso, V. (eds.) Principles of Waveform Diversity and Design, pp. 152–162. SciTech Publishing, Rayleigh (2011)

39. Luke, S.: Essentials of Metaheuristics, `http://cs.gmu.edu/~sean/book/metaheuristics`
40. Michael, J.J.: Sparse Frequency Transmit and Receive Waveform Design. IEEE Trans. On Aerospace and Electronic Systems 40(3), 851–861 (2004)
41. Nathanson, F.E., Riley, J.P., Cohen, M.N.: Radar Design Principles: Signal Processing and the Environment. SciTech Publishing, Mendham (1999)
42. Natarajan, B., Das, S., Stevens, D.: Design of Optimal Complex Spreading Codes for DSCDMA using an Evolutionary Approach. In: IEEE Global Telecommunications Conference, Dallas, USA, pp. 3882–3886 (2004)
43. Natarajan, B., Das, S., Stevens, D.: An Evolutionary Approach to Desginging Complex Spreading Codes for DS-CDMA. IEEE Trans. on Wireless Communications 4(5), 2051–2056 (2005)
44. Nelder, J.A., Mead, R.: A Simplex Method for Function Minimization. Computer Journal 7, 308–313 (1965)
45. Oppermann, I., Vucetic, B.S.: Complex spreading sequences with a wide range of correlation properties. IEEE Trans. on Commun. 45(3), 365–375 (1997)
46. Pace, P.E.: Detecting and Classifying Low Probability of Intercept Radar. Artech House, London (2004)
47. Powell, M.J.D.: An Efficient Method for Finding the Minimum of a Function of Several Variables Without Calculating Derivatives. Computer Journal 7, 152–162 (1964)
48. Rapajic, P.B., Kennedy, R.A.: Merit factor based comparison of new polyphase sequences. IEEE Commun. Letters 2(10), 269–270 (1998)
49. Roberton, M., Brown, E.R.: Integrated Radar and Communications based on Chirped Spread- Spectrum Techniques. In: IEEE International Microwave Symposium, Philadelphia, USA, pp. 611–614 (2003)
50. Rosenbrock, H.H.: An Automatic Method for Finding the Greatest or Least Value of a Function. Computer Journal 3(3), 175–184 (1960)
51. Sumathi, S., Surekha, P.: Computational Intelligence Paradigms: Theory and Application with Matlab. CRC Press, Boca Raton (2010)
52. Tantar, A.-A., Melab, N., Talbi, E.-G.: A Grid-Based Hybrid Hierarchical Genetic Algorithm for Protein Structure Prediction. In: de Vega, F.F., Cantú-Paz, E. (eds.) Parallel and Distributed Computational Intelligence. Studies in Computational Intelligence, vol. 269, pp. 291–319. Springer, Heidelberg (2010)
53. Tavik, G.C., et al.: The Advanced Multifunction RF Concept. IEEE Trans. Microw. Theory and Techn. 53(3), 1009–1020 (2005)
54. Venkataraman, P.: Applied Optimization with Matlab Programming. John Wiley & Sons, Chichester (2009)
55. Walsh, J.L.: A closed set of normal orthogonal functions. Am. J. Math. 45, 5–24 (1923)
56. Weise, T.: Global Optimization Algorithms - Theory and Applications (Self-Published), `http://www.it-weise.de`
57. Woodward, P.M.: Probability and Information Theory with Applications to Radar. Artech House, London (1980)
58. Wysocki, B.J., Wysocki, T.A.: Optimization of Orthognonal Polyphase Spreading Sequences for Wireless Data Applications. In: IEEE Vehicular Technology Conference, Atlantic City, USA, pp. 1894–1898 (2001)
59. Xu, S.J., Chen, Y., Zhang, P.: Integrated Radar and Communication based on DS-UWB. In: IEEE Ultrawideband and Ultrashort Impulse Signals, Sevastopol, Ukraine, pp. 142–144 (2006)
60. Xu, S.J., Chen, B., Zhang, B.P.,, P.: Radar-Communication Integration based on DSSS.Techniques. In: IEEE International Confonference on Signal Processing, Beijing, China, pp. 16–20 (2006)

61. Yang, X.-S.: Introduction to Mathematical Optimization: From Linear Programming to Metaheuristics. Cambridge International Science Publishing, Cambridge (2008)
62. Yang, X.-S.: Engineering Optimization: An Introduction with Metaheuristics Applications. John Wiley & Sons, Chichester (2010)
63. Zepernick, H.-J., Finger, A.: Pseudo Random Signal Processing: Theory and Application. John Wiley & Sons, Chichester (2005)

Chapter 8
Parameter Estimation from Laser Flash Experiment Data

Louise Wright, Xin-She Yang, Clare Matthews, Lindsay Chapman, and Simon Roberts

Abstract. Optimisation techniques are commonly used for parameter estimation in a wide variety of applications. The application described here is a laser flash thermal diffusivity experiment on a layered sample where the thermal properties of some of the layers are unknown. The aim is to estimate the unknown properties by minimising, in a least squares sense, the difference between model predictions and measured data. Two optimisation techniques have been applied to the problem. Results suggest that the classical nonlinear least-squares optimiser is more efficient than particle swarm optimisation (PSO) for this type of problem. Results have also highlighted the importance of defining a suitable objective function and choosing appropriate model parameters.

8.1 Introduction

Many components that operate in a high-temperature corrosive environment, such as engine parts and turbine blades, use coatings to increase their operational lifetime. In some cases these coatings are grown on the component by reaction (e.g. oxide layers), and in other cases they are separate substances applied to the surface of the component before it is put into operation. It is often difficult to obtain samples of the coating on its own, since the coating is often too thin and too fragile to be removed from the component in pieces of a usable size.

Accurate prediction of the behaviour and, in particular, the lifetime of such components in operation can avoid unexpected component failure and hence reduce downtime and maintenance costs. Models for prediction of component lifetime

Louise Wright · Xin-She Yang · Clare Matthews · Lindsay Chapman · Simon Roberts
Mathematics and Scientific Computing,
National Physical Laboratory,
Teddington, Middlesex TW11 0LW, UK

X.-S. Yang & S. Koziel (Eds.): Computational Optimization & Applications, SCI 359, pp. 205–220.
springerlink.com

generally require a coupled thermal-mechanical analysis to predict stresses caused by differential thermal expansion and oxide layer growth. The thermal part of this analysis requires knowledge of the thermal conductivity of each material present within the component, including coatings.

This chapter describes the application of an optimisation process to a finite volume model of the laser flash experiment using a layered sample. The optimisation minimises the difference between the measured data and model predictions by adjusting model parameters, including the thermal conductivity of one of the layers. The aim is to demonstrate that the thermal conductivity of a layer within a sample can be obtained using optimisation techniques.

The process described shows the steps required for application of optimisation techniques to a real-world problem: data preparation, model development, choice of objective function and parameters, and choice of an appropriate optimisation method. The work reported also illustrates that each of these steps may be revisited repeatedly before a fit-for-purpose model is achieved.

The laser flash experiment will be summarised in section 8.2, and the initial model used to simulate the experiment will be defined in section 8.3. The initial optimisation results obtained will be discussed in section 8.4, and subsequent alterations to the model will be explained in section 8.4.2. The final optimisation results will be discussed in section 8.4.3. Our concluding remarks are given in section 8.5.

8.2 The Laser Flash Experiment

The laser flash experiment measures the thermal diffusivity of materials. Thermal diffusivity is a measure of how quickly heat travels through a material and has units of $m^2\ s^{-1}$. Thermal diffusivity α is related to density ρ, thermal conductivity λ, and specific heat capacity c_p by the equation

$$\alpha = \frac{\lambda}{\rho c_p}, \tag{8.1}$$

and so if the density and the specific heat capacity of a material can be obtained from other experiments, the thermal conductivity can be calculated from the thermal diffusivity.

The laser flash experiment generates a set of temperature measurements gathered over time. The model used to determine the thermal diffusivity from the measurements is based on a number of assumptions, including the assumption that the material is uniform and isotropic. These assumptions are clearly not true for layered and coated materials such as the components described above. Since it is generally difficult to obtain samples of the coating that are sufficiently large to use in the laser flash experiment, a method of obtaining the thermal properties of each layer within a layered sample would enable the properties of the coating to be determined.

The experiment exposes one circular face of a cylindrical sample of material to a pulse of laser light, and measures the temperature rise of the centre of the opposite circular face. The sample is placed in a furnace so that measurements can be carried out at well-controlled temperatures. The sample is put in the furnace before the experiment starts and the experiment is not started until the furnace temperature and sample temperature are judged to be equal. The sample is supported by three small pins to minimise conductive losses, and the furnace is held in near-vacuum conditions to minimise convective losses.

The laser flash is a pulse of laser light lasting less than 1 ms. The laser power is adjusted so that the maximum temperature rise in the sample caused by the laser flash is typically between 3 K and 4 K. This temperature rise gives a good signal-to-noise ratio on the detected signal, but is sufficiently small that radiative losses can be approximated well by linearisation and can be taken into account in a straightforward manner when calculating the thermal diffusivity. The laser power used is not known by the user and cannot be obtained from the equipment.

The temperature change of the rear face of the sample is measured throughout the experiment by an infra-red (IR) temperature sensor. The temperature sensor has a finite spot size and so the measurement is an average over an area rather than a value at a single point.

In order to shield the temperature sensor from the laser flash, a guard cap with a window in it is placed over the end of the sample. The guard cap should not be in contact with the material sample since the conductive heat losses from the sample to the cap will affect the measured temperature and hence the calculated thermal diffusivity value.

The measurement data set used in this work is shown in Fig. 8.1. This data set is used as target data in the work reported here, meaning that the aim of the optimisation work was to generate model results that fit these data well. The measurement was carried out at a furnace temperature of 947.15 K, and Fig. 8.1 shows the change in temperature relative to the firnace temperature. This data set was chosen because i) the ambient temperature was sufficiently high that radiative losses would be significant, and so determination of emissivity would be a possibility, and ii) the same data set had been studied previously [1], giving values to which the calculated results could be compared.

The measurement data set consists of temperature change measurements every 1.568 ms. The measurements are continued for 2.373952 s after the laser flash, giving a total of 1515 measurements for time $t \geq 0$ (the first measurement being at $t = 0$). It is assumed that the sample temperature has fully stabilised by the time that the laser is fired, and so the temperature measured at $t = 0$ is taken to be the ambient (furnace) temperature. The time axis is scaled such that the laser was fired at $t = 0$. The small peak shortly after $t = 0$ is caused by energy from the laser flash that has not been absorbed by the sample being measured by the temperature sensor.

The simplest form of data analysis of these data [9] is based on an analytical solution to the transient heat flow equation that assumes a uniform sample, an instantaneous uniform laser flash, and no heat losses from the sample. This approach leads to a 1-D model for the heat flow, and solution of this model gives an expression of the form

$$\Delta T = \Delta T_m \left(1 + 2 \sum_{n=1}^{\infty} (-1)^n \exp(-n^2 \pi^2 \alpha t / L^2) \right) \tag{8.2}$$

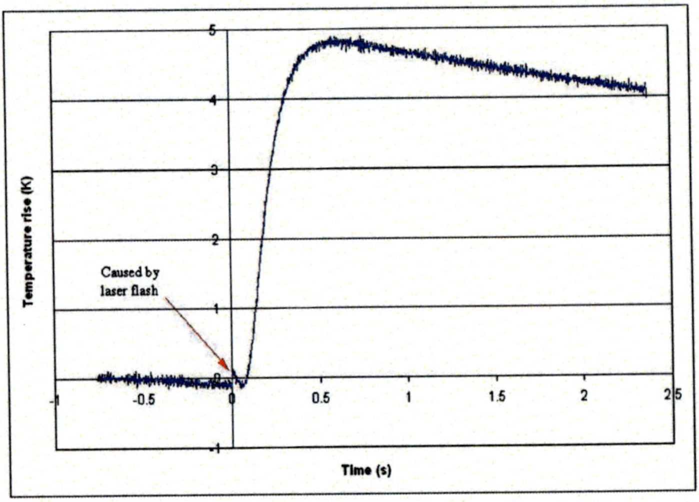

Fig. 8.1 Measured data set used for the work reported here

where $\Delta T(t)$ is the temperature rise of the rear face at time t, ΔT_m is the maximum temperature rise, and L is the thickness of the sample. Defining $t_{1/2}$ as the time taken for the temperature rise to reach half of its maximum value, which can be determined from the measured temperature values, gives

$$\sum_{n=1}^{\infty}(-1)^n \exp(-n^2\pi^2\alpha t_{1/2}/L^2)+\frac{1}{4}=0. \tag{8.3}$$

This is a nonlinear equation that gives α in terms of known values. Solving the equation gives

$$\alpha = 0.138785\frac{L^2}{t_{1/2}} \tag{8.4}$$

Subsequent work [2, 5, 7, 11] has developed corrections to the simple one-dimensional model to allow for the finite duration of the laser pulse and for radiant heat losses (including those from the curved faces). The methods of data processing that include corrections still make a number of assumptions, including spatial uniformity and isotropy of sample properties, insignificant temperature-dependency of material properties during the experiment, spatial uniformity of the laser flash, and absence of conductive and convective heat losses. The first of these assumptions is clearly not the case for the layered samples of interest in this work. It will be shown in section 8.3 that the final assumption is not valid either.

For the purposes of the modelling work, the sample is assumed to be perfectly cylindrical with plane parallel circular faces of radius 6 mm. The data shown in Fig. 8.1 were gathered during the measurement of a layered sample. It is assumed that the sample consists of two distinct layers. Each layer is assumed to be uniform and isotropic. The known sizes and properties of the layers in the sample are listed in table 8.1. The emissivity, ε, is required for implementation of radiative boundary conditions. A dash in a cell indicates that the property is unknown and is to be determined using optimisation methods to minimise the difference between the measured data and the model predictions. As has been mentioned above, the power of the laser that generates the flash is unknown and also must be determined using optimisation.

Table 8.1 Thicknesses and thermal properties of the layers. A dash in a cell indicates that the property is unknown and is to be determined using optimisation

Material	P92 steel	Oxide
Thickness (mm)	2.0942	0.2265
Density (kg m^{-3})	7871	5015
Specific heat capacity (J kg^{-1} K^{-1})	1473.2	934.8
Thermal conductivity (W m^{-1} K^{-1})	45.181	-
Emissivity	0.8	-

Previous analysis and simulation relating to this sample have been described in an NPL report [1]. All properties used in the work reported here have been taken from that report or from the references therein. The full chemical composition of the steel is given in the earlier report. The oxide layer consists of two components, magnetite and iron/chromium spinel, but they have been treated as a single uniform substance in order to provide a simpler model for initial investigations. The model could easily be extended to account for more complex layered structures.

8.3 Mathematical Model

8.3.1 Governing Equations

The model considers heat flow within the sample and assumes that the heat flow within the rest of the equipment is either irrelevant or can be taken into account via an appropriate choice of boundary conditions. Cylindrical polar coordinates $\mathbf{r} = \{r, \theta, z\}$ and total temperature (rather than temperature change relative to furnace temperature) will be used throughout.

The temperature distribution within the sample obeys the transient heat flow equation

$$\rho(\mathbf{r})c_p(\mathbf{r})\frac{\partial T(\mathbf{r},t)}{\partial t} = \nabla.(\lambda(\mathbf{r})\nabla T(\mathbf{r},t)) + Q(\mathbf{r},t), \tag{8.5}$$

where $\mathbf{r}$ denotes a position within the sample, $T(\mathbf{r},t)$ is the temperature at a point $\mathbf{r}$ and time t, and $Q(\mathbf{r},t)$ is a heat source term that is used to account for the laser flash.

The domain is defined as $0 \leq r \leq R, 0 \leq \theta \leq 2\pi, 0 \leq z \leq L$ where R is the radius of the sample and L is its thickness. It is assumed that the problem is axisymmetric so that variation with θ can be neglected. This reduces the model to two dimensions, making it simpler and quicker to solve.

The domain is split into two layers in the z direction. The properties of each layer are isotropic and uniform and the two layers are assumed to have a perfect thermal bond. The layers are of thickness z_1 (P92) and z_2 (oxide), with $z_1 + z_2 = L$. Then the material properties of the two layers are dependent only on z and are given by

$$\lambda(z) = \begin{cases} \lambda_1 & 0 \leq z \leq z_1 \\ \lambda_2 & z_1 < z \leq L \end{cases} \tag{8.6}$$

$$\rho(z) = \begin{cases} \rho_1 & 0 \leq z \leq z_1 \\ \rho_2 & z_1 < z \leq L \end{cases} \tag{8.7}$$

$$\varepsilon(z) = \begin{cases} \varepsilon_1 & 0 \leq z \leq z_1 \\ \varepsilon_2 & z_1 < z \leq L \end{cases} \tag{8.8}$$

$$c_p(z) = \begin{cases} c_{p1} & 0 \leq z \leq z_1 \\ c_{p2} & z_1 < z \leq L \end{cases} \tag{8.9}$$

The source term Q is assumed to affect a uniform layer (thickness Δz) at the front of the sample directly. It is assumed that the flash is of equal intensity over its finite duration. If the duration is t_0 and the intensity is I then

$$Q(z,t) = \begin{cases} I & 0 \leq z \leq \Delta z, \ 0 \leq t \leq t_0 \\ 0 & \text{otherwise} \end{cases} \tag{8.10}$$

A value for t_0 is known from the experiment and a value for Δz will be assigned when the numerical solution method is described. The value of t_0 used in this work was 0.8 ms.

The initial conditions assume that the sample is uniformly at the ambient temperature T_0 at time 0, so that

$$T(r,z,0) = T_0, \ 0 \leq r \leq R, \ 0 \leq z \leq L. \tag{8.11}$$

8.3.2 Boundary Conditions

The simplest modelling assumption for boundary conditions is that all cooling is due to radiation only. A straight-line fit to the cooling section of the curve shown in Fig. 8.1 shows a cooling rate of approximately 0.42 K s^{-1}. A simplified model assuming instantaneous uniform temperature change throughout the sample during cooling shows that the maximum radiative heat loss for a sample of this type at an ambient temperature of 947.15 K and a sample temperature of 952.15 K is approximately 0.02 K s^{-1} (this value is likely to be an overestimate due to the model assumptions). The difference of an order of magnitude in cooling rates suggests that the sample must be losing heat via some mechanism in addition to radiation.

Considering the experimental set-up, the most likely source of extra heat loss from the sample is contact between the guard cap and the sample. This contact could be caused by thermal expansion of the sample, the sample being too large for the holder, or poor positioning of the sample within the holder.

The cross section of the guard cap is shown in Fig. 8.2, including dimensions. The cap has a window at its centre through which the sample temperature is measured and which is transparent to infrared radiation. The radiative losses of the sample pass through this window. It is assumed that there is a perfect thermal bond between the surface of the guard cap marked with a heavy line and the equivalent portion of the sample, and that the guard cap is uniformly at the ambient temperature. It is assumed that the window is not in contact with the sample since in reality it is slightly offset from the main part of the guard cap. These assumptions avoid the need to include the heat flow within the guard cap in the model and enable the conductive heat losses to be modelled as a boundary condition.

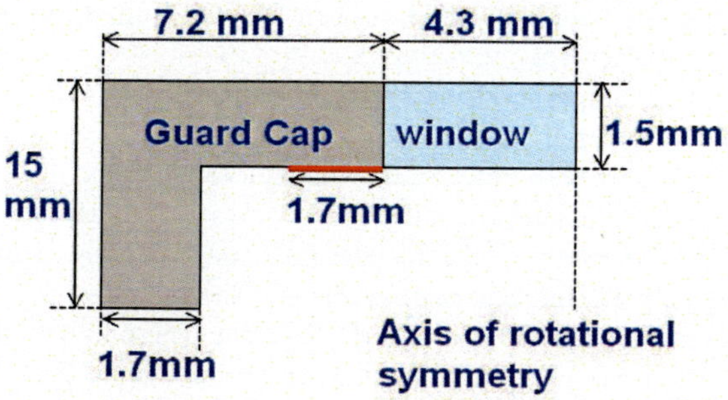

Fig. 8.2 Sketch of the guard cap geometry (not to scale). The dark area marks the region of contact between guard cap and sample

The curved surfaces of the sample are assumed to be perfectly insulated. The temperature gradient along the axis of symmetry must be zero for axisymmetry to be valid. The flat face exposed to the laser is assumed to lose heat radiatively. These boundary conditions can be expressed as

$$\left.\frac{\partial T}{\partial r}\right|_{r=0} = 0, \quad 0 \le z \le L, \tag{8.12}$$

$$\left.\frac{\partial T}{\partial r}\right|_{r=R} = 0, \quad 0 \le z \le L, \tag{8.13}$$

$$\left.\frac{\partial T}{\partial z}\right|_{z=0} = \varepsilon\sigma(T(^4r,0,t) - T_0^4), \quad 0 \le r \le R \tag{8.14}$$

$$\left.\frac{\partial T}{\partial z}\right|_{z=L} = -\varepsilon\sigma(T^4(r,L,t) - T_0^4), \quad 0 \le r \le r_w \tag{8.15}$$

$$T(r, L, t) = T_0 \quad r_w < r \leq R \tag{8.16}$$

where r_w is the radius of the window in the guard cap and σ is the Stefan-Boltzmann constant.

8.3.3 Numerical Methods

These equations, boundary conditions, initial conditions, and material properties fully define the two-dimensional model. The model cannot be solved analytically. A numerical approximation technique must be used instead.

The technique used to solve the model numerically is based on the finite volume method. The structure and approach are described in detail elsewhere [1, 6]. The work reported here has used a version of TherMol [1, 6], an NPL software package for multiphysics applications focussing on the diffusion equation, as the basis for the model. The software used has been adapted from a three-dimensional implementation of TherMol.

The finite volume mesh uses two different volume sizes Δz in the z direction, one for each material. The oxide layer had $\Delta z = 0.0453$ mm and the P92 steel had $\Delta z = 0.0419$ mm. The latter value was used as the value of Δz in the definition of $Q(\mathbf{r}, t)$ in (8.10). A uniform volume size of $\Delta r = 0.1$ mm was used in the r direction.

The finite volume model calculates the change in temperature relative to the initial temperature. This approach means that the rounding errors caused by the use of finite-precision arithmetic have little effect. The only feature of the model that requires the use of the true temperature is the calculation of radiative losses, and the apropriate formulation is used in that section of the software.

An explicit time integration method has been used for the transient calculations for simplicity and ease of implementation. The time step was chosen by trial and error for a typical set of parameter values, and was then divided by 10 to ensure that the model would run for more extreme parameter value choices. No numerical stability problems have been encountered during the work.

The results of interest from the calculations were the temperature changes of the rear face averaged over the spot size of the temperature sensor. It was assumed that the temperature sensor spot size was the same size as the window of the guard cap. The results were output at the same time intervals as the measurements to enable direct comparison.

The software TherMOL has been used before with an optimisation routine to determine unknown properties of the laser flash experiment [1]. The work reported used a one-dimensional model with radiative cooling only and an extension of the Nelder-Mead algorithm, COBYLA [10], able to handle constraints on the parameter values. When applied to the data shown in Fig. 8.1, the optimisation process gave a thermal conductivity of 2.1 W m^{-1} K^{-1} for the oxide layer, and a laser power intensity of 1.6×10^8 W m^{-2}, but the model results did not fit the cooling part of the temperature curve at all well.

8.4 Optimisation Results

The initial optimisation aimed to minimise the relative difference between the measured temperatures and the model predictions by varying the thermal conductivity of the oxide layer, λ_2, the laser power I, and the emissivity of the oxide layer ε_2. For any optimisation problem, the formulation of the right objective function is important. Here we intend to minimise errors, but errors can be defined as relative errors and absolute errors, which implies two different ways of defining the objective function.

8.4.1 *Initial Optimisation Results*

The objective function was initially defined as the root mean square average of the relative differences between the measured data and the calculated values:

$$\sqrt{\frac{1}{N}\sum_{n=1}^{N}\left(1-\frac{\bar{T}(t_n;\lambda_2,I,\varepsilon_2)}{T_n}\right)^2}, \tag{8.17}$$

where $N = 1515$ is the number of data points. T_n is the measured temperature rise (i.e. $T - T_0$) at time t_n, and $\bar{T}(t_n;\lambda_2,I,\varepsilon_2)$ is the calculated averaged surface temperature rise over the spot size of the temperature sensor for a given set of values of the model parameters $\lambda_2, I, \varepsilon_2$.

Two optimisation algorithms were used: the Levenburg-Marquardt algorithm within a trust-region [3, 4] and a particle swarm optimisation algorithm (PSO) [8]. The Levenburg-Marquardt algorithm is an efficient local optimiser and will find global minima for smooth unimodal surfaces. The work reported here uses the Matlab Optimisation Toolbox function `lsqnonlin`, an implementation of the Levenburg-Marquardt algorithm designed for minimisation of least-squares functions such as (8.17). The PSO implementation used was developed at NPL. The technique is a global optimiser and can be used to check that the local optimiser's results are globally optimal. The PSO is expected to take more time to converge to the optimal solution than the local optimiser.

The initial optimisation results gave a poor fit to the measurement data, particularly at later times. A typical set of results, generated by parameter values identified as optimal by both algorithms, is shown in Fig. 8.3. The lower of the plots shows a close-up of the first 0.15 seconds. The measured temperature changes here are very close to zero, and the model is a very good fit to these values. The small values of T_i for these initial times mean that the relative errors are very large, and so the fit of the model to the data at small times dominates the overall fit.

This dominance of small measured temperature change values suggests that the root mean square of differences,

$$\sqrt{\frac{1}{N}\sum_{n=1}^{N}\left(T_n-\bar{T}(t_n;\lambda_2,I,\varepsilon_2)\right)^2}, \tag{8.18}$$

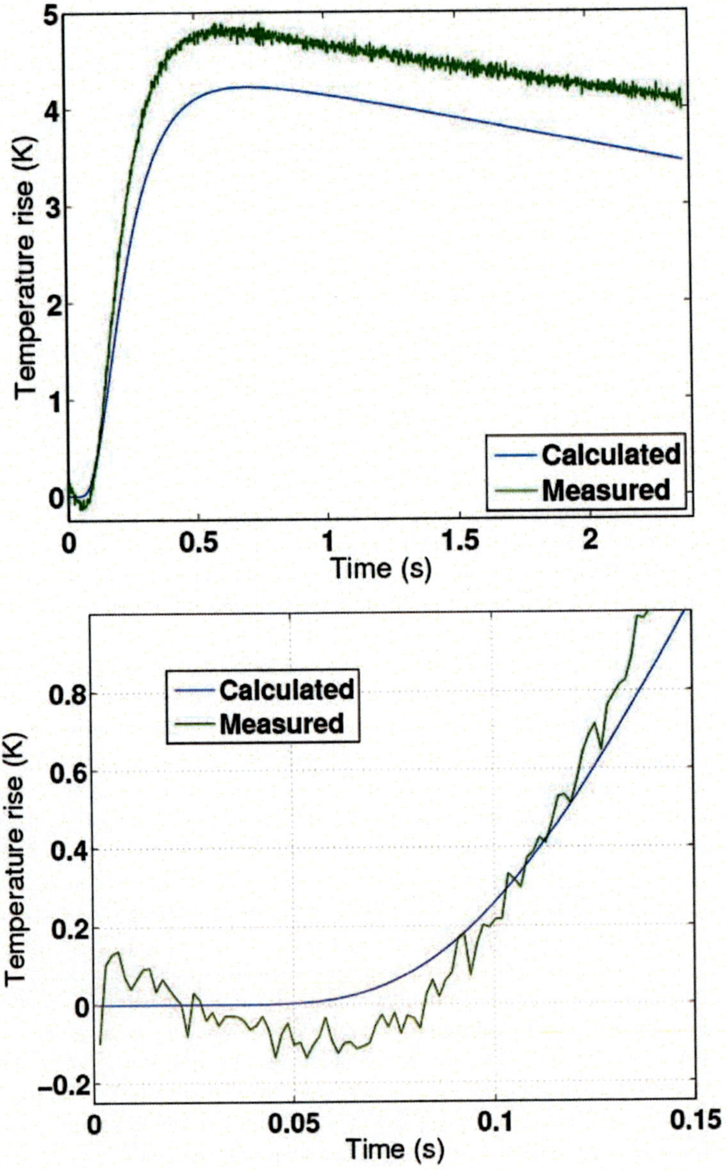

Fig. 8.3 An inappropriate objective function leads to a poor fit of model results to measured data (upper figure), but an unnecessarily good fit to values around 0 (lower figure)

would be a better choice of objective function. This function has been used to generate all results in the rest of this chapter. Normally this sum would be weighted according to the uncertainties associated with the measurements, but here it is assumed that the uncertaintes were the same for all measurements.

In general, if the target data contains values close to zero, absolute differences may be a better choice of objective function than relative differences. Whilst minimisation of relative errors can be a good way to combine results of different types, the measured data may not be the best choice of scaling factor when measured values are close to zero. The problems experienced illustrate the importance of considering the construction of the objective function carefully.

In order to check the sensitivity of the converged solution to the initial parameter estimates, five sets of optimisation runs were carried out using each optimisation algorithm. The runs started from randomly-generated points within the parameter search space. The optimal solutions found and the number of function evaluations required to find them are summarized in Table 8.2.

The results in Table 8.2 show that both algorithms converge to the same optimal solution repeatedly. The PSO showed more variation within the five runs than `lsqnonlin`, leading to a higher standard deviation for each of the parameters. The size of the standard deviation is strongly linked to the stopping criteria of the optimisation algorithms. The deviations listed below are consistent with the algorithms effectively arriving at the same solution. As expected, PSO required more function evaluations (about 327) to converge than the efficient local optimiser (about 32). As stated in section 8.3, earlier work had found values of $\lambda_2 = 2.1$ W m^{-1} K^{-1} and $I = 1.6 \times 10^8$ W m^{-2}, which is a change of about 25% in the value of λ_2.

Table 8.2 Summary of optimisation results. Means and standard deviations of parameters calculated from 5 runs. Note that intervals specified here are ±one standard deviation

	PSO	LSQnonlin
Evaluations	327 ± 56	32 ± 3
Mean/Std	$\lambda_2 = 2.83 \pm 0.23$ W m^{-1} K^{-1}	2.82 ± 0.005 W m^{-1} K^{-1}
	$I = [1.69 \pm 0.12] \times 10^8$ W m^{-2}	$[1.69 \pm 0.04] \times 10^8$ W m^{-2}
	$\varepsilon_2 = 0.092 \pm 0.073$	0.0000 ± 0.002

The covariance matrix, V_a, associated with these parameter estimates has been calculated from the goodness of fit and Jacobian matrix, using the equation

$$V_a = (J^{\mathrm{T}} J)^{-1} \frac{1}{N-m} \sum_{n=1}^{N} (T_n - \bar{T}(t_n; \lambda_2, I, \varepsilon_2))^2 \tag{8.19}$$

where J is the Jacobian matrix and $m = 3$ since there are three parameters. The Jacobian matrix has been estimated using finite difference approximations since the objective function of this model is a black box. The standard uncertainties associated with the parameter estimates are given by the square root of the diagonal entries of the resulting matrix. The standard uncertainties were, in the order λ_2, I, ε_2, 0.047 W m^{-1} K^{-1}, 7.2×10^5 W m^{-2}, and 0.17. This reflects the high degree of uncertainty about the emissivity. The associated estimate of the goodness of fit was 8.3×10^{-3} K.

The results of the model obtained by using the optimal parameter values are shown in Fig. 8.4. These model results are clearly a better fit to the measured data than those shown in Fig. 8.3, illustrating the benefit of changing the objective function. The results are also a better fit to the measured values than the results obtained during the work described in [1], illustrating that the new model simulates the experiment better than the original version.

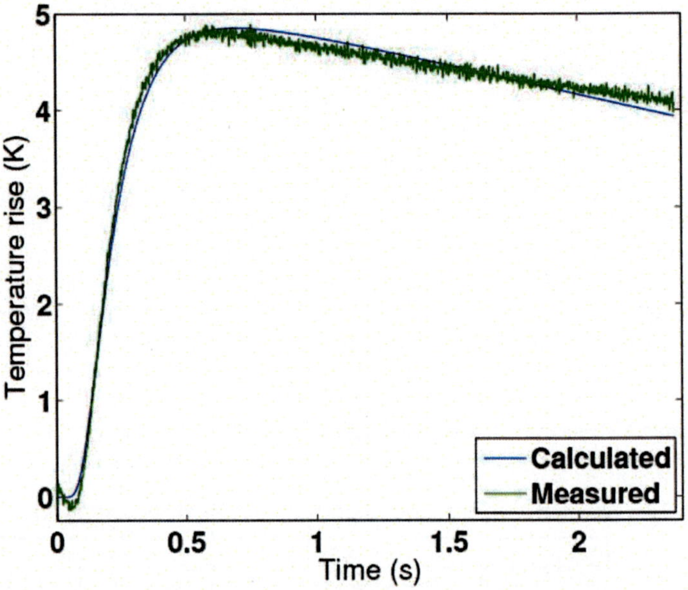

Fig. 8.4 Model results calculated using the best solutions found ($\lambda_2 = 2.8229$, $I = 1.6903 \times 10^8$ and $\varepsilon = 0.0$)

There is still a discrepancy in the cooling part of the curve: the model results appear to cool too fast. In addition, the value of ε that has been found is unexpectedly low (physically the value must be between 0 and 1, and was expected to be close to 0.8). These observations suggest that the conductive losses through the guard cap over-estimate the true cooling, and that the conductive losses dominate the radiative losses to the point where emissivity cannot be determined.

8.4.2 Revision of the Model

The results of the initial model suggest that whilst the inclusion of the effects of the guard cap improves the fit to the measured data, the model is still not predicting the cooling part of the measured data correctly. In order to improve the fitting of the cooling part of the data, the perfect bond between the guard cap and the sample is changed to an imperfect thermal bond parameterised by the unknown thermal bond quality parameter β with units W m^{-2} K^{-1}.

The imperfect bond is defined as an extra layer between the guard cap and the surface of the sample. The bond quality parameter is effectively the thermal conductivity of the extra layer divided by its thickness. The boundary conditions describing the imperfect bond are generated by solving a one-dimensional steady state heat flux equation analytically at each point on the sample surface that is in contact with the guard cap, and imposing continuity of temperature and flux at the boundaries of the extra layer. This approach assumes that the only heat flow within the extra layer only occurs in the z direction, which is valid because the extra layer is not real and is only a simulation of a poor thermal bond.

The imperfect thermal bond boundary condition is implemented as

$$\lambda \frac{\partial T(r,z,t)}{\partial z}\bigg|_{z=L} = \frac{\beta\lambda/(\Delta z/2)(T_0 - T(r,L-\Delta z/2,t))}{\beta + \lambda/(\Delta z/2)}, \tag{8.20}$$

where $r_w \leq r \leq R$. From this implementation it is clear that $\beta = 0$ is a perfectly insulating boundary, and that as $\beta \to \infty$ the condition tends towards a perfect thermal bond with $T(r,L,t) = T_0$ as in the initial model.

8.4.3 Optimisation Results with the New Model

Following the successful application of the Levenbug-Marquardt algorithm to the initial version of the model, the optimisations using the new model with the imperfect bond are carried out using `lsqnonlin` only. The new model uses four parameters and so it is expected that more function evaluations will be required to find a converged solution. The optimisation identified the best parameter values as

- λ_2=3.55 W m^{-1}K^{-1},
- I=1.67 $\times 10^8$ W m^{-2},
- ε_2=1.0,
- β=1.92 $\times 10^4$ W m^{-2}K^{-1}.

These values were obtained from five runs started from randomly-generated initial parameter estimates. The average number of function evaluations required for convergence was 126. The standard deviations of each of the parameter values across the five runs were less than 10^{-3}, suggesting good repeatability. The covariance matrix was calculated using equation (8.19). This calculation was complicated by the Jacobian estimates suggesting that the derivative of the objective function with respect to ε_2 at $\varepsilon_2 = 1$ was zero. This zero sensitivity meant that the emissivity could not be included in the covariance calculation, so the covariance matrix only considered λ_2, I, and β. The standard uncertainties found from the matrix were, in the order λ_2, I, and β, 0.045 W m^{-1} K^{-1}, 1.9 $\times 10^5$ W m^{-2}, and 620 W m^{-2}K^{-1}. The associated estimate of the goodness of fit was 5.2×10^{-3} K, an improvement on the previous model.

The model results obtained by using these parameter values are plotted in Fig. 8.5, along with the measured data and the model results shown in Fig. 8.4 (dashed line). These plots show that the use of an imperfect thermal bond improves the fit of the model results to the measured data, particularly for $t > 0.5$. The old and new models are in close agreement for $t < 0.3$, which is the time period where the energy absorbed by the sample from the laser flash is likely to dominate the heat flow and differences in the value of λ_2 are less likely to have an effect as the oxide layer is comparatively thin.

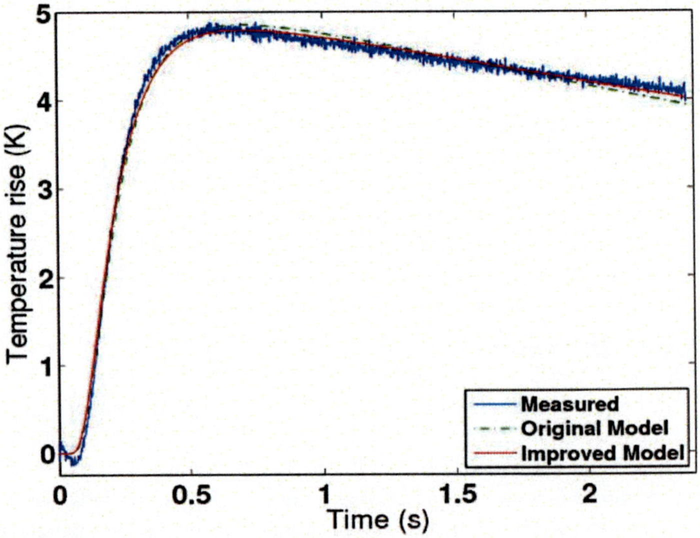

Fig. 8.5 Comparison of new model and original model and their best-fit curves

8.5 Discussion

Using the revised model, the value of λ_2 has increased by about 20 %, and that of ε_2 has gone from 0 to 1, whereas the value of I has not changed significantly. The parameter I defines how much energy goes into the sample during the laser flash, and the good fit of the model predictions to the peak temperature rise suggests that this value has been determined accurately. The value of λ_2 affects how the heat flows within the sample, so it is expected that a change in the boundary conditions would affect the optimal value of λ_2. Whilst the new value of ε_2 is closer to the expected value of 0.8, it is by no means certain that this value is a good estimate of the true value. It is likely that the cooling due to the contact with the guard cap still dominates the heat loss, making determination of ε_2 difficult to determine.

It is worth pointing out that the fit, though improved, is still not perfect. Ideally the differences between the model predictions and measured values would lie below the level of the measurement noise, but this level of agreement has not been achieved. There are differences between the model results and the measured values for $0.3 \leq t \leq 0.5$ which suggest that further improvements could be made to the model. Possibilities for improvements include adding circumferential heat losses, considering an imperfect thermal bond between the sample and the oxide, adding extra layers to allow for the multi-phase nature of the oxide, and including a full model of the guard cap so that the conductive losses are modelled more accurately.

It is clear from the results shown here that choosing the best optimisation algorithm can significantly reduce the number of function evaluations, leading to a reduction in computational time. There was no difference between the algorithms in terms of the accuracy of the estimates. The results suggest that the accuracy of the parameter estimates is constrained by the quality of the underpinning model and (were the model to be improved) the uncertainties associated with the measurement data.

Recent trends suggest that metaheuristic algorithms such as PSO are increasingly widely used [8, 12], but popularity does not mean the algorithm is the best choice. In this case study, both PSO and nonlinear least squares provided very good results, but the classical, well-tested nonlinear squares required significantly fewer function evaluations to reach a converged solution. The objective function in this case study was formulated in the least-squares sense, and the results suggest that a unique global optimum exists, which makes the least-squares optimiser more suitable than a metaheuristic algorithm.

Experience gained in this case study and in other applications suggests that it is a good idea to use well-established algorithms when first solving a new optimisation problem. If the well-established algorithms fail, it is worth trying metaheuristic algorithms [12]. This approach can avoid unnecessary and time-consuming trial and error.

References

1. Chapman, L.A., Fry, A.T., Roberts, S.J.: Thermal diffusivity and thermal conductivity measurements on oxide scales. Technical Report MATC(D) 188, National Physical Laboratory (2004)
2. Clark, L.M., Taylor, R.E.: Radiation loss in the flash method for thermal diffusivity. J. Appl. Phys. 46, 714–719 (1975)
3. Coleman, T.F., Li, Y.: On the convergence of reflective Newton methods for large-scale nonlinear minimization subject to bounds. Mathematical Programming 67, 189–224 (1994)
4. Coleman, T.F., Li, Y.: An interior trust-region approach for nonlinear minimization subject to bounds. SIAM J. Optim. 6, 418–485 (1996)

5. Cowan, R.D.: Pulse method for measuring thermal diffusivity at high temperatures. J. Appl. Phys. 34, 926–927 (1963)
6. Duncan, B.C., Pilkington, G., Nottay, J.S., Allen, C.R.G., Lawrence, K., Urquhart, J., Roberts, S.J.: Diffusion of moisture in adhesive bonds. Technical Report DEPC-MPR 062, National Physical Laboratory (2007)
7. Dusza, L.: Combined solution of the simultaneous heat loss and finite pulse corrections with the laser flash method. High Temp. High Press. 27/28, 467–473 (1995/1996)
8. Kennedy, J., Eberhardt, R.C.: Particle swarm optimization. In: Proc. IEEE Int. Conf. Neural Networks, pp. 1942–1948. IEEE Press, Los Alamitos (1995)
9. Parker, W.J., Jenkins, R.J., Butle, C.P., Abbott, G.L.: Flash method for determining thermal diffusivity, heat capacity and thermal conductivity. J. Appl. Phys. 32, 1679–1684 (1961)
10. Powell, M.J.D.: A direct search optimization method that models the objective and constraint functions by linear interpolation. In: Gomez, S., Hennart, J.P. (eds.) Avdances in Optimization and Numerical Anlaysis. Kluwer Academic, Dordrecht (1994)
11. Taylor, R.E., Clark, L.M.: Finite pulse time effects in flash diffusivity method. High Temp. High Press. 6, 65–72 (1974)
12. Talbi, E.G.: Metaheuristics: From Design to Implementation. Wiley, Chichester (2009)

Chapter 9
Applications of Computational Intelligence in Behavior Simulation of Concrete Materials

Amir Hossein Gandomi and Amir Hossein Alavi

Abstract. The application of Computational Intelligence (CI) to structural engineering design problems is relatively new. This chapter presents the use of the CI techniques, and specifically Genetic Programming (GP) and Artificial Neural Network (ANN) techniques, in behavior modeling of concrete materials. We first introduce two main branches of GP, namely Tree-based Genetic Programming (TGP) and Linear Genetic Programming (LGP), and two variants of ANNs, called Multi Layer Perceptron (MLP) and Radial Basis Function (RBF). The simulation capabilities of these techniques are further demonstrated by applying them to two conventional concrete material cases. The first case is simulation of concrete compressive strength using mix properties and the second problem is prediction of elastic modulus of concrete using its compressive strength.

9.1 Introduction

Modeling of structural engineering nonlinear systems is a diverse research area where different kinds of methods can be utilized. Due to the large variety of this field, no method can impose itself as the best solution. Estimating both the structure and the parameters of the structural engineering problems makes their modeling process a difficult task. Different criteria for model classification can be characterized while dealing with a system modeling task [1]. A model can be

Amir Hossein Gandomi
College of Civil Engineering, Tafresh University, Tafresh, Iran
e-mail: a.h.gandomi@gmail.com

Amir Hossein Alavi
School of Civil Engineering,
Iran University of Science and Technology, Tehran, Iran
e-mail: ah_alavi@hotmail.com

X.-S. Yang & S. Koziel (Eds.): Computational Optimization & Applications, SCI 359, pp. 221–243.
springerlink.com

classified as phenomenological or behavioral [2]. A phenomenological model is derived by taking into account the physical relations governing the system. As a result, the structure of the model is selected according to the prior knowledge about the system. It is not always possible to design phenomenological models for many of the structural engineering systems due to their complexity. In order to overcome such a problem, the behavioral models are commonly employed. Such models approximate the relationships between the inputs and outputs based on a measured set of data without a need to prior knowledge about the mechanism that produced the experimental data. The behavioral models can provide very good results with a minimal effort [2]. Traditional statistical regression techniques are commonly used for the behavioral modeling purposes. The regression analysis can have large uncertainties. It has major drawbacks for idealization of complex processes, approximation, and averaging widely varying prototype conditions. Another important issue is due to the limitation of this method. The regression analysis tries to model the nature of the corresponding problem by a pre-defined linear or nonlinear equation. Another major constraint in application of the regression analysis is the assumption of normality of residuals.

In the case of the behavioral models, several alternative computer-aided pattern recognition and data classification approaches have been developed. Computational intelligence (CI) [3] techniques are well-known pattern recognition methods. Developments in the computer hardware during the last two decades have made it much easier for these techniques to grow into more efficient frameworks. In addition, various CI-based approaches may be used as efficient tools in problems where conventional approaches fail or perform poorly. Artificial neural networks (ANNs) are the most widely-used CI methods. ANNs have been used for a wide range of structural engineering problems (e.g. [4]). In spite of the successful performance of ANNs, they usually do not give a deep insight into the process which they use the available information to obtain a solution. In the present study, the approximation ability of two of the most widely used ANN architectures, namely Multi Layer Perceptron (MLP) and Radial Basis Function (RBF) are investigated.

Genetic algorithm (GA) is a powerful stochastic search and optimization method based on the principles of genetics and natural selection. GA has been shown to be suitably robust for a wide variety of complex civil engineering problems (e.g. [5]). Genetic programming (GP) [6] is an alternative approach for behavior modeling of geotechnical engineering tasks. GP is a developing subarea of evolutionary algorithms inspired from the Darwin's evolution theory. It may generally be defined as a specialization of GA where the solutions are computer programs rather than fixed-length binary strings. The programs generated by traditional GP are represented as tree structures and expressed in the functional programming language [6]. This classical GP approach is referred to as Tree-Based GP (TGP). For the last ten years, traditional GP has been pronounced as an alternative method for simulating the behavior of civil engineering problems (e.g. [7]). Linear genetic programming (LGP) [8] is a new subset of GP with a linear structure similar to the DNA molecule in biological genomes. More specifically, LGP operates on programs that are represented as linear sequences of instructions of an imperative programming language [8, 9]. In contrast with traditional GP and ANNs, application of LGP in the field of civil engineering is totally new and original. (e.g., [10, 11]).

This chapter presents the feasibility of using TGP, LGP, MLP, and RBF for behavior simulation of concrete materials. To verify the capabilities of these techniques, they are applied to the modeling of compressive strength and elastic modulus of concrete. The chapter is organized as follows: Section 9.2 presents the basic aspects and the characteristics of the employed algorithms. The modeling process and parameters settings for the methods are given in Section 9.3. Numerical examples and the obtained results are presented in Section 9.4. Section 9.5 presents a discussion of the capabilities of the methods. Finally, some concluding remarks are provided in Section 9.6.

9.2 Computational Intelligence

Evolutionary algorithms (EAs) [12] are a subset of evolutionary computations. They use biology-inspired mechanisms to optimize a solution with regard to desired result. Computational intelligence (CI) [3] includes EAs and all of their different branches with artificial neural networks and fuzzy logic. The CI techniques have wide ranging applications for approximating the nonlinearities. A survey of the literature reveals the growing interest of the research community in the relatively new field of computational intelligence. In the following subsections, different branches of the CI techniques employed in this research are briefly introduced.

9.2.1 Genetic Programming

GP is a symbolic optimization technique that creates computer programs to solve a problem using the principle of Darwinian natural selection [6]. The breakthrough in GP then came in the late 1980s with the experiments on symbolic regression. GP was introduced by Koza [6] as an extension of GA. Most of the genetic operators used in GA can also be implemented in GP with minor changes. The main difference between GP and GA is the representation of the solution. GA creates a string of numbers that represent the solution. The GP solutions are computer programs represented as tree structures and expressed in a functional programming language (like LISP) [6, 10, 20]. In other words, in GP, the evolving programs (individuals) are parse trees than can vary in length throughout the run rather than fixed-length binary strings. Essentially, this is the beginning of computer programs that program themselves [6]. Since GP often evolves computer programs, the solutions can be executed without post-processing, while coded binary strings typically evolved by GA require post-processing. The traditional optimization techniques, like GA, are generally used in parameter optimization to evolve the best values for a given set of model parameters. GP, on the other hand, gives the basic structure of the approximation model together with the values of its parameters [13]. GP optimizes a population of computer programs according to a fitness landscape determined by a program ability to perform a given computational task. The fitness of each program in the population is evaluated using a fitness function. Thus, the fitness function is the objective function GP aims to optimize [14].

The classical GP technique is also referred to as tree-based GP (TGP) [6]. In TGP, a random population of individuals (computer programs) is created to achieve high diversity. A population member in TGP is a hierarchically structured tree comprising functions and terminals. The functions and terminals are selected from a set of functions and a set of terminals. For example, the function set F can contain the basic arithmetic operations (+, -, ×, /, etc.), Boolean logic functions (AND, OR, NOT, etc.), or any other mathematical functions [10,20]. The terminal set T contains the arguments for the functions and can consist of numerical constants, logical constants, variables, etc. The functions and terminals are chosen at random and constructed together to form a computer model in a tree-like structure with a root point with branches extending from each function and ending in a terminal. An example of a simple tree representation of a TGP model is illustrated in Fig. 9.1.

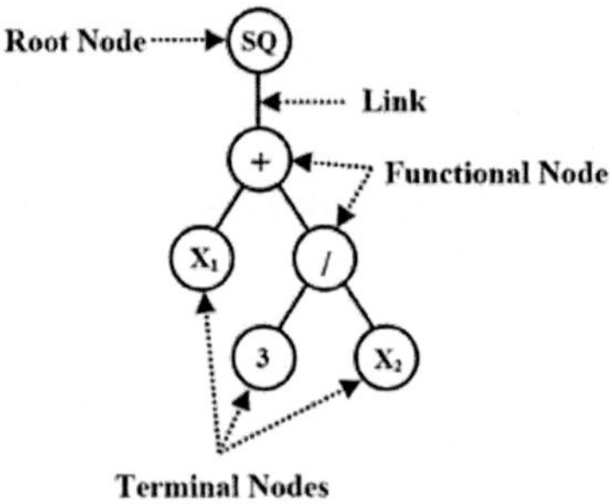

Fig. 9.1 The tree representation of a TGP model $(X_1 + 3/X_2)^2$ (After [10])

Once a population of individuals (models) has been created at random, the TGP algorithm evaluates the fitness of individuals, selects individuals for reproduction, and generates new individuals by reproduction, crossover and mutation [6]. The reproduction operation gives a higher probability of selection to more successful individuals. They are copied into the next generation without any change. The crossover operation ensures the exchange of genetic material between the evolved programs. During the crossover procedure, a point on a branch of each solution (program) is selected at random and the set of terminals and/or functions from each program are then swapped to create two new programs. Fig. 9.2 shows a typical crossover operation of two computer programs consisting of several function and terminal genes. Two new child computer programs (Child I, Child II) are generated from two parental computer programs (Parent I, Parent II). In Fig. 9.2, the randomly generated crossover points are shown by dotted lines. It can be seen that both child organisms include the genetic material from their parents. It is necessary to preserve syntactic structure of the programs during the crossover process.

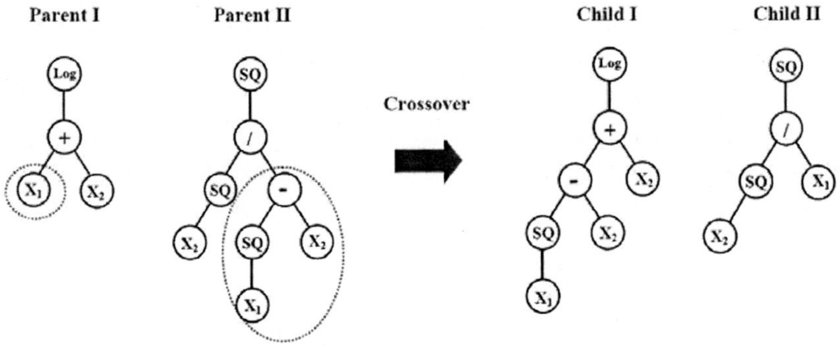

Fig. 9.2 Typical crossover operation in TGP

During this mutation process, the TGP algorithm occasionally selects a function or terminal from a model at random and mutates it. The mutation operation can be applied to the function or terminal nodes. A node in the tree is selected at random. If the selected node is a terminal, it is replaced by another terminal. If the node is a function and point mutation is to be applied, it is replaced by a new function with the same parity. If a tree mutation is to be performed, a new function node, which is not necessarily with the same parity, is chosen. Then, the original node together with its relative sub-tree is replaced by a new randomly created sub-tree. Fig. 9.3 illustrates a typical mutation operation in TGP. The best program that appeared in any generation, the best-so-far solution, defines the output of the GP algorithm [6].

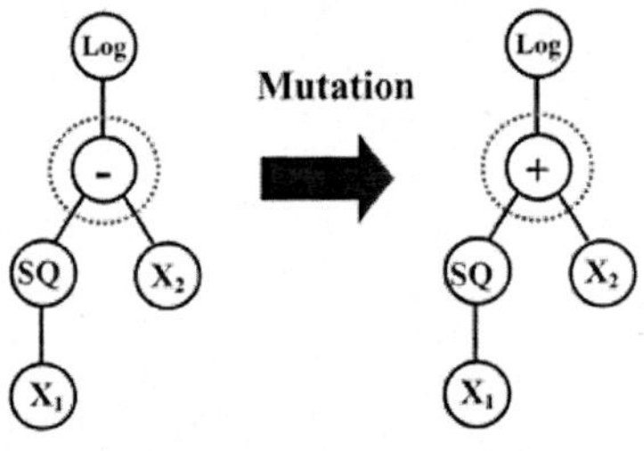

Fig. 9.3 Typical mutation operation in GP

In addition to traditional tree-based GP, there are other types of GP where programs are represented in different ways. These are linear and graph-based GP [15]. The emphasis of the present study is placed on the linear GP techniques. Several linear variants of GP have recently been proposed such as linear genetic programming (LGP) and multi-expression programming (MEP). The linear variants of GP make a clear distinction between the genotype and phenotype of an individual. In these variants, individuals are represented as linear strings [11,16]. Such linear programs can have a complex control flow similar to the trees of standard GP when executed. There are some main reasons for using linear GP. Basic computer architectures are fundamentally the same now as they were twenty years ago, when GP began. Almost all architectures represent computer programs in a linear fashion. In other words, computers do not naturally run tree-shaped programs. Hence, slow interpreters have to be used as part of tree-based GP. Conversely, by evolving the binary bit patterns, the use of an expensive interpreter (or compiler) is avoided. Consequently, the linear GP methods can run several orders of magnitude faster than comparable interpreting systems [11,17]. The enhanced speed of the linear variants of GP (e.g., LGP) permits conducting many runs in realistic timeframes. This leads to deriving consistent, high-precision models with little customization [18].

9.2.1.1 Linear Genetic Programming

LGP is a subset of GP with a linear representation of individuals. The main characteristic of LGP in comparison with traditional tree-based GP is that expressions of a functional programming language (like LISP) are substituted by programs of an imperative language (like C/C++) [8, 9]. Fig. 9.4 presents a comparison of the program structures in LGP and tree-based GP. As shown in Fig. 9.4(a), a linear genetic program can be seen as a data flow graph generated by multiple usage of register content. That is, on the functional level the evolved imperative structure denotes a special directed graph. As can be observed from Fig. 9.4(b), in tree-based GP, the data flow is more rigidly determined by the tree structure of the program [9,11].

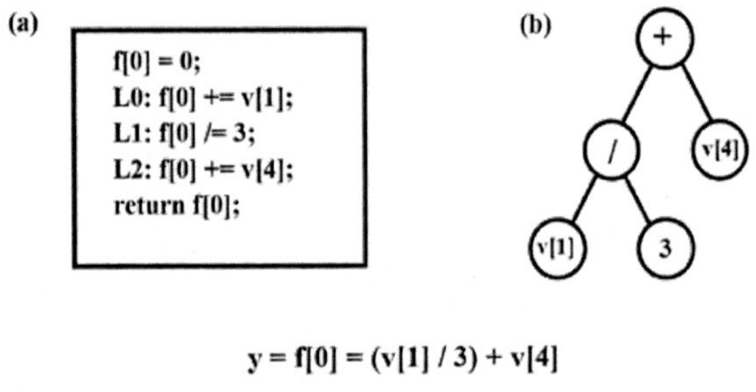

Fig. 9.4 Comparison of the GP program structures. (a) LGP (b) Tree-based GP (after [19])

In the LGP system described here, an individual program is interpreted as a variable-length sequence of simple C instructions. The instruction set or function set of LGP consists of arithmetic operations, conditional branches, and function calls. The terminal set of the system is composed of variables and constants. The instructions are restricted to operations that accept a minimum number of constants or memory variables, called registers (r), and assign the result to a destination register, e.g., $r_0 := r_1 + 1$. A part of a linear genetic program in C code is represented in Fig. 9.5. In this figure, register r[0] holds the final program output.

```
void LGP (double r[5])
{ ...
r[0] = r[5] + 70;
r[5] = r[0] - 50;
if (r[1] > 0)
if (r[5] > 2)
r[4] = r[2] × r[1];
r[2] = r[5] + r[4];
r[0] = sin(r[2]);
}
```

Fig. 9.5 An excerpt of a linear genetic program

Here are the steps the LGP system follows for a single run [8,11,20]:

1. Initializing a population of randomly generated programs and calculating their fitness values.
2. Running a Tournament. In this step four programs are selected from the population randomly. They are compared and based on their fitness, two programs are picked as the winners and two as the losers.
3. Transforming the winner programs. After that, two winner programs are copied and transformed probabilistically into two new programs via crossover and mutation operators.
4. Replacing the loser programs in the tournament with the transformed winner programs. The winners of the tournament remain without change.
5. Repeating steps two through four until termination or convergence conditions are satisfied.

Crossover occurs between instruction blocks. Fig. 9.6 demonstrates a two-point linear crossover used in LGP for recombining two tournament winners. As it is seen, a segment of random position and arbitrary length is selected in each of the two parents and exchanged. If one of the two children would exceed the maximum length, crossover is aborted and restarted with exchanging equally sized segments. The mutation operation occurs on a single instruction. Two types of standard

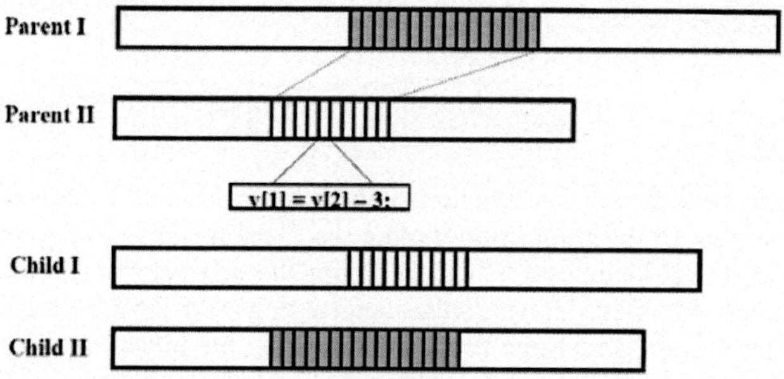

Fig. 9.6 Crossover in LGP [9]

LGP mutations are commonly used: micro and macro mutation. Micro mutation changes an operand or an operator of an instruction. The macro mutation operation inserts or deletes a random instruction [9]. Comprehensive descriptions of the basic parameters used to direct a search for a linear genetic program can be found in [8, 20].

9.2.2 Artificial Neural Network

ANNs have emerged as a result of simulation of biological nervous system. The ANN method was founded in the early 1940s by McCulloch and co-workers [21]. The first researches were focused on building simple neural networks to model simple logic functions. At the present time, ANNs can be applied to problems that do not have algorithmic solutions or problems with complex solutions. ANN formulates a mathematical model for a system in which no clear relationship is available between inputs and outputs. ANNs use the data alone to determine the structure of the model and unknown model parameters. The ability of ANNs to learn by example makes them very flexible and powerful techniques. Thus, this approach has widely been applied to solving regression and classification problems in many fields. In this study, the approximation ability of two of the most well-known ANN architectures, MLP and RBF, are investigated.

9.2.2.1 Multilayer Perceptron Network

MLPs are a class of ANN structures using feed forward architecture. The MLP networks are usually applied to perform supervised learning tasks, which involve iterative training methods to adjust the connection weights within the network. MLPs are universal approximators, that is, they are capable of approximating essentially any continuous function to an arbitrary degree of accuracy. They are often trained with back propagation (BP) [22] algorithm. Fig. 9.7 shows a schematic representation of an MLP network. MLP consist of an input layer, at least one hidden layer of neurons and an output layer. Each of these layers has several processing units and each unit is fully interconnected with weighted connections to units in the subsequent layer. Each layer contains a number of nodes. Every input is multiplied by the interconnection weights of the nodes. The output (h_j) is obtained by passing the sum of the product through an activation function as follows:

$$h_j = f\left(\sum_i x_i w_{ij} + b\right) \tag{9.1}$$

where $f()$ is activation function; x_i is the activation of ith hidden layer node; w_{ij} is the weight of the connection joining the jth neuron in a layer with the ith neuron in the previous layer, and b is the bias for the neuron. For nonlinear problems, the sigmoid functions (Hyperbolic tangent sigmoid or Log-sigmoid) are usually adopted as the activation function. Adjusting the interconnections between layers will reduce the following error function:

$$E = \frac{1}{2}\sum_{n}\sum_{k}\left(t_k^n - h_k^n\right)^2 \tag{9.2}$$

where t_k^n and h_k^n are respectively the calculated output and the actual output value, n is the number of sample and k is the number of output nodes. Further details of MLPs can be found in [23, 24].

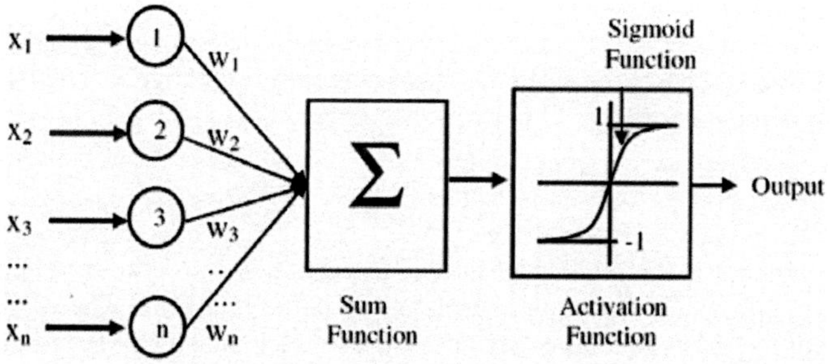

Fig. 9.7 A schematic representation of an MLP network

9.2.2.2 Radial Basis Function

RBFs have feed forward architectures. Compared to other ANN structures such as MLPs, the RBFs procedure to find complex relationships is generally faster and their training is much less computationally intensive. The schematic representation of the RBF network is illustrated in Fig. 9.8. The structure of the RBF network consists of an input layer, a hidden layer with a non-linear RBF activation function, and a linear output layer. Input vectors are transformed into radial basis functions by means of the hidden layer.

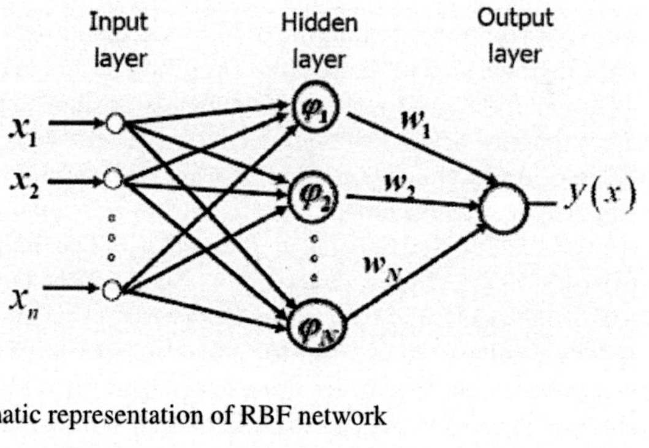

Fig. 9.8 A schematic representation of RBF network

The transformation functions used are based on a Gaussian distribution as an activation function. Center and width are two important parameters that are related to the Gaussian basis function. As the distance, that is usually Euclidean distance, between the input vector and its center increases the output given by the activation function decays to zero. The rate of decrease in the output is controlled by the width of RBF. The Gaussian basis function (c) is given in the following form:

$$c_j(x) = \exp\left(-\frac{\left\| x - \mu_j \right\|^2}{2\sigma_j^2}\right) \tag{9.3}$$

where ||•|| is the Euclidian norm, x is the input pattern, and μ_j and σ_j are the center and the spread of the Gaussian basis function respectively. The output of kth neuron in the output layer of network is computed as:

$$y_k(x) = \Sigma_{j=1}^{n} w_{jk} h_j(x) + b_k \tag{9.4}$$

in which n is the number of the hidden neurons, w_{jk} is the weight between j^{th} hidden neuron and k^{th} output neuron and b_k is the bias term. The RBF networks with Gaussian basis functions have been shown to be universal function approximators with high pointwise convergency [25].

9.3 Modeling Process and Parameters Setting

9.3.1 Model Development Using GP-Based Methods

Various parameters are involved in the TGP and LGP predictive algorithm. Several runs were conducted to come up with a parameterization of TGP and LGP that provided enough robustness and generalization to solve the problems. In this study, basic arithmetic operators and mathematical functions were utilized to get the optimum GP models. The number of programs in the population that TGP and LGP will evolve is set by the population size. A run will take longer with a larger population size. The maximum number of tournaments sets the outer limit of the tournaments that will occur before the program terminates the run. The proper number of population and tournaments depends on the number of possible solutions and complexity of the problem. Different levels were tested for the number of population and tournaments to find models with minimum error. The program was run until there was no longer significant improvement in the performance of the models or the runs terminated automatically. The mutation rate was set to 90%. At the low level the crossover rate is 50% and at the high level it is 95%. The values of the other involved parameters were selected based on some previously suggested values [6, 11, 20, 26] and also after a trial and error approach. Different parameter combinations were tested and 10 replications for each were carried out. For each of the problems, the overall number of runs was equal to 60 for each of the TGP and LGP algorithms. For the TGP analysis, a MATLAB toolbox, namely GPLAB [27] was used. The LGP algorithm was implemented using the Discipulus software [28].

Overfitting is one of the principal problems in machine learning generalization. An efficient approach to prevent overfitting is to test other individuals from the run on a validation set to find a better generalization [15]. This technique was used in this study for improving the generalization of the models. For this purpose, the available data sets were randomly divided into learning, validation and testing subsets. The learning data were used for training (genetic evolution). The validation data were used to specify the generalization capability of the evolved programs on data they did not train on (model selection). In other words, the learning and validation data sets were used to select the best evolved programs and included in the training process. Thus, they were categorized into one group referred to as "training data". The testing data were finally used to measure the performance of the models obtained by TGP and LGP on data that played no role in building the models. A trial study was conducted to find a consistent data division. The selection was such that the statistical properties (e.g. mean and standard deviation) of the training and testing subsets were similar.

9.3.2 Model Development Using ANN-Based Methods

For the development of the MLP and RBF models, two scripts were written in the MATLAB environment using Neural Network Toolbox 5.1 [29]. The performance of an ANN model mainly depends on the network architecture and parameter settings. For the traditional MLP, a single hidden layer network is sufficient to uniformly approximate any continuous and nonlinear function according to a universal approximation theorem [23]. Choice of the number of the hidden layer nodes, learning rate, epochs and types of activation function plays an important role in the construction of the MLP and RBF models. Hence, several network models with different settings for the mentioned characters were trained to reach the optimal configurations with the desired precision [30]. The written program automatically tries various numbers of neurons in the hidden layer and reports the R and MAE values. The data division for the MLP and RBF analyses was similar to that considered for TGP and LGP.

9.3.3 Finding the Optimum Models

The best models were chosen on the basis of a multi-objective strategy as follows:

- The simplicity of the models, although this was not a predominant factor.
- Providing the best fitness value on the learning set of data.
- Providing the best fitness value on a validation set of data.

The first objective can be controlled by the user through the parameter settings (e.g., program size for GP and hidden layer neurons for ANN). For the other objectives, the following objective function (OBJ) was constructed as a measure of how well the model predicted output agrees with the experimentally measured output. The selection of the best models was deduced by the minimization of the following function:

$$OBJ = \left(\frac{No._{Learning} - No._{Validation}}{No._{Training}}\right)\frac{MAE_{Learning}}{R^2_{Learning}} + \frac{2No._{Validation}}{No._{Training}}\frac{MAE_{Validation}}{R^2_{Validation}} \quad (9.5)$$

where $No._{Learning}$, $No._{Validation}$ and $No._{Training}$ are respectively the number of learning, validation and training data; R and MAE are respectively correlation coefficient and mean absolute error given in the form of formulas as follows:

$$R = \frac{\sum_{i=1}^{n}\left(h_i - \overline{h_i}\right)\left(t_i - \overline{t_i}\right)}{\sqrt{\sum_{i=1}^{n}\left(h_i - \overline{h_i}\right)^2 \sum_{i=1}^{n}\left(t_i - \overline{t_i}\right)^2}} \quad (9.6)$$

$$MAE = \frac{\sum_{i=1}^{n}\left|h_i - t_i\right|}{n} \quad (9.7)$$

in which h_i and t_i are respectively actual and calculated outputs for the ith output, $\overline{h_i}$ is the average of the actual outputs, and n is the number of sample. It is well known that the R value alone is not a good indicator of prediction accuracy of a model. This is because that by shifting the output values of a model equally, the R value will not change. The constructed objective function takes into account the changes of R and MAE together. Higher R values and lower MAE values result in lower OBJ and, consequently, indicate a more precise model. In addition, the above function considers the effects of different data divisions for the learning and validation data.

9.4 Case Studies

9.4.1 Compressive Strength of Concrete

The performance characteristics of concrete are major concerns in construction of civil engineering applications. The enhanced performance characteristics of concrete are generally achieved by addition of various cementitious materials and chemical and mineral admixtures to the conventional concrete mix designs. According to the Abrams' well-known rule, the correlation of the strength of concrete with the water to cement ratio is negative. This rule indicates that only the quality of the cement paste controls the strength of comparable cement. Based on a variety of experimental studies, this is not quite true. For example, if two comparable concrete mixtures have the same water to cement ratio, the strength of the concrete with the higher cement content is lower [31]. Several studies have independently shown that concrete strength development is determined not only by the water to cement ratio, but that it is also influenced by the content of other ingredients [32]. Advances in recent years have been assisted by the use and understanding of chemical admixtures, notably super plasticizers, and cement replacement materials, notably fly ash, blast furnace slag, etc. The use of fly ash and slag plays an important role in contributing to a better workability and low slump loss rates of concrete. This is due to the mutual containment with surface lubrication and the ball-bearing effects among the fly ash and micro fine materials. In many cases, there is also the economic benefit of the price differential between cement and the supplementary cementitious material. Additionally, partial replacement of cement nearly always allows a significant reduction in the dosage of the super plasticizer, which is a particularly expensive ingredient [33].

9.4.1.1 Modeling

In its current state, behavior modeling of the compressive strength of concrete containing additives is inherently more difficult than for the conventional concrete. In order to provide accurate assessment of the performance characteristics of the concrete mix, the effects of several parameters should be incorporated into the model development. Therefore, in this study, the GP and ANN-based approaches were utilized to obtain meaningful relationships between the compressive strength (f_c) of concrete mixes and the predictor variables as follows:

$$f_c = f\left(\frac{\mathrm{W}}{\mathrm{C}}, \mathrm{B}, \mathrm{F}, \mathrm{S}, \frac{\mathrm{Ca}}{\mathrm{Fa}}, \mathrm{Ln(A)}\right) \tag{9.8}$$

where,

W/C: Water to cement ratio
%B: Blast furnace slag content
%F: Fly ash content
%S: Superplasticizer content
Ca/Fa: Coarse aggregate to fine aggregate ratio
Ln(A): Natural logarithm of age

The above variables were chosen as the input variables on the basis of a literature review [33-35].

It is known that the models derived using the GP, ANNs or other CI approaches, in most cases, have a predictive capability within the data range used for their development. Thus, the amount of data used for the training of these algorithms is an important issue, as it heavily bears on the reliability of the final models. The only way to overcome this limitation is to employ comprehensive data sets for training their algorithms. Hence, a reliable database consisting of tests on mixtures with a wide range of aggregate gradation and properties was obtained from the literature to develop the generalized models. The database contains 1133 compressive strength of concrete test results presented by Yeh [34, 35]. It includes measurements of water (W), cement (C), blast furnace slag (B), fly ash (F), superplasticizer (S), coarse aggregate (CA), fine aggregate (FA), age of specimens (A) and f_c of concrete mixes. To visualize the distribution of the samples, the data are presented by frequency histograms (Fig. 9.9).

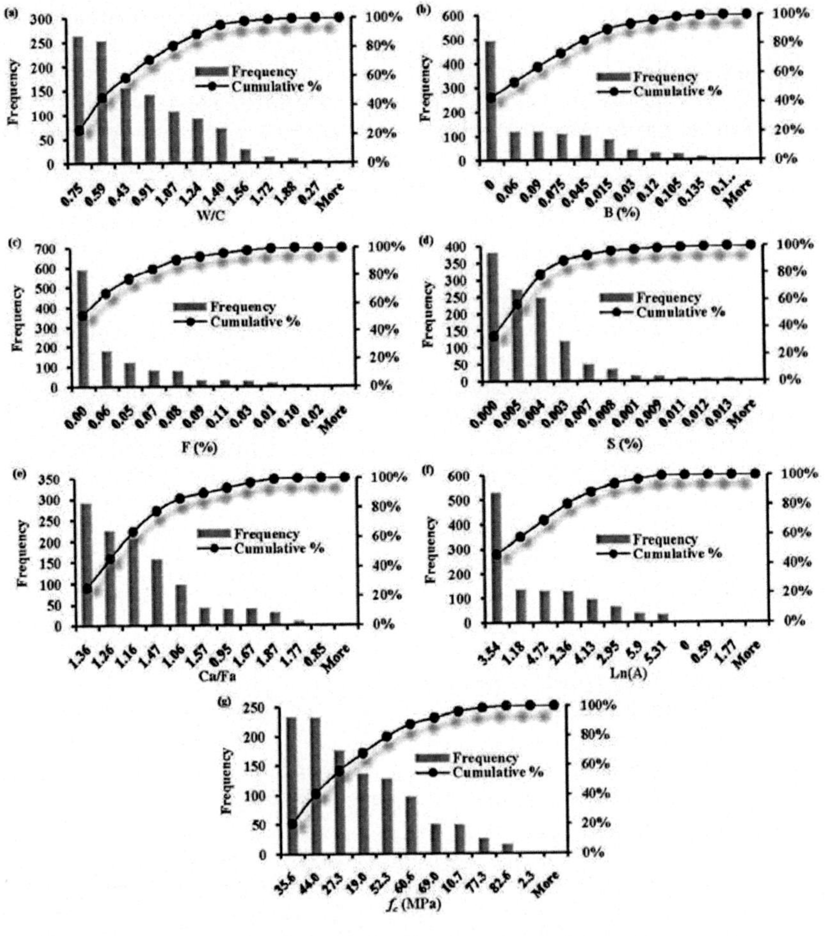

Fig. 9.9 Histograms of the variables used in the model development

Some of the HPC property variables may be fundamentally interdependent. The first step in the analysis of interdependency of the data is to make a careful study of what it is that these variables are measuring, noting any highly correlated pairs. High positive or negative correlation coefficients between the pairs may lead to poor performance of the models and difficulty in interpreting the effects of the explanatory variables on the response. This interdependency can cause problems in analysis as it will tend to exaggerate the strength of relationships between variables. This is a simple case commonly known as the problem of multicollinearity [36]. Thus, the correlation coefficients between all possible pairs were determined and shown in Table 9.1. As it is seen, there are not high correlations between the predictor variables. For the analysis, 907 values (80%) of the data were taken for

Table 9.1 Correlation coefficients between all pairs of the explanatory variables

Variable	W/C	B (%)	F (%)	S (%)	CA/FA	Ln(Age) (day)
W/C	1.000	0.341	0.337	-0.141	-0.091	0.044
B (%)	0.341	1.000	-0.275	0.045	0.065	-0.026
F (%)	0.337	-0.275	1.000	0.393	-0.078	0.000
S (%)	-0.141	0.045	0.393	1.000	-0.255	-0.060
CA/FA	-0.091	0.065	-0.078	-0.255	1.000	0.056
Ln(Age) (day)	0.044	-0.026	0.000	-0.060	0.056	1.000

the training process (807 sets for learning and 100 sets for validation). The rest of the data were used for the testing of the generalization capability of the models.

The maximum tree depth and program size for the optimal TGP and LGP modes were respectively equal to 30 and 256. Various training algorithms were implemented for the training of the MLP networks. The best results were obtained by Quasi-Newton back-propagation method. Also, hyperbolic tangent sigmoid was adopted as the transfer function between the input and hidden layer. The transfer function between the hidden layer and output layer was a linear transfer function. The best MLP model was built with one hidden layer with 18 hidden neurons, a learning rate of 0.05 and was trained for 1000 epochs. For the RBF analysis different spreads were checked and the optimum one was equal to 4.4.

9.4.1.2 Comparison of the Results

The compressive strength prediction equations for the best results of the TGP and LGP algorithms are given as follows:

$$f_{c,TGP}(MPa)=\frac{C}{W}\left((F+4)\left(\frac{Ca}{Fa}+Ln(A)+\left(F+B\frac{W}{C}\right)\left(Ln(A)(Ln(A)+5)+13B\left(\frac{W}{C}+F+Ln(A)\right)\right)\right)\right)+1 \quad (9.9)$$

$$f_{c,LGP}(MPa)=\frac{C}{W}Ln(A)\left(36F+36B-\frac{2Ln(A)}{3}\left(1-\frac{2W}{3C}\right)+\frac{Ca}{Fa}+5\right)-\frac{W}{C}-Ln(A)-S+7 \quad (9.10)$$

Comparisons of experimental versus predicted compressive strength values using TGP, LGP, MLP and RBF are illustrated in Fig. 9.10. The other performance statistics of these models is presented in Table 9.2. It is notable that no rational model to predict the compressive strength of HPC mixes has yet been developed that would encompass the influencing variables considered in this study. Therefore, it was not possible to conduct a more comprehensive comparative study herein. The results indicate that the TGP, LGP, MLP and RBF models are able to predict the compressive strength with high degree of accuracy. Comparing the performance of the GP-based methods, it can be observed from Fig. 9.10 and Table 9.2 that LGP has produced better outcomes than TGP. The ANN-based techniques have provided better results than TGP and LGP. The best results for the training and testing data are respectively obtained by RBF and MLP.

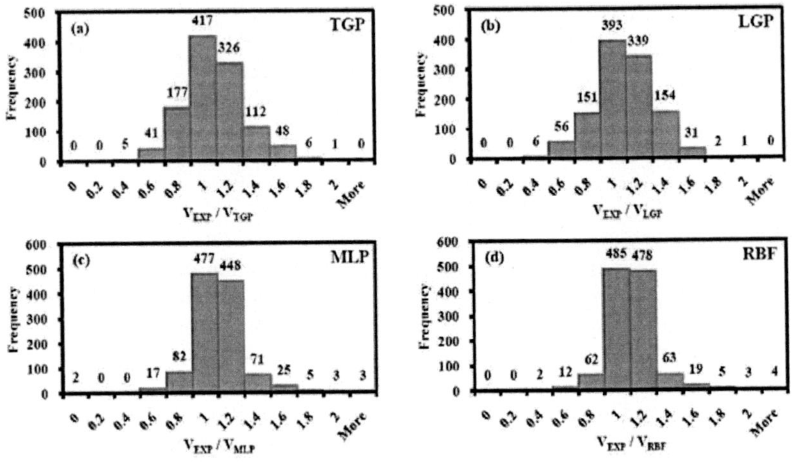

Fig. 9.10 Histograms of the CI models (a) TGP, (b) LGP, (c) MLP, and (d) RBF

Table 9.2 Statistical performances of the CI models

Model	Training data				Testing data			
	R	MAE	Ave.	Std.	R	MAE	Ave.	Std.
GP techniques								
TGP	0.873	5.97	0.980	0.223	0.881	6.14	0.979	0.233
LGP	0.894	5.70	0.969	0.218	0.906	5.71	0.983	0.232
ANN techniques								
MLP	0.943	3.99	1.018	0.260	0.935	4.31	0.992	0.278
RBF	0.956	3.44	1.005	0.207	0.930	4.37	1.043	0.232

9.4.2 *Elastic Modulus of Concrete*

The elastic modulus of concrete is a key factor in structural and material engineering. Designers need the elastic modulus for estimating immediate and time-dependant deformation, determining modular ratio, and evaluating the stiffness of buildings and members. The modulus of elasticity is also important in reinforced and pre-stressed concrete for creep and shrinkage evaluation, as well as in crack control, especially at an early age [26,37]. The modulus of elasticity can be derived from the stress–strain responses of concrete under compression. The modulus of elasticity is defined in the region in which Hooke's law is obeyed for the material as the ratio of stress over strain [38]. In mechanics, Hooke's law of elasticity is an estimation that states that the amount of strain is linearly related to the stress. This can be determined from the slope of compressive stress-strain curves. As shown in Fig. 9.11, in a typical stress-strain diagram of concrete, the first part of the curve is nearly a straight line with some curvature at σ, which is equal to half of the maximum value, σ_u. The initial slope of the stress-strain curve defines the initial or tangent modulus used with the parabolic stress method. The slope of the chord connecting the origin of the coordinate system to $0.5\sigma_u$ determines the secant modulus of elasticity, which is generally used in straight-line stress calculation [26,39].

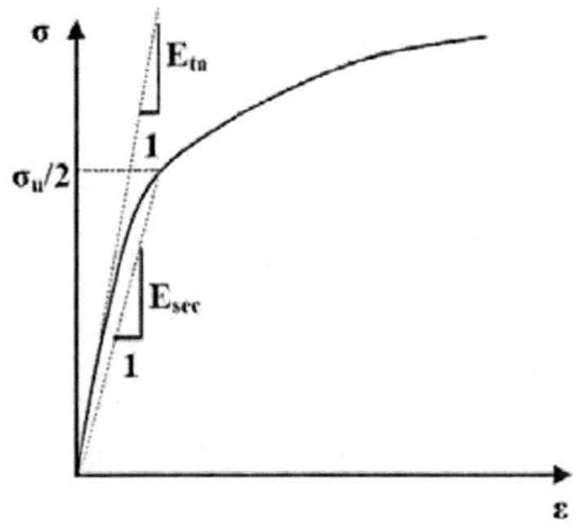

Fig. 9.11 Typical stress-strain diagram of concrete

Despite its importance, tensile strength (and elastic modulus) is not usually measured in the site for compliance purposes. It is often estimated from the measured compressive strength based on the empirical relationships proposed by various codes of practice. This is mainly to avoid performing laborious and time-consuming direct measurements from load-deformation curve [40].

9.4.2.1 Modeling

The GP and ANN-based approaches were employed to formulate the elastic modulus (E_c) of NSC and HSC in terms of compressive strength (f_c) as follows:

$$E_c = f(f_c) \tag{9.11}$$

An experimental database of previously published test results [41] was utilized to develop the models. The database has previously been employed by Gandomi et al. [26] and Demir [42] to develop the LGP and ANN models, respectively. The database contains 89 and 70 test results respectively for the elastic modulus of normal-strength concrete (NSC) and high-strength concrete (HSC). Of the total 159 data sets for HSC and NSC, 126 values were taken for the training of the algorithm (112 sets for learning and 15 sets for validation). The remaining 33 values were used for the testing of the derived models. To visualize the distribution of the samples, the data are presented by frequency histograms (Fig. 9.12).

For the TGP analysis, the maximum tree depth was set to 10. The parameters setting of the LGP algorithm can be found in [26]. The characteristics of the best

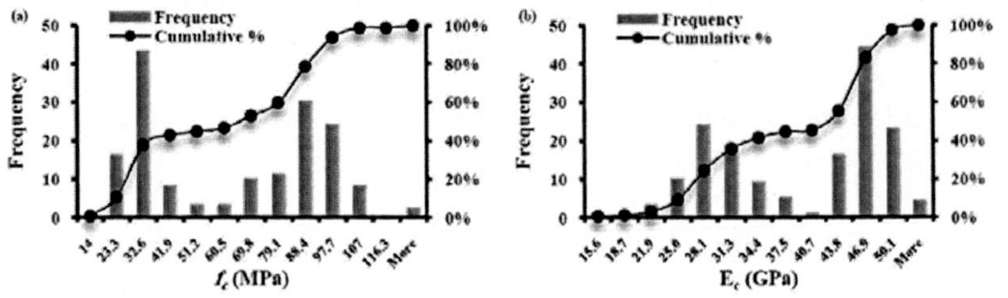

Fig. 9.12 Histograms of: (a) compressive strength and (b) elastic modulus for all data

MLP structure are given in [42]. For the RBF analysis, different spreads were checked and the optimum one was equal to 35.

9.4.2.2 Comparison of the Results

The TGP and LGP-based formulation of the E_c of concrete in terms of f_c are as given below:

$$E_{c,TGP}(GPa) = 4\sqrt{f_c + f_c/5} + 8 \tag{9.12}$$

$$E_{c,LGP}(GPa) = 4\sqrt{f_c - 3} + 9 \tag{9.13}$$

The elastic modulus predictions obtained by TGP, LGP, MLP and RBF are shown in Fig. 9.13. Statistical performance of different models in terms of their prediction capabilities is summarized in Table 9.3. As it seen, TGP, LGP, MLP and RBF give precise estimates of the target values. The performance of these techniques is fairly similar to each other. Overall, RBF has provided better results than other methods.

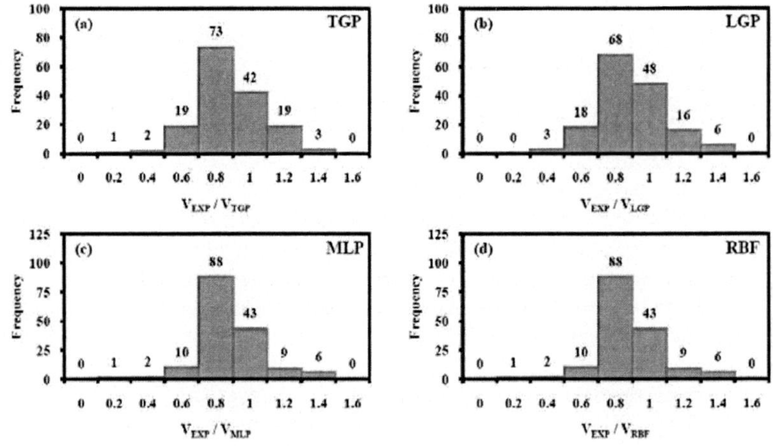

Fig. 9.13 Histograms of the CI models (a) TGP, (b) LGP, (c) MLP, and (d) RBF

Table 9.3 Statistical performances of the CI models

Model	Training data				Testing data			
	R	MAE	Ave.	Std.	R	MAE	Ave.	Std.
GP techniques								
TGP	0.946	2.382	0.997	0.096	0.956	2.254	0.954	0.174
LGP	0.946	2.354	1.001	0.096	0.957	2.181	0.957	0.173
ANN techniques								
MLP	0.952	2.215	1.000	0.094	0.966	2.110	0.951	0.170
RBF	0.961	2.031	0.997	0.083	0.962	2.276	0.945	0.171

9.5 Discussion

Based on a logical hypothesis [43], if a model gives $R > 0.8$, and the error values (e.g., MAE) are at the minimum, there is a strong correlation between the predicted and measured values. The model can therefore be judged as very good. It can be observed from Figs. 9.10 and 9.13 and Tables 9.2 and 9.3 that all the GP and ANN-based models with very high R and low MAE values can accurately predict the target values. Meanwhile, it is noteworthy that the MAE values are not only low but also as similar as possible for the training and testing sets. This suggests that the proposed models have both predictive ability (low values) and generalization performance (similar values).

The task faced by the GP-based approaches is mainly the same as that faced by the ANN-based methods. GP and ANNs are machine learning techniques that can effectively be applied to the classification and approximation problems. They directly learn from raw experimental (or field) data presented to them in order to extract the subtle functional relationships among the data, even if the underlying relationships are unknown or the physical meaning is difficult to be explained. Contrary to these methods, most conventional empirical and statistical methods like finite element method need prior knowledge about the nature of the relationships among the data [11]. Classical constitutive models rely on assuming the structure of the model in advance, which may be suboptimal. Therefore, the GP and ANN-based approaches are well-suited to modeling the complex behavior of most geotechnical engineering problems with extreme variability in their nature [44]. In spite of similarities, there are some important differences between GP and ANNs. ANNs suffer from some shortcomings including lack of transparency and knowledge extraction. That is, they do not explicitly explain the underlying physical processes. The knowledge extracted by ANNs is stored in a set of weights that cannot properly be interpreted. Due to the large complexity of the network structure, ANNs do not give a transparent function relating the inputs to the corresponding outputs. The main advantage of GP over ANNs is that GP generates a transparent and structured representation of the system being studied. An additional advantage of GP over ANNs is that determining the ANN architecture is a difficult task. The structure and network parameters of ANNs (e.g. number of inputs, transfer functions, number of hidden layers and their number of nodes, etc.) should be identified a priori, which is usually done through a time consuming trial and error procedure [11]. In GP, the number and combination of terms are automatically evolved during model calibration [13, 44]. A notable limitation of GP and its variants is that these methods are parameter sensitive. The performance of the TGP and LGP algorithms employed herein can be improved by using any form of optimally controlling the parameters of the run (e.g., GAs).

However, one of the goals of introducing the expert systems, such as the GP and ANN-based approaches, into the design processes is better handling of the information in the pre-design phase. In the initial steps of design, information about the features and properties of targeted output or process are often imprecise and incomplete [11,45]. Nevertheless, it is idealistic to have some initial estimates of the outcome before performing any extensive laboratory or field work. The approaches employed in this research are based on the data alone to determine the structure and parameters of the models. Thus, the derived models can particularly be valuable in the preliminary design stages. For more reliability, the results of the analyses are suggested to be treated as a complement to conventional computing techniques. In any case, the importance of engineering judgment in interpretation of the obtained results should not be underestimated. In order to develop a sophisticated prediction tool, TGP, LGP, MLP, and RBF can be combined with advanced deterministic models. Assuming the deterministic model captures the key physical mechanisms, it needs appropriate initial conditions and carefully calibrated parameters to make accurate predictions. An idea could be to calibrate the required parameters by the use of TGP, LGP, MLP, and RBF which take into account historic data sets as well as the laboratory or field test results. This allows integrating the uncertainties related to in-situ conditions which the deterministic model does not explicitly account for. TGP and LGP provide a structured representation for the constitutive material model that can readily be incorporated into the finite element or finite difference analyses. In this case, it is possible to use a suitably trained GP-based material model instead of a conventional (analytical) constitutive model in a numerical analysis tool such as finite element code or finite difference software (like FLAC) [11]. It is notable that the numerical implementation of ANNs in the finite element analyses has already been presented by several researchers (e.g. [46]). This strategy has led to some qualitative improvement in the application of finite element method in engineering practice [13].

9.6 Conclusions

In this study, the TGP, LGP, MLP, and RBF paradigms were utilized to assess different characteristics of concretes. The following conclusions can be derived from the results presented in this research:

1. TGP, LGP, MLP, and RBF are effectively capable of predicting the compressive strength and elastic modulus of concrete. The validity of the derived models was tested for a part of test results beyond the training data domain. In all cases, LGP gives more accurate predictions than TGP. RBF and MLP generally provide better results than TGP and LGP.

2. The proposed TGP, LGP, MLP, and RBF models simultaneously take into account the role of several important factors representing the concrete behavior. The proposed TGP and LGP simplified formulations can reliably be employed for the pre-design purposes or may be also used as quick checks on solutions developed by more time consuming and in-depth deterministic analyses.
3. Utilizing the models derived via the GP and ANN methods, the concrete compressive strength can easily be estimated from the design mixture basic properties and subsequently the elastic modulus can be assessed using the compressive strength. Thus, there is no need to go through sophisticated and time-consuming laboratory tests.
4. A substantial distinction of GP and ANN to the statistical techniques lies in their powerful abilities to model the complex behavior of the concrete without any need to pre-defined equations or simplifications.
5. Although ANNs are successful in prediction, they usually do not give a certain function to calculate the outcome using the input values. Furthermore, they require the structure of the neural network (e.g. number of inputs, transfer functions, number of hidden layers, etc.) to be identified a priori. On the other hand, the GP-based techniques provide greatly simplified prediction equations.
6. The constitutive models derived using TGP, LGP, MLP, and RBF are basically different from the conventional constitutive models based on the first principles (e.g., elasticity and plasticity theories). One of the distinctive features of GP and ANN-based constitutive models is that they are based on the experimental data rather than on the assumptions made in developing the conventional models [11]. Consequently, as more data becomes available, these material models can be improved by re-training the TGP, LGP, MLP, and RBF algorithms.

References

1. Walter, E., Pronzato, L.: Identification of parametric models from experimental data. Springer, London (1997)
2. Metenidis, M.F., Witczak, M., Korbicz, J.: A novel genetic programming approach to nonlinear system modelling: application to the DAMADICS benchmark problem, Eng. Appl. Art. Int. 17, 363–370 (2004)
3. Schwefel, H.P., Wegener, I., Weinert, K.: Advances in Computational Intelligence – Theory and Practice. Springer, Berlin (2002)
4. Sakla, S.S., Ashour, A.F.: Prediction of tensile capacity of single adhesive anchors using neural networks. Comput. Struct. 83(21–22), 1792–1803 (2005)
5. Levasseur, S., Malécot, Y., Boulon, M., Flavigny, E.: Statistical inverse analysis based on genetic algorithm and principal component analysis: Method and developments using synthetic data'. Int. J. Num. Anal. Meth. Geomech. 33(12), 1485–1511 (2009)
6. Koza, J.R.: Genetic programming: On the programming of computers by means of natural selection. MIT Press, Cambridge (1992)
7. Johari, A., Habibagahi, G., Ghahramani, A.: Prediction of Soil-Water Characteristic Curve Using Genetic Programming. J. Geotech. Geoenvir. Eng. 132(5), 661–665 (2006)
8. Brameier, M., Banzhaf, W.: Linear genetic programming. Springer Science + Business Media, New York (2007)

9. Brameier, M., Banzhaf, W.: A comparison of linear genetic programming and neural networks in medical data mining. IEEE Tran. on Evol. Comput. 5(1), 17–26 (2001)
10. Gandomi, A.H., Alavi, A.H., Mirzahosseini, M.R., Moqaddas Nejad, F.: Nonlinear Genetic-Based Models for Prediction of Flow Number of Asphalt Mixtures. J. Mater. Civil Eng. ASCE 23(3), 248–263 (2011)
11. Alavi, A.H., Gandomi, A.H.: A robust data mining approach for formulation of geotechnical engineering systems. Eng. Computations 28(3), 242–274 (2011)
12. Bäck, T.: Evolutionary Algorithms in Theory and Practice: Evolution Strategies, Evolutionary Programming, Genetic Algorithms. Oxford University Press, USA (1996)
13. Javadi, A.A., Rezania, M.: Applications of artificial intelligence and data mining techniques in soil modeling. Geomech. Eng. 1(1), 53–74 (2009)
14. Torres, R.S., Falcão, A.X., Gonçalves, M.A., Papa, J.P., Zhang, B., Fan, W., Fox, E.A.: A genetic programming framework for content-based image retrieval. Pattern Recogn. 42(2), 283–292 (2009)
15. Banzhaf, W., Nordin, P., Keller, R., Francone, F.: Genetic programming - an introduction. In: On the automatic evolution of computer programs and its application. Morgan Kaufmann, San Francisco (1998)
16. Oltean, M., Grossan, C.: A comparison of several linear genetic programming techniques. Adv. Comp. Sys. 14(4), 1–29 (2003)
17. Poli, R., Langdon, W.B., McPhee, N.F., Koza, J.R.: Genetic programming: An introductory tutorial and a survey of techniques and applications. Technical report [CES–475], University of Essex, UK (2007)
18. Francone, F.D., Deschaine, L.M.: Extending the boundaries of design optimization by integrating fast optimization techniques with machine–code–based, linear genetic programming. Inf. Sci. 161, 99–120 (2004)
19. Alavi, A.H., Gandomi, A.H., Heshmati, A.A.R.: Discussion on soft computing approach for real–time estimation of missing wave heights. Ocean Engineering 37, 1239–1240 (2010)
20. Gandomi, A.H., Alavi, A.H., Sahab, M.G.: New formulation for compressive strength of CFRP confined concrete cylinders using linear genetic programming. Mater. Struc. 43(7), 963–983 (2010)
21. Perlovsky, L.I.: Neural networks and intellect. Oxford University Press, Oxford (2001)
22. Rumelhart, D.E., Hinton, G.E., Williams, R.J.: Learning internal representations by error propagation. MIT Press, Cambridge (1986)
23. Cybenko, J.: Approximations by Superpositions of a Sigmoidal Function. Math. Cont. Sign. Syst. 2, 303–314 (1989)
24. Haykin, S.: Neural networks – A comprehensive foundation, 2nd edn. Prentice Hall Inc., Englewood Cliffs (1999)
25. Girosi, F., Poggio, T.: Networks and the best approximation property. Biological Cybernetics 63(3), 169–176 (1990)
26. Gandomi, A.H., Alavi, A.H., Sahab, M.G., Arjmandi, P.: Formulation of Elastic Modulus of Concrete Using Linear Genetic Programming. Journal of Mechanical Science and Technology 24(6), 1011–1017 (2010)
27. Silva, S.: GPLAB, a genetic programming toolbox for MATLAB (2007), `http://gplab.sourceforge.net`
28. Conrads, M., Dolezal, O., Francone, F.D., Nordin, P.: Discipulus–fast genetic programming based on AIM learning technology. Register Machine Learning Technologies Inc., Littleton (2004)

29. MathWorks, Inc. MATLAB the language of technical computing, Version 7.4, Natick, MA, USA (2007)
30. Eberhart, R.C., Dobbins, R.W.: Neural Network PC Tools, A Practical Guide. Academic Press, San Diego (1990)
31. Popovics, S.: Analysis of the concrete strength versus water-cement ratio relationship. ACI Mater. J. 87(5), 517–529 (1990)
32. Oluokun, F.A.: Fly ash concrete mix design and the water–cement ratio law. ACI Mater. J. 91(4), 362–371 (1994)
33. Yeh, I.C.: Modeling of strength of high-performance concrete using artificial neural networks. Cement Concrete Res. 28(12), 1797–1808 (1998)
34. Yeh, I.C.: Exploring concrete slump model using artificial neural networks. J. Comput. Civil Eng. 20(3), 217–221 (2006)
35. Yeh, I.C.: Analysis of strength of concrete using design of experiments and neural networks. ASCE, J. Mater. Civil Eng. 18(4), 597–604 (2006)
36. Dunlop, P., Smith, S.: Estimating key characteristics of the concrete delivery and placement process using linear regression analysis. Civil Eng. Environ. Syst. 20, 273–290 (2003)
37. Khan, A., Cook, W.D., Mitchell, D.: Early age compressive stress– strain properties of low, medium and high strength concretes. ACI Mater. J. 92(6), 617–624 (1995)
38. McNaught, A.D., Wilkinson, A.: Compendium of Chemical Terminology. In: IUPAC, 2nd edn., Blackwell Scientific Publications, Oxford (1997)
39. Ferguson, P.M., Breen, J.E., Jirsa, J.O.: Reinforced Concrete Fundamentals, 5th edn. John Wiley & Sons, Chichester (1988)
40. ASTM-C469. Standard test method for static modulus of elasticity and poisson's ratio of concrete in compression. Annual Book of ASTM standards (1994)
41. Demir, F.: A new way of prediction elastic modulus of normal and high strength concrete–fuzzy logic. Cem. Concr. Res. 35, 1531–1538 (2005)
42. Demir, F.: Prediction of elastic modulus of normal and high strength concrete by artificial neural networks. Constr. Build. Mater. 22, 1428–1435 (2008)
43. Smith, G.N.: Probability and statistics in civil engineering. Collins, London (1986)
44. Shahin, M.A., Jaksa, M.B., Maier, H.R.: Recent advances and future challenges for artificial neural systems in geotechnical engineering applications. Adv. Artif. Neur. Syst., Article ID 308239 (2009)
45. Kraslawski, A., Pedrycz, W., Nyström, L.: Fuzzy neural network as instance generator for case–based reasoning system: an example of selection of heat exchange equipment in mixing. Neural Computing and Applications 8(2), 106–113 (1999)
46. Javadi, A.A., Tan, T.P., Elkassas, A.S.I.: Intelligent finite element method. In: the 3rd MIT Conference on Computational Fluid and Solid Mechanics, Cambridge, Massachusetts, USA (2005)

Chapter 10
A New Approach to Network Optimization Using Chaos-Genetic Algorithm

Golnar Gharooni-fard and Fahime Moein-darbari

Abstract. Genetic Algorithms (GAs) have been widely used to solve network optimization problems with varying degrees of success. Part of the problem with GAs lies in the premature convergence when dealing with large-scale and complex problems; Caught in local optima, the algorithm might fail to reach the global optimum even after a large number of iterations. In order to overcome the problems with traditional GAs, a method is proposed to integrate Chaos Optimization Algorithms (COAs) with GA to fully exploit their respective searching advantages. The basic idea of COA is to transform the problem variables, by way of a map, from the solution space to a chaos space and to perform a search that benefits from the randomness, orderliness and ergodicity of chaos variable. In this chapter, we will first discuss network optimization in general, and then focus on how chaos theory can be incorporated into the GA in order to enhance its optimization capacities. We will also examine the efficiency of the proposed Chaos-Genetic algorithm in the context of two different types of network optimization problems, Grid scheduling and Network-on-Chip mapping problem.

Keywords: network optimization, Genetic Algorithm, Chaos theory, Grid scheduling, Network-on-Chip mapping problem.

10.1 Introduction

Network theory basically deals with problems that have a graph structure. Graphs are mathematical structures used to model pair wise relations between objects. They consist of points, and lines connecting pairs of points. The points are called

---

Golnar Gharooni-fard · Fahime Moein-darbari
Computer Department of Islamic Azad University,
Mashhad Branch, Young Researchers Club, Iran
e-mail: golnar.ghf@gmail.com, fahime.md61@gmail.com

X.-S. Yang & S. Koziel (Eds.): Computational Optimization & Applications, SCI 359, pp. 245–270.
springerlink.com

nodes or vertices and the lines are called arcs. The arcs may have a direction on them, in which case they are called directed arcs. If an arc has no direction, it is often called an edge. If all the arcs in a graph are directed, the graph is said to be directed (digraph). Graphs are among the most ubiquitous models of both natural and human-made structures. They can be used to model many types of relations and process dynamics in physical, biological and social systems. Many problems of practical interest can be represented by graphs [1]. In computer science, graphs are used to represent networks of communication, data organization, computational devices, the flow of computation, etc. Fig. 10.1 is an example of a network modeled with graphs. At any given time, a message may take a certain amount of time to traverse each line (due to congestion effects, switching delays, etc.). The expended time can vary greatly and telecommunication companies dedicate a significant amount of their resources tracking these delays. Assuming a centralized switcher knows these delays, there remains the problem of routing a call so as to minimize the delays. This is an example of a particular type of network model, called the *shortest path* which includes a network with weighted edges and two special nodes: a source and a destination. The goal is to find a path from the source to the destination with the minimum total weight.

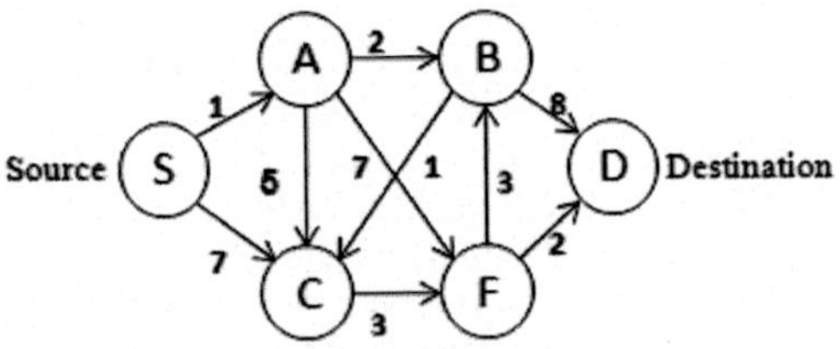

Fig. 10.1 A Phone network modeled by a graph

Network problems that involve finding the least-cost solution to a problem where each solution is associated with a numerical cost are generally studied under combinatorial optimization which concerns the efficient allocation of limited resources to meet desired objectives when the values of some or all of the variables are restricted to be integral [2]. Still, in most such problems, there are many possible alternatives to consider and one overall goal determines which of these alternatives is best.

Different approaches have been used to solve network optimization problems [3] among which are a large family of algorithms collectively labeled metaheuristics. A metaheuristic designates a computational method that optimizes a problem by iteratively trying to improve a candidate solution with regard to a given measure of quality. Metaheuristics make few or no assumptions about the problem being optimized and can search very large spaces of candidate solutions [4], [5]. Metaheuristics can be used for the purpose of combinatorial optimization where an optimal solution is sought over a discrete search-space. Popular metaheuristics for combinatorial problems include Simulated Annealing (SA) [6], Genetic Algorithm (GA) [7]' Particle Swarm Optimization (PSO) [8], Ant Colony Optimization

(ACO) [9] and Tabu Search (TS) [10]. Since our focus here is on GAs, in the next section we will discuss them as one of the most popular metaheuristics used for optimization purposes. The interested reader is referred to [11] and [12] for more general surveys on the metaheuristics.

10.2 Genetic Algorithms

Genetic algorithms are inspired by the evolutionary theory of the origin of species which explains how weak and unfit species in nature face extinction by way of natural selection. Natural selection is the process by which traits become more or less common in a population due to consistent effects upon the survival or reproduction of their bearers, the strong species. In the long run, species carrying the correct combination in their genes become dominant in their population. Sometimes, during the slow process of evolution, random changes may occur in the genes. If these changes provide additional advantages in the challenge for survival, new species evolve from the old ones. Unsuccessful changes are eliminated by natural selection.

The concept of Genetic Algorithms (GAs) was introduced by John Holland in the early seventies as a special technique for function optimization [7]. In GA terminology, a solution vector is called an individual or a *chromosome*. Chromosomes are made of discrete units called *genes*. Each gene controls one or more features of the chromosome. In the original implementation of GA by Holland, genes are assumed to be binary numbers. In later implementations, more varied gene types have been introduced. Normally, a chromosome corresponds to a unique solution in the solution space. The GA operates with a collection of chromosomes, called a *population*. The population is normally randomly initialized. As the search goes on, populations evolve to include fitter and fitter solutions, and eventually converge, to a single solution.

The basic idea of a GA is that the genetic pool of a given population potentially contains the best solution, to a given adaptive problem, although this solution might not have been realized yet. The algorithm operates in an iterative manner and evolves a new generation from the current generation by applying genetic operators [13]. Given a clearly defined problem to be solved and strings of candidate solutions, a simple GA works as follows:

1. Initialize the population.
2. Calculate the fitness value for each individual in the population.
3. Reproduce selected individuals to form a new population.
4. Perform crossover and mutation on the population.
5. Loop to step 2 until some termination condition is met.

In some GA implementations, operations other than crossover and mutation are carried out in step 4. Crossover is considered by many to be an essential operation of all GAs. It plays an important role in distributing the individuals over the space of interest through the GA. Termination of the algorithm is usually based either on achieving a population member with some specified fitness or on running the algorithm for a given number of generations. Like many other metaheuristics, GAs do not guarantee an optimal solution is ever found. They often show a very fast initial convergence followed by progressive slower improvement. Therefore different techniques have been used to improve the results obtained from the GAs [14]. By introducing Chaos theory in the next section, we will explain how to integrate this concept with GA, in order to enhance the quality of the solutions.

10.3 Chaos Theory

Chaos theory is the study of the behavior of dynamical systems that are highly sensitive to initial conditions. In common usage, "chaos" means "a state of disorder", but the adjective "chaotic" is defined more precisely in chaos theory. Although there is no universally accepted mathematical definition of chaos, a commonly used definition describes, chaos as a non-periodic, long-term behavior in a deterministic system that exhibits sensitive dependence on initial conditions [15]. None-periodic long-term behavior means that the system's trajectory in phase space does not settle down to any fixed points or periodic orbits, as time tends to infinity. Deterministic systems can have no random (or probabilitistic) parameters. It is a common misconception that chaotic systems are noisy systems driven by random processes. The irregular behavior of chaotic systems arises from intrinsic nonlinearities rather than noise. Sensitive dependence on initial conditions, the proverbial "the butterfly effect", requires that trajectories originating from nearly identical initial conditions diverge exponentially. Despite what the name suggests, chaos is not the absence of order; it is a subtle state that is poised between order and randomness, with both aspects intermingled.

If a chaotic system's behavior is plotted in a graph over an extended period, obscure patterns might emerge. When a bounded chaotic system does have some long term pattern, but not a simple periodic oscillation or orbit, it is said to have a strange attractor [16]. In other words, strange attractor is the natural shape of chaos. It is called strange because of its complex geometry, and it is an attractor because the system that it describes is always drawn to the behavior that it represents as if attracted to it. The mathematical model developed, called the "Lorenz system[1] has been used as a paradigm for chaotic systems that satisfy the above definition. The Lorenz system consists of three first-order coupled differential equations as follows

$$\begin{cases} \dfrac{dx}{dt} = \sigma(y - x) \\ \dfrac{dy}{dt} = x(\rho - z) - y \\ \dfrac{dz}{dt} = xy - \beta z \end{cases} \tag{10.1}$$

[1] The "Lorenz system" is named after the American meteorologist Edward N. Lorenz, who in 1963 discovered chaotic behavior in a computer study of weather.

where all σ, ρ, β > 0, but usually σ = 10, β = 8/3 and ρ is varied. The system exhibits chaotic behavior when ρ = 28 [15]. The Lorenz system has three dynamic variables, and consequently the state-space picture of such a system is three-dimensional. Plotting the trajectory of the Lorenz system in state space, shown in Fig. 10.2, reveals what was earlier defined as a strange attractor (the Lorenz chaotic attractor). The map shows how the state of a dynamical system (the three variables of a three-dimensional system) evolves over time in a complex, non-repeating pattern.

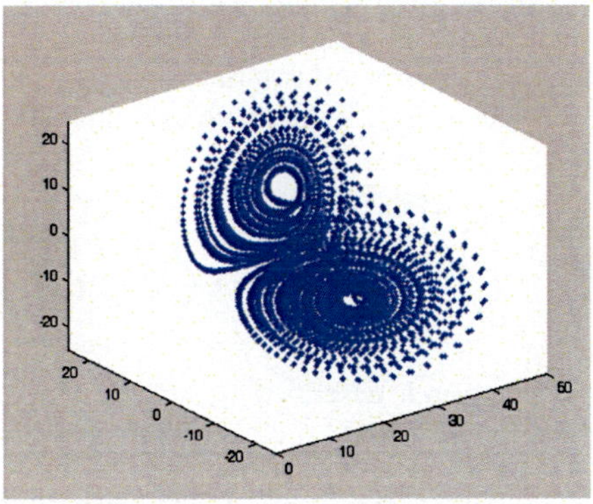

Fig. 10.2 The Lorenz attractor

One-dimensional noninvertible maps are the simplest systems capable of generating chaotic motion. As such, they serve as a convenient starting point for the study of chaos [17]. Here, we introduce some well known one-dimensional maps.

Logistic Map. The logistic map proposed by Robert May is a polynomial map and is often cited as an example of how complex behavior can arise from a very simple nonlinear dynamical equation [15]. This map is defined as

$$x_{n+1} = f(\mu, x_n) = \mu x_n(1 - x_n) \ , \ 0 < \mu \leq 4 \tag{10.2}$$

where μ is a control parameter, and x is a variable. Since the equation represents a deterministic dynamic system, it might seem like its long-term behavior can be predicted, but that is in fact not the case since its behavior is heavily dependent on the variations of μ. The value of the control parameter, determines whether x converges to a constant point, oscillates between two or more values, or behaves chaotically in an unpredictable pattern [18].

Tent Map. In mathematics, the tent map is an iterated function, in the shape of a tent, forming a discrete-time dynamical system. It takes a point x_n on the real line and maps it to another point as

$$x_{n+1} = \begin{cases} \mu x_n & , \quad x_n < \frac{1}{2} \\ \mu(1 - x_n) \ , & \frac{1}{2} \leq x_n \end{cases} \tag{10.3}$$

where μ is a positive real constant [19]. The tent map and the logistic map are topologically conjugate and thus their behavior under iteration is identical in this sense. Depending on the value of μ, the tent map demonstrates a range of dynamical behavior ranging from predictable to chaotic.

Bernoulli Shift Map. The Bernoulli shift map belongs to a class of piecewise linear maps which consist of a number of piecewise linear segments. This map is a particularly simple, consisting of two linear segments to model the active and passive states of the source [20]. It is defined as follows

$$x_{n+1} = \begin{cases} \frac{x_n}{(1-\lambda)} & , \quad 0 < x_n < (1 - \lambda) \\ \frac{x_n - (1-\lambda)}{\lambda} & , \quad (d \equiv 1 - \lambda) < x_n < 1 \end{cases} \tag{10.4}$$

Sine Map. The sine map is described by the following equation

$$x_{n+1} = \frac{a}{4} \sin(\pi x_n) \tag{10.5}$$

where $0 < a \leq 4$. Qualitatively this map has the same shape as the logistic map.

ICMIC Map. The iterative chaotic map with infinite collapses (ICMIC) has infinite fixed points in comparison with finite collapses one-dimensional maps [21], [22]. The ICMIC map is described by following equation

$$x_{n+1} = \sin \frac{a}{x_n} \tag{10.6}$$

where $a \in (0, \infty)$ is an adjustable parameter.

10.4 Chaos Optimization Algorithm (COA)

In random-based optimization algorithms, the methods using chaotic variables instead of random variables are called Chaotic Optimization Algorithm (COA) [23], [24]. Originally proposed by Li and Jiang, COA searches the solution space based on the regularity of chaotic variables and more easily escapes local minima compared with stochastic optimization algorithm [25]. By means of ergodicity, regularity and semi-stochastic properties of chaos, the optimal solution migrates in a chaotic way among the local minima and finally converges to the global optimal solution [26]. Experimental studies assert that the benefits of using chaotic signals instead of random signals are often evident although it is not mathematically proved yet [27]. The procedure of COA is demonstrated as followings:

1. Set k = 0 and $f(x_i^k)$ as a random solution in the problem domain and a chaotic variable $0 < r_i < 1,\ (i = 1,2,\cdots,n)$.
2. Map the chaotic sequences r_i^k to x^* according to the characteristics of the particular problem.
3. Compare the function value of $f(x^*)$ with $f(x_i^k)$, pick the better value and replace it with $f(x_i^k)$. Then replace x^* with x_i^k.
4. Apply one of the aforementioned chaotic equations (denoted by *M*)

$$r_i^{k+1} = M(r_i^k) \tag{10.7}$$

   Note that the interval of chaotic sequences is between 0 and 1.
5. Set $k = k + 1$ and loop back to step 2 until the termination condition is reached.

Numerical results show that COA takes less iteration to reach to an optimum solution than most global optimization methods [25]. However, COA has the deficiency of taking much time to get to the optimum value, which affects the speed of convergence [28]. To overcome this limitation, an improved chaos optimization method that combines COA and GA is presented in the next section.

10.4.1 Chaos-Genetic Algorithm (CGA)

The idea of using chaotic systems instead of random processes has recently been noticed in several fields, including optimization theory. The basic idea is to transform the variables of a problem from the solution space to chaos space and then perform a search to find a solution by virtue of the randomness, orderliness and ergodicity of the chaos variable. Although the COA has many advantages, it makes no use of the experiential information previously acquired [29]. Furthermore, in GAs there is no guaranteed convergence even to a local minimum [30]. Since the genes from a few highly fit (but not optimal) individuals may rapidly come to dominate the population, causing it to converge on local minima and once the population has converged, the ability of the GA to continue to search for better solutions is largely compromised.

In order to overcome the shortcomings of both COA and GA, one option is to integrate the two in order to bring together the searching advantages of both algorithms. The concept of Chaos-genetic algorithms (CGA), first introduced in [30], has the following characteristics: Firstly, CGA benefits from the characteristics of the chaotic variables to make the individuals of subgenerations distributed ergodically in the defined space and thus to avoid premature convergence in the subgenerations. Secondly, according to its evolutionary nature, CGA maintains the fittest individuals in each run and hence increases the probability of finding the global optimal solution. CGA can be implemented by simply adding a chaotic mapping operator to the standard GA operators, namely crossover and mutation.

As an example of a chaotic equation, the logistic map has been extensively analyzed in the past decade. The evolution of the chaotic variables could be defined through the following equation [31],

$$r_i^{k+1} = 4r_i^k\left(1 - r_i^k\right), \quad i = 1, 2, \ldots, n. \tag{10.8}$$

In principle, this is the same as the equation introduced for logistic map in Section 10.3. The value of the parameter $\mu = 4$ is chosen in order for the system to act chaotically. Here r_i is the i-th chaotic variable and k denotes the number of iterations. The value of r_i , is distributed in the range of $[0\,, 1]$ and n denotes the number of genes in each chromosome. In order to perform the chaotic mapping, the following procedure is proposed.

1. Divide the interval $[0\,, 1]$ to n equal sub-intervals, of which the lower limit $[a_1, a_2, \ldots, a_n]$ is represented by vector a, and the upper limit $[b_1, b_2, \ldots, b_n]$ by vector b.
2. The real value of each x_i in the first randomly produced population is linearly mapped to new values of $1 < r_i < 0$, using

$$r_i = \frac{1}{b_i - a_i}(x_i - a_i). \tag{10.9}$$

3. The next iteration chaotic variables $r_i^{(2)}$, will be produced through applying the logistic map equation to $r_i^{(1)}$ values, generated in the previous section.
4. The chaotic variables $r_i^{(2)}$, are then used to produce $x_i^{(2)}$, using

$$x_i^{(2)} = a_i + r_i^{(2)}(b_i - a_i)\,, \; i = 1, 2, \ldots, n\,. \tag{10.10}$$

We can repeat the process in order to produce the next values of $x_i^{(k)}$. Although chaos variables are usually generated by the logistic map, there's no reason not to try any of the previously defined one-dimensional maps in order to form a chaotic mapping operator. Fig. 10.3 demonstrates a flowchart of the overall process of Chaos-genetic algorithm using the logistic map as a chaotic mapping operator to produce the chaotic population P_2 from the randomly produced initial population P_1. In the next section we will examine the performance of CGA in two types of network optimization problems, namely Grid scheduling and Network-on-Chip mapping problem.

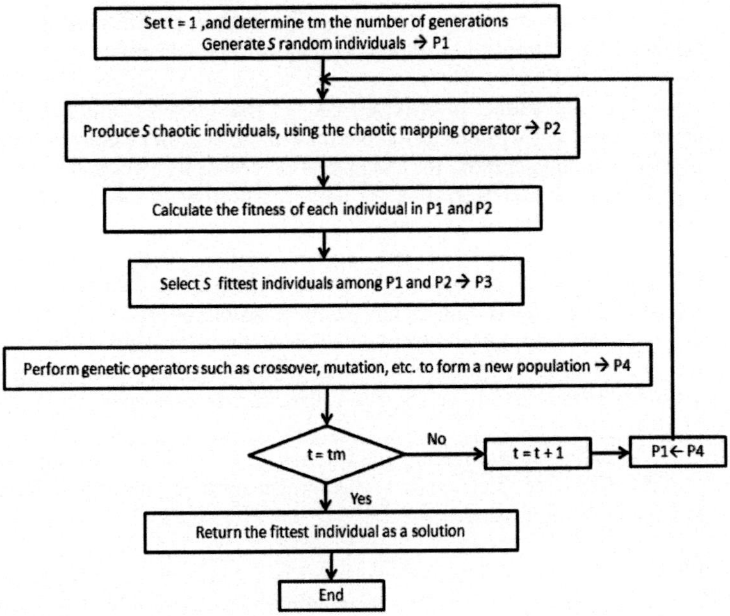

Fig. 10.3 Chaos-Genetic Algorithm procedure

10.5 Grid Scheduling: Case Study # 1

A grid is a hardware and software infrastructure that provides dependable, consistent, pervasive, and inexpensive access to high-end computational capabilities [30]. It is a shared environment, implemented via the deployment of a persistent, standards-based service infrastructure that supports the creation and sharing of the resource within distributed communities. The resources might be computers, storage space, instruments, software applications, and data, all connected through the Internet and a middleware software layer that provides basic services for security, monitoring, resource management, and etc. Resources owned by various administrative organizations are shared under locally defined policies that specify what is shared, who is allowed to access what, and under what conditions [32].

From the point of view of scheduling systems, a higher level abstraction for the Grid can be applied by ignoring some infrastructure components such as authentication, authorization, resource discovery and access control. Thus, the following definition for the term *Grid* is adopted in our study: "A type of parallel and distributed system that enables the sharing, selection, and aggregation of geographically distributed autonomous and heterogeneous resources dynamically at runtime depending on their availability, capability, performance, cost, and users' quality-of-service requirements" [33].

To facilitate the discussion on grid scheduling, we need to define some frequently used terms; *tasks* are atomic units to be scheduled by the scheduler and assigned to resources. The *properties* of a task are parameters like CPU/memory requirement, deadline, priority, etc. A *job* (*metatask* or *application*) is a set of atomic tasks that will be carried out on a set of resources. *Resources* are required to carry out an operation, for example: a processor for data processing, a data storage device, or a network link for data transporting. A *site* (or *node*) is an autonomous entity composed of one or multiple resources.

Based on the definitions above, *task scheduling* can be defined as the mapping of tasks to a selected group of resources which may be distributed in multiple administrative domains. Although, a grid is a system of high diversity, which is rendered by various applications, middleware components, and resources, we can still find a logical architecture of the task scheduling subsystem in the grid that as noted by Schopf in [34], can be generalized into three stages:

1. Resource discovering and filtering,
2. Resource selecting and scheduling according to certain objectives,
3. Job submission.

Since the study of scheduling algorithms is our primary concern, we mainly focus on the second step. Scheduling of interdependent tasks in distributed heterogeneous computing environments is well known to be an NP-hard problem [35]. Several heuristic algorithms have been applied to solve the scheduling problem. These can be classified into two major groups, in view of their main objectives. First, a group of works that only attempt to minimize workflow execution time, without considering user's budget. *Min-Min*, which sets the highest priority to tasks with the shortest execution time, and *Max-Min*, which sets the high priority to the tasks with the long execution times are two major heuristic algorithms employed for scheduling workflows on grids [36]. *Sufferage*, is another heuristic algorithm which sets high scheduling priority to tasks whose completion time by the second best resource is far from that of the best resource [36]. Another workflow scheduling algorithm developed by the authors of [37], is based on a Greedy Randomized Adaptive Search Procedure (*GRASP*). Another workflow level heuristic is a Heterogeneous-Earliest-Finish-Time (*HEFT*) algorithm proposed by Wieczorek et al. [38]. Second, a group of works which address scheduling problems based on user's budget constraints. Nimrod-G [39] schedules independent tasks for parameter-sweep applications to meet user's budget. More recently, *LOSS* and *GAIN* scheduling approaches were developed, to adjust a schedule which is generated by a time-optimized heuristic and cost optimized heuristic to meet the user's budget constraints [40].

10.5.1 Challenges of Scheduling Algorithms in Grid Computing

Although previous research in this area is of great value, traditional scheduling models generally produce poor grid schedules in practice [32]. To remedy this let us go through the assumptions underlying traditional systems:

- All resources reside within a single administrative domain.
- To provide a single system image, the scheduler controls all of the resources.
- The resource pool is invariant.
- Contention caused by incoming applications can be managed by the scheduler according to some policies, so that its impact on the performance that the site can provide to each application can be well predicted.
- Computations and their data reside in the same site.

Unfortunately, not all of these assumptions hold in grid circumstances. There are unique characteristics in grid computing, listed by the authors of [41], which make the design of scheduling algorithms more challenging:

• *Heterogeneity and Autonomy.* In grid computing, because resources are distributed in multiple domains on the Internet, heterogeneity is a characteristic not only of computational and storage nodes but also of the underlying networks connecting them. This results in different capabilities for job processing and data access. The autonomy also gives way to aa diverse array of local resource management techniques and access control policies, such as, priority settings for different applications and resource reservation methods. Thus, a grid scheduler is required to be adaptive to different local policies. The heterogeneity and autonomy on the grid user side are represented by various parameters, including application types, resource requirements, performance models, and optimization objectives.

• *Performance Dynamism.* Making a feasible scheduling usually depends on the performance estimate that candidate resources can provide, especially when the algorithms are static. Grid schedulers work in a dynamic environment where performance of available resources is constantly changing. The change comes from site autonomy and competition for resources by various applications.

• *Resource Selection and Computation. Data Separation* In traditional systems, executable codes of applications and input/output data are usually in the same site, or the input sources and output destinations are determined before the application is submitted. Thus the cost for data staging can either be neglected or is a constant determined before execution, and scheduling algorithms need not consider it. But in a grid which consists of a large number of heterogeneous computing sites (from supercomputers to desktops) and storage sites connected via wide area networks, the computation sites of an application are usually selected by the grid scheduler according to resource status and certain performance models. Additionally, in a grid, the communication bandwidth of the underlying network is limited and shared by a host of background loads, so the inter-domain communication cost cannot be neglected.

Many grid applications are data intensive, so the data staging cost is considerable. This situation brings about the computation-data separation problem: the advantage brought by selecting a computational resource that can provide low computational cost may be neutralized by its high access cost to the storage site. These challenges depict unique characteristics of grid computing, and put significant obstacles to design and implement efficient and effective grid scheduling systems. It is believed, however, that research achievements on traditional scheduling problems can still provide stepping-stones for a new generation of scheduling systems.

In order to introduce the Chaos-genetic algorithm to solve the workflow scheduling problem, we need to define an appropriate problem representation, fitness assignment, and genetic operators. These will be discussed in the following subsections.

10.5.2 Problem Description

As mentioned in the previous section, the scheduling problem becomes more challenging because of some unique characteristics of grid computing. The grid scheduling problem can be defined as follows: A workflow application can be modeled as a Directed Acyclic Graph (DAG). There is a finite set of tasks T_i $(i = 1,2, \ldots, n)$ and a set of directed arcs of the form (T_i, T_j), where T_i is the parent task of T_j, and T_j is the child of T_i. A child task can never be executed unless all of its parent tasks have been completed. Let B be the cost constraint (budget) and D the time constraint (deadline), specified by the user's workflow execution. The total number of available services is shown by m. There's a set of services S_j $(j = 1,2, \ldots, m)$ capable of executing task T_i, but each task can only be assigned for execution to one of these services. Services have varied processing capabilities delivered at different prices. We denote t_i^j as the processing time, and c_i^j as the service price for processing T_i on service S_j. The scheduling problem is to map every T_i onto a suitable S_j in order to get the best trade-off between execution time and cost in a workflow considering the user's budget and deadline.

10.5.3 The Chaos-Genetic Scheduling Algorithm (CGA)

For a workflow scheduling problem, a feasible solution is required to meet several conditions:

1. A task can only be started after all its predecessors have completed.
2. Every task appears once and only once in the schedule.
3. Each task must be allocated to one available time slot of a service capable of executing the task.

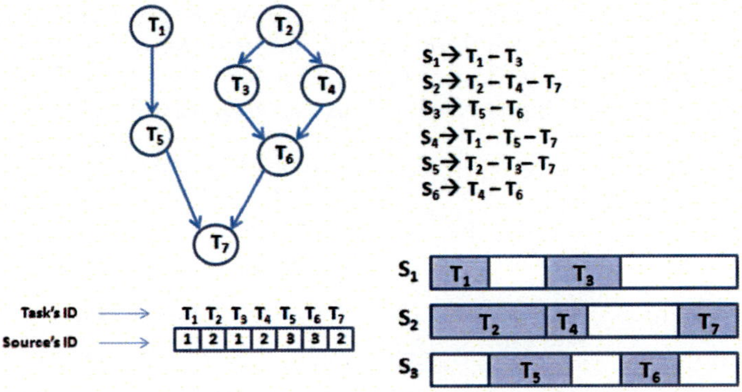

Fig. 10.4 A sample workflow followed by a set of source-to-task assignments

Each individual in the population represents a feasible solution to the problem, and consists of a vector of task assignments. Each task assignment includes four elements (task ID, service ID, start time, end time) [42]. The first two parameters identify to which service each task is assigned. Since involving time frames during the genetic operation may lead to a very complicated situation [43], we ignore the time frames here. Therefore, the operation strings (chromosomes) encode only the service allocation for each task and the order of the tasks allocated to each service. Different execution priorities of such parallel tasks within the workflow may impact the performance of workflow execution significantly. For this reason, the solution representation strings are required to show the order of task assignments on each service in addition to service allocation of each task. As suggested by Buyya [43], we create an array to represent a schedule as illustrated in Fig.10.4. Each element of this array represents a service and the indexes refer to the task number.

As stated earlier, the problem is to schedule a workflow execution considering both time and user budget constraints. The first decision to be made is how to represent the solution, which was shown in Fig.10.4. Initializing the population is done randomly using a random generator to produce values between 1 to n. For each task, these random values are chosen from sources that are capable of executing that task. The length of the chromosome depends on the number of tasks in the workflow. A chaotic mapping operator is then applied to the initial population, generating a new chaotic population.

At this stage, the fitness of the individuals of the entire population is evaluated. The fitness value is often proportional to the output value of the function being optimized according to the given objectives. As the goal of scheduling is to get the best trade-off between the time and cost of the workflow execution, the fitness function divides the evaluation into two parts [43]: cost-fitness and time-fitness. For budget constrained scheduling, the cost-fitness component produces results with less cost. The cost fitness function of an individual I is defined by

$$F_{cost}(I) = \frac{c(I)}{B} \tag{10.11}$$

where *c(I)* is the sum of the task execution cost and data transmission cost of *I* and *B* is the budget of the workflow. For budget constrained scheduling, the time-fitness component is designed to produce individuals that satisfy the deadline constraint. The time-fitness function of an individual *I* is defined by

$$F_{time}(I) = \frac{t(I)}{D} \qquad (10.12)$$

where *t(I)* is the completion time of *I*, *D* is the deadline of the workflow. The final fitness function combines the two parts and it is expressed as:

$$F(I) = \begin{cases} F_{cost}(I) + F_{time}(I), if\ F_{cost}(I) > 1\ or\ F_{time}(I) > 1 \\ \frac{c(I)}{maxcost} \times \frac{t(I)}{maxtime} \qquad otherwise \end{cases} \qquad (10.13)$$

where *maxcost* is the most expensive solution of the current population and *maxtime* denotes the largest completion time in the current population.

Elitism is incorporated into the algorithm by transferring the single fittest individual directly to the next generation. Crossover is used to create new solutions by rearranging parts of the existing solutions in the current population. The idea behind the crossover operation is that a higher quality solution may result from the combination of two of the current fittest solutions [44]. We have implemented a two-point crossover which is illustrated in Fig. 10.5. For population based algorithms, mutation occasionally occurs in order to allow a child to obtain features that are not possessed by either of its parents. This process helps the algorithm explore new and possibly better genetic material than has been previously considered. The process of mutation is shown in fig. 10.6.

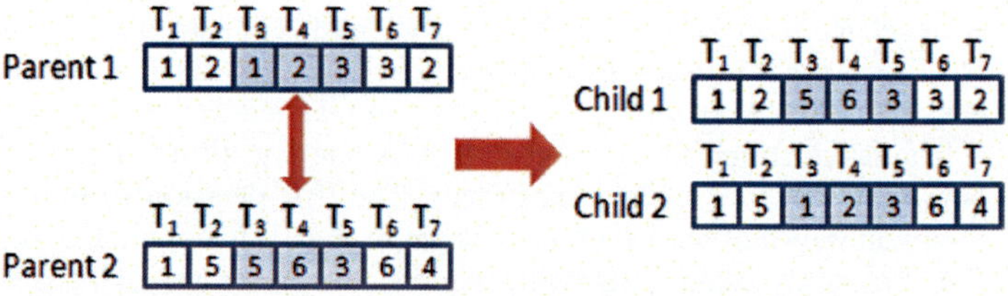

Fig. 10.5 The Crossover operation: First, two random parents are chosen from the current population. Then two random points are selected from the schedule order of both parents. The locations of all tasks between the two parents are exchanged. Two new offsprings are generated by combining task assignments taken from two parents.

Fig. 10.6 The Mutation operation: A task is randomly selected in a chromosome. An alternative service which is also capable of executing the task is randomly selected to replace the current task allocation

The new population is now ready for another round of chaotic mapping, crossover, and mutation, producing yet another generation. So the initial population is replaced by the newly generated individuals. More generations are produced until the stopping condition (a maximum number of generations) is met. The fittest chromosome is thus returned as a solution.

10.5.4 Experimental Results

Given that different workflow applications may have different impact on the performance of the scheduling algorithms, we have evaluated algorithms on different workflow structures. According to many grid workflow projects [45], workflow applications can be categorized into *balanced structures* and *unbalanced structures*. Fig. 10.7 shows balanced and unbalanced-structure applications used in our experiments. As shown in Fig. 10.7(a), the balanced-structure application consists of several parallel pipelines, which require the same types of services but process different data sets. As can be seen in Fig. 10.7(b), the structure of the unbalanced application is more complex. Unlike the balanced-structure application, many parallel tasks in the unbalanced structure require different types of services, and their workload and I/O data varies significantly.

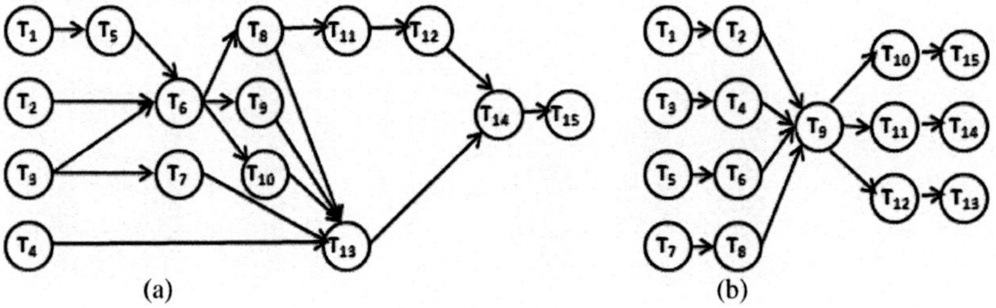

Fig. 10.7 Workflow structures: (a) Balanced workflow (fMRI). (b) Unbalanced workflow (DNA)

A Chaos-Genetic scheduling Algorithm (CGA) is introduced to solve the workflow execution planning problem. Our goal is to simultaneously minimize two conflicting objectives; execution time and execution price while meeting users' maximum time constraint (deadline) and price constraint (budget). We have simulated 15 types of services with various price levels. The parameter settings used as a default configuration for the algorithms are listed in Table 10.1. The behaviors of algorithms are also observed at three constraint levels, namely relaxed constraint, medium constraint, and tight constraint. The relaxed constraint level assumes that users require relatively large deadline and budget, while the tight constraint level assumes that users require small deadline and budget. In other words, the relaxed/tight deadlines and budgets of an application are determined by the maximum/minimum time and cost for the workflow execution.

Table 10.1 Parameter settings for workflow scheduling problem

Parameter	Value/type
Population size	10
Initial population	randomly generated solution
Maximum Generation	100
Crossover Probability	0.98
Mutation Probability	0.05
Maximum Iteration	10

As it is illustrated in Fig. 10.8, neither of GA and CGA satisfy the low budget constraint (about G$3500), however CGA shows better results in both applications. Results are gradually improved under medium budget constraints. Obviously, the descending trend in the diagram shows that as the budget increases, it'll be easier for the algorithms to meet the user budget constraints. On the other hand, considering the differences between the two approaches, it is clear that GA takes longer to complete even under relaxed constraints. Therefore, CGA shows better performance compared to GA in both applications.

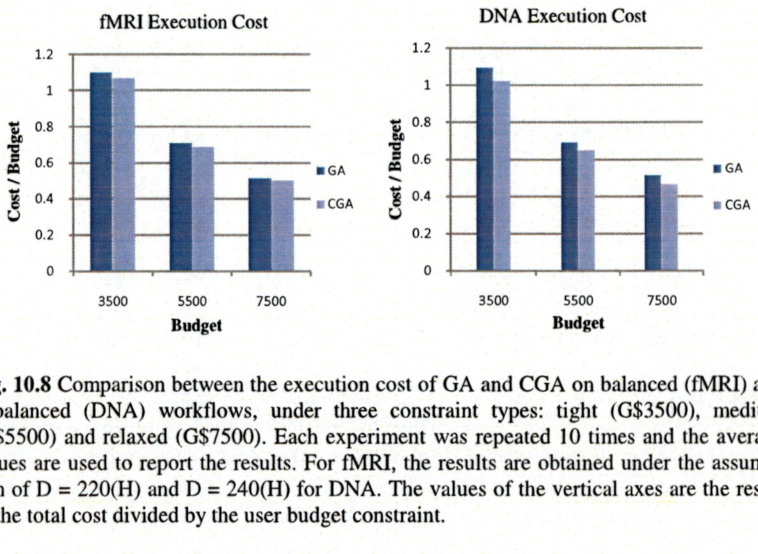

Fig. 10.8 Comparison between the execution cost of GA and CGA on balanced (fMRI) and unbalanced (DNA) workflows, under three constraint types: tight (G$3500), medium (G$5500) and relaxed (G$7500). Each experiment was repeated 10 times and the average values are used to report the results. For fMRI, the results are obtained under the assumption of D = 220(H) and D = 240(H) for DNA. The values of the vertical axes are the result of the total cost divided by the user budget constraint.

In Fig. 10.9 a comparison between the execution times of the two algorithms on fMRI and DNA workflows is illustrated. Here we change the user deadline values from 190(H) to 290(H) for DNA and from 170(H) to 270(H) for fMRI, since the latter is a balanced workflow and takes less time to complete. It can be seen that GA takes longer to complete in most of the conditions. The differences are obviously better observed in the unbalanced workflow structure.

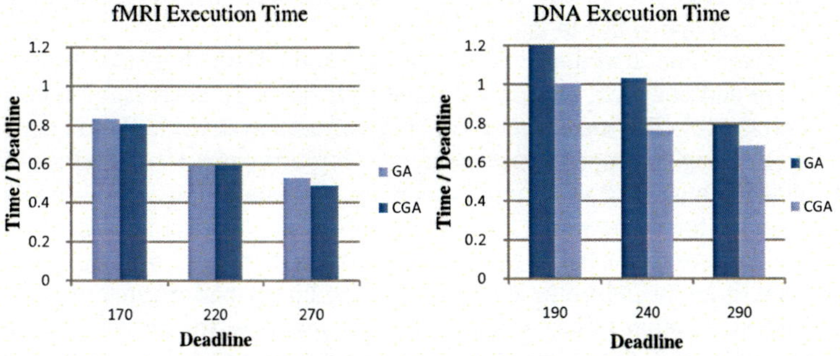

Fig. 10.9 Comparison between the execution time of GA and CGA on balanced (fMRI) and unbalanced (DNA) workflows, under three constraints: tight (around 180H), medium (around 230H) and relaxed (around 280H) with a medium budget of G$5000. Each experiment was repeated 10 times and the average values are used to report the results.

In all of the above diagrams, there are conditions where CGA and GA show similar results (for instance in Fig. 10.9 for fMRI, under medium constraint). These are the conditions where GA solutions were not trapped in a local optimum, resulting in similar performance patterns for the two algorithms. In those conditions, CGA does not do any good in keeping the suitable solutions. In the rest of the states though, GA, is stuck somewhere in a local optimum (as it usually is), which prevents it from producing possible better results. In other words, CGA takes advantage of the characteristics of the chaotic variable to make the individuals of subgenerations distributed ergodically in the defined space and thus to avoid premature convergence [30]. It also takes advantage of the convergence characteristic of GA to overcome the randomness of the chaotic process and hence to increase the probability of finding the global optimal solution.

10.6 Network-on-Chip (NoC): Case Study # 2

System-on-Chip (SoC) is a chip design method where all of the components of an electronic system are integrated into a single chip. Benefits of this integration compared with traditional multi-chip design include a size and energy reduction. An important concept in chip design is the *core*. A *core* is basically a separate and reusable unit of logic. Examples of *cores* include processors, memory banks and external communication components. These cores may be licensed from a number of vendors, under the common label *Intellectual Property-cores* (*IP-cores*). A System on Chip can include many *IP-cores* that need to communicate with each other. This is traditionally done by shared buses and ad-hoc core to core links. Using such traditional communication structures, functions well without creating communication bottlenecks when a system has few *cores* [46]. But as the number of *IP-cores* increases, the number of potential connections between them increases exponentially to a point where assigning the same bus to many *IP-cores* is not a practical option due to latency issues.

Over time, traditional SoC communication methods gradually became inefficient and complex, and do not scale well for large SoCs (say, more than 20 IP-cores) [46]. The increasing complexity of such systems leads to some difficulties in creating a proper communications infrastructure for the chip. When time-division buses and custom point to point communications are no longer sufficient, more elaborate networks are the obvious choice. By going beyond current buses and custom communication designs for the higher levels of interconnection on the chip, it might indeed be possible to reach higher performance with lower design and verification costs. A scalable communication architecture that avoids these problems is required and this is where creating a global network on the chip becomes a viable option.

10.6.1 Network on Chip

Network on Chip (NoC) is an emerging communication method for a System on Chip [47]. NoC attempts to solve the communication problems mentioned in the previous section by creating an inter-chip network consisting of network adapters, routers/switches and links between them. Each IP-core is connected to a network adapter which converts the transaction data from the IP-core to the flow digits (flits) transmitted across the network. Fig. 10.10 illustrates the basic concepts in a NoC.

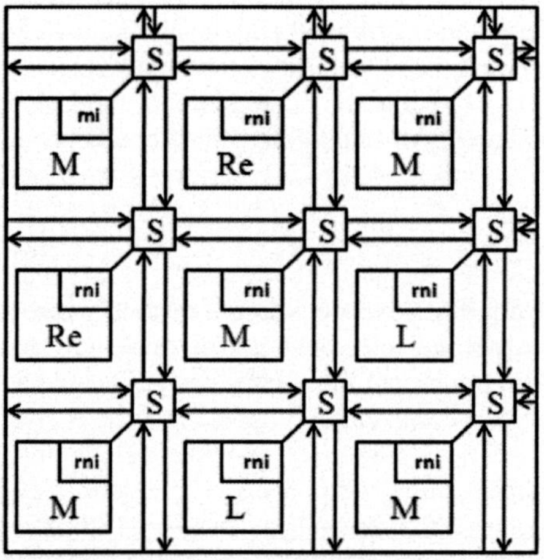

Fig. 10.10 An example of the Network-on-Chip architecture with 'S' for Switches, 'M' for Memory, 'Re' for Reconfigurable logic, 'rni' for resource-network interface and 'L' for dedicated hardware.

The individual IP-cores do not need to be aware of how the data is transmitted on the network. This decoupling of the processing from communication is an important benefit of NoC. This means that IP-cores with different transaction standards can easily communicate with each other, simplifying the design process. Other benefits include shorter, simplified wiring and lower energy usage. There are also some potential drawbacks including increased delay/latency especially if the network is congested, and the extra space used on the chip for the routers and network adapters. However the overhead is estimated to be fairly small [48] and space is usually not the bottleneck in chip design especially with the continuing shrinking microchip technologies.

The future for NoC looks promising, but many problems need to be addressed in order for it to find more widespread application. One of the problems is how to connect systems of IP-cores that vary in size and/or communication requirements (a heterogeneous SoC) defined as the network layout or topology selection. Three main factors that have to be taken into account when evaluating a NoC design are latency, energy usage and size (area overhead). Another important matter is application mapping which deals with finding the best node arrangement with the aim of improving the quality of service parameters. The mapping process can be described as follows: first select a set of IP-cores to distribute the data processing on and secondly construct a topology that connects the IP-cores and minimizes communication costs. The selected set of IP-cores and the data transmission between them constitutes what is defined as the Core Graph.

10.6.2 Problem Description

The investigation of different network topologies pointed to a two-dimensional mesh as the most suitable topology for most on-chip networks. This is also the common topology proposed by most researchers [48], [49], [50]. The main reasons for selecting the two-dimensional mesh instead of other topologies such as hypercubes, butterflies, or trees are that a two-dimensional mesh has an acceptable wire cost, reasonably high bandwidth, and a nice mapping onto a chip. Routing either refers to the problem of connecting a topology or choosing data transmission routes through a constructed topology. The transmission routes can either be static or dynamic. Dynamic routing is definitely more flexible [46] but requires more complex routers and larger buffers in the network. The static routing scheme chosen in the implementation of the algorithm is a shortest path routing algorithm.

The input of our problem is a directed task graph $G(V, E)$, in which every $v_i \in V$ denotes a processing element or a memory unit (generally an IP core), and a directed edge $e_k = (v_i, v_j)$ denotes a communication trace from the source node v_i to the destination node v_j. The weights of the edges $w(e_k)$ usually refer to the communication cost between two corresponding nodes. A mesh based topology of NoC is defined by (U, L), where each vertex $u_i \in U$ denotes a node in the topology and each $l_i \in L$ denotes a physical link between two vertices. The weight of a link $w\ (l_k)$ represents the bandwidth available across the link l_k. Fig. 10.11 exhibits the mapping process of a sample task graph onto a tile-based mesh structure.

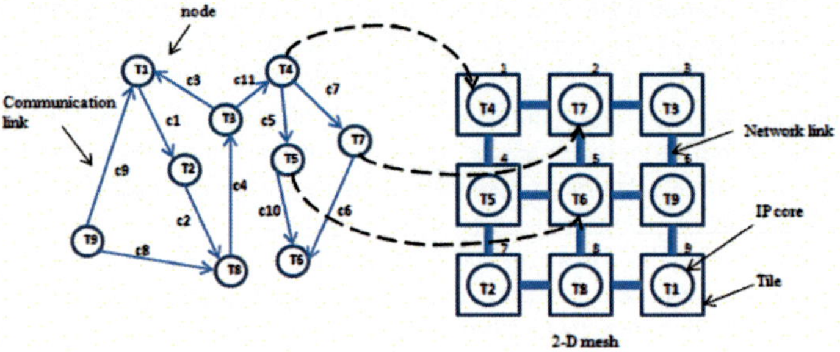

Fig. 10.11 The mapping process

In order to optimize the results of the mapping process, various authors have tried to enhance the results considering different performance elements. Lei and Kumar, proposed a two-step GA for mapping task graphs to the NoC architecture in [51], with the objective of minimizing the average communication delay of the network. Following the same objective, Murali and De Micheli, proposed NMAP, as a fast algorithm that maps the cores onto mesh NoC architecture under bandwidth constraints, in [52]. BMAP a binomial mapping and optimization algorithm that reduce the hardware cost of on-chip network infrastructure [53].

In Chaos-Genetic Mapping (CGMAP) approach [54], determining solution representation is the first priority. The values of the genes in this problem can only be integer values between 1 and n (the value of n is proportional to the number of tiles in the mesh). The length of each chromosome depends on the number of nodes in the communication task graph. Population size is another important parameter. In an actual application, it would be common to have somewhere between a few dozen and a few hundred individuals. For the purposes of this problem, we assume that the first population consists of 100 individuals. The initialization of the first population is done randomly by means of a random number generator which assigns values between 1 and n, to each of the n positions in every one of 100 individuals. Then the chaotic mapping operator is applied to each individual in the initial population and creates the chaotic population. At this stage, the fitness of all 200 individuals is evaluated. The fitness value is often proportional to the output value of the function being optimized. Since data always take the shortest distance in the network and often more than one such path exists for data going from node $v_i = (x_i, y_i)$ to $v_j = (x_j, y_j)$, we estimated this hop distance as

$$hd(e_k) = (|x_i - x_j| + (|y_i - y_j|), \tag{10.14}$$

and defined the fitness function as follows

$$F = \sum\nolimits_{\forall e_k} w(e_k) hd(e_k)\,. \tag{10.15}$$

Elitism is incorporated into the algorithm by transferring the single fittest individual directly to the next generation. Crossover and mutation are also performed on randomly selected individuals. The initial population is replaced by these newly generated individuals. Obviously, more generations are produced until the stopping condition (a maximum number of generations) is met. The fittest chromosome is thus returned as a solution.

10.6.3 Simulation Results

The results of the execution of CGMAP on two benchmark applications are demonstrated in this section; a Video Object Plane Decoder (VOPD) with 16 IP-cores and 20 links and an MPEG-4 decoder with 12 nodes and 13 links. Fig. 10.12 is an instance of a VOPD task graph mapped onto a two-dimensional mesh using CGMAP. Afterwards the results are compared with those of previous mapping algorithms such as NMAP [52], BMAP [53], PBB [55], etc. using the same routing and scheduling characteristics.

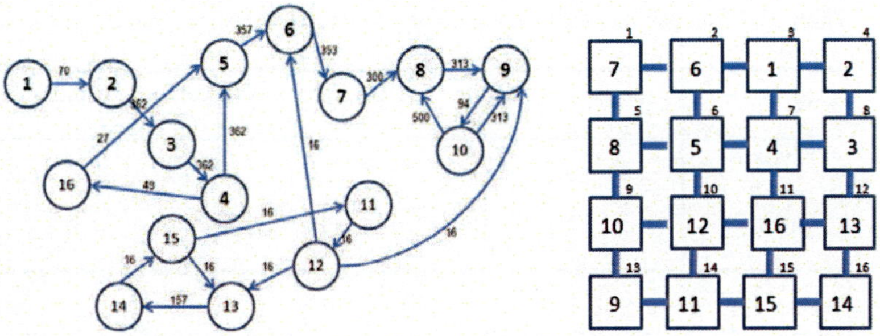

Fig. 10.12 VOPD task graph and the place of each associated IP core in a 2-dimensional mesh

Fig. 10.13 demonstrates the results of CGMAP compared with five other mapping algorithms in both applications, considering the communication costs. As it is clear in the figure, CGMAP performs well in both applications. Table 10.2, shows a comparison between the hop counts of the three most efficient mapping algorithms for two benchmark applications. The hop count is a measure of distance across an IP-based network which keeps track of the number of intermediate devices (like routers) an IP packet has to pass through in order to reach its destination. Generally speaking, the more hops data must traverse to reach their destination, the greater the transmission delay incurred. Assuming the average hop count in NMAP is 1, the table proves that using CGMAP decreases the hop number to an average of 0.97 in the first application (MPEG-4) and to 0.99, in the case of the second application (VOPD).

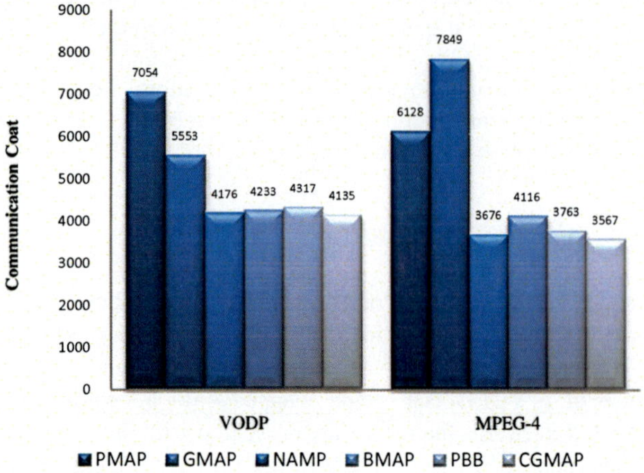

Fig. 10.13 Comparison between the communication costs of six mapping algorithm in NoC

Table 10.2 Comparision between the average hop count of mapping algorithms in NoC

	NMAP	BMAP	CGMAP	GMAP	PMAP	PBB
VOPD	1	1.01	0.99	1.33	1.69	1.03
MPEG - 4	1	1.71	0.95	2.13	1.67	1.02

10.6.4 Performance Analysis of One-Dimensional Chaotic Maps

In this section, the efficiency of the discussed chaotic maps is compared when used as an operator in GA [56]. Fig. 10.14 shows a convergence rate comparison

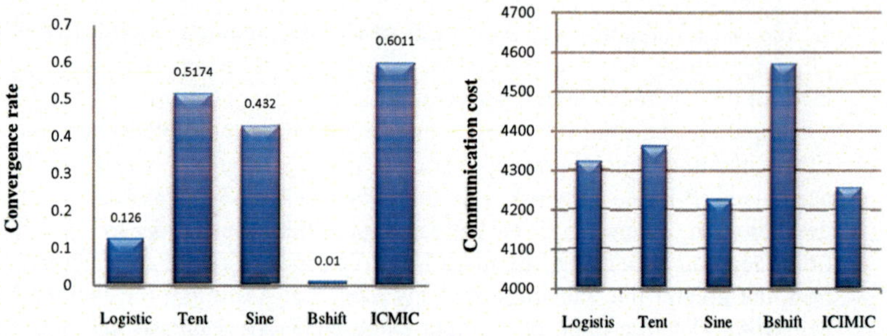

Fig. 10.14 Convergence rate and communication cost comparison between chaotic maps. The algorithm is run10 times with each chaotic operator and the average results are depicted. The benchmark application used in this section is VOPD.

between five one-dimensional chaotic maps. It is noticeable that, ICMIC and Tent map have the greatest convergence rates and the lowest convergence rate belongs to the Bernoulli Shift and the Logistic map. This means that in order to get the best results for this specific application, CGMAP should be implemented with the ICMIC map as a chaotic operator. This way the algorithm reaches an optimum solution within the shortest time period.

Communication costs of executing CGMAP are also compared with each of the discussed chaotic maps and the average results are demonstrated in Fig. 10.14. The Sine map achieves the lowest communication cost among all and the Bernoulli Shift costs a lot to complete the application. The main aim of this experiment was to prove that the choice of the most effective chaotic map is a function of the benchmark problem on the one hand and the main objective of the problem on the other.

10.7 Concluding Remarks

The chaos optimization algorithm adopts chaos variable to search, and the search goes on according to the regularity characteristics of the chaotic variables. Chaos variable's traversal property ensures that a true optimum solution can be found if allowed to run for sufficient time. Even if the optimization calculation time is limited we can get approximate solution with extremely good precision.

Grid Scheduling and Network-on-Chip mapping problems both belong to the group of NP-complete problems, which are traditionally solved using metaheuristic algorithms such as GA. In this chapter a Chaos-Genetic Algorithm (CGA) was used in order to take advantage of the properties of the chaotic variables to make the search of optimal values in GA more effective and faster. This is done by designing a chaotic mapping operator, using one-dimensional chaotic maps and applying it to the GA along with the common genetic operators, namely crossover and mutation. Experimental results were highly dependent upon the chaotic map that was used. Therefore, by prioritizing the favorites that one seeks, a chaotic equation may be selected that is the most congruent with ones will.

References

1. Bondy, J.A., Murty, U.S.R.: Graph Theory. Springer, Heidelberg (2008)
2. Korte, B., Vygen, J.: Combinatorial Optimization: Theory and Algorithms, Algorithms and Combinatorics, 4th edn. Springer, Heidelberg (2008)
3. Weise, M.: Global Optimization – Theory and Application, 2nd edn. Thomas Weise (2009)
4. Yang, X.-S.: Nature-Inspired Metaheuristic Algorithms. Luniver Press (2008)
5. Yang, X.-S.: Engineering Optimization: An Introduction with Metaheuristic Applications. Wiley, Chichester (2010)
6. Kirkpatrick, S., Gelatt Jr., C.D., Vecchi, M.P.: Optimization by Simulated Annealing. Science 220(4598), 671–680 (1983)

7. Holland, J.H.: Adaptation in Natural and Artificial Systems. University of Michigan Press, Ann Arbor (1975)
8. Kennedy, J., Eberhart, R.C., Shi, Y.: Swarm Intelligence. Morgan Kaufmann, New York (2001)
9. Dorigo, M.: Optimization, Learning and Natural Algorithms (Phd Thesis), Politecnico di Milano, Italy (1992)
10. Glover, F., Laguna, M.: Tabu Search. USA Norwell. Kluwer Academic Publishers, Dordrecht (1997)
11. Blum, C., Roli, A.: Metaheuristics in Combinatorial Optimization: Overview and conceptual comparison. ACM Computing Surveys 35(3), 268–308 (2003)
12. Ribeiro, C., Hansen, P.: Essays and Surveys in Metaheuristics. Kluwer Academic Publishers, Norwell (2002)
13. Melanie, M.: An Introduction to Genetic Algorithm. A Bradford book MIT press, London (1998)
14. Haupt, R.L., Haupt, S.E.: Practical Genetic Algorithms, 2nd edn. Wiley-Interscience Publication, Hoboken (1998)
15. Stavroulakis, P.: Chaos Application in Telecommunications. CRC Press Taylor and Francis group, New York (2006)
16. Strogatz, S.: Nonlinear dynamics and chaos: with applications to physics, biology, chemistry, and engineering. In: Perseus Books (1994)
17. Tavazoei, M.S., Haeri, M.: Comparison of Different One-Dimensional Maps as Chaotic Search in Chaos Optimization Algorithms. Applied Mathematics and Computation 187(2), 1076–1085 (2007)
18. He, Y.Y., Zhou, J.Z., Xiang, X.Q.: Comparison of different chaotic maps in particle swarm optimization algorithm for long term cascaded hydroelectric system scheduling. Chaos Solitons Fractals 42(5), 3169–3176 (2009)
19. Ott, E.: Chaos in Dynamical Systems. Cambridge University Press, U.K (2002)
20. Erramili, A., Singh, R.P., Pruthi, P.: Modeling Packet Traffic with Chaotic Maps. Royal Institute of Technology, Sweden (1994), ISRN KTH/IT/R-94/18-SE
21. He, D., He, C., Jiang, L.G., Zhu, H.W., Hu, G.R.: A chaotic map with infinite collapses. In: Proc IEEE tencon., Kuala Lumpur, Malaysia, vol. 3(9), pp. 95–99 (2000)
22. He, D., He, C., Jiang, L.G., Zhu, H.W., Hu, G.: Chaotic characteristics of a one-dimensional iterative map with infinite collapses. IEEE Trans. 48(7), 900–906 (2001)
23. Lu, Z., Shieh, L.S., Chen, G.R.: On robust control of uncertain chaotic systems: a sliding-mode synthesis via chaotic optimization. Chaos Solitons & Fractals 18(4), 819–836 (2003)
24. Yang, J.J., Zhou, J.Z., Wu, W., Liu, F.: A chaos algorithm based on progressive optimality and tabu search algorithm. In: IEEE Proc. 4th International Conf. Machine Learning and Cybernetics, vol. 5, pp. 2977–2981 (2005)
25. Li, B., Jiang, W.S.: Chaos Optimization Method and Its Application. Control Theory and Application 14, 613–615 (1997)
26. Gao, L., Liu, X.: A Resilient Particle Swarm Optimization Algorithm based on chaos and applying it to optimize the fermentation process. International Journal of Information and Systems Sciences 5(3-4), 380–391 (2009)
27. Bucolo, M., Caponetto, R., Fortune, L., Frasca, M., Rizzo, A.: Does chaos work better than noise? IEEE Circuits and Systems Magazine, 4–19 (2002)
28. Hongkai, W., Zhiming, C., Pingbo, W., Yinfeng, F.: Study of Intelligent Optimization Methods Applied in Fractional Fourier Transform. International Journal of Computer Theory and Engineering 2(4), 1793–8201 (2010)

29. Cheng, C.: Optimizing hydropower reservoir operation using hybrid genetic algorithm and chaos. Water Resources Management 22(7), 895–909 (2008)
30. Yan, X.F., Chen, D.Z., Hu, X.S.: Chaos-genetic algorithms for optimizing the operating conditions based on RBF-PLS model. Elsevier Computers and Chemical Engineering, 1390–1404 (2003)
31. Moein-Darbari, F., Khademzaheh, A., Gharoonifard, G.: CGMAP: A new Approach to Network-on-Chip Mapping Problem. IEICE Electronic Express 6(1), 27–34 (2009)
32. Foster, I., Kesselman, C.: Computational Grids. In: The Grid: Blueprint for New Computing Infrastructure, pp. 15–52. Morgan Kaufmann, San Francisco (1998)
33. Dong, F., Akl, S.G.: Scheduling Algorithms for Grid Computing: State of the Art and Open Problems. In: School of Computing, Queen's University Kingston, Ontario, pp. 1–55 (2006)
34. Schopf, J.M.: Ten Actions When SuperScheduling, document of Scheduling Working Group. In: Global Grid Forum (2001)
35. Yang, Y., Casanova, H.: NP-complete Scheduling Problems. Journal of Computer and System Sciences 10, 434–439 (1975)
36. Mandal, A., Kennedy, K., Koelbel, C., Martin, G., Mellor-Crummey, J., Liu, B., Johnsson, L.: Scheduling Strategies for Mapping Application Workflows onto the Grid. In: IEEE International Symposium on High Performance Distributed Computing (HPDC 2005), Research Triangle Park, NC, pp. 125–134 (2005)
37. Eilam, T., Appleby, K., Breh, J., Breiter, G., Daur, H., Fakhouri, S.A., Hunt, G.D.H., Lu, T., Miller, S.D., Mummert, L.B., Pershing, J.A., Wagner, H.: Using a utility computing framework to develop utility systems. IBM System Journal 43(1), 97–120 (2004)
38. Laszewski, G.V.: Java CoG Kit Workflow Concepts for Scientific Experiments. Argonne National Laboratory, Argonne, IL, USA Technique Report (2005)
39. Buyya, R., Giddy, J., Abramson, D.: An Evaluation of Economy-based Resource Trading and Scheduling on Computational Power Grids for Parameter Sweep Applications. In: 2nd Workshop on Active Middleware Services (AMS 2000). Kluwer Academic Press, Dordrecht (2000)
40. Sakellariou, R., Zhao, H., Tsiakkouri, E., Dikaiakos, M.: Scheduling workflows with budget constraints. In: Gorlatch, S., Danelutto, M. (eds.) Integrated Research in GRID Computing, ser., pp. 189–202. Springer, Heidelberg (2007)
41. Zhu, Y.: A Survey on Grid Scheduling System. Department of Computer Science, Hong Kong University of Science and Technology (2003)
42. Gharooni-fard, G., Moein-darbari, F., Deldari, H., Morvaridi, A.: Scheduling of Scientific Workflows Using a Chaos-Genetic Algorithm. In: Procedia Computer Science vol. 1(1), pp. 1439–1448 (2010)
43. Yu, J., Buyya, R.: Scheduling Scientific Workflow Applications with Deadline and Budget Constraints using Genetic Algorithms. Scientific Programming, 217–230 (2006)
44. Yu, J., Kirley, M., Buyya, R.: Multi-objective Planning for Workflow Execution on Grids. In: 8th IEEE ACM International Conference on Grid Computing, Singapore, pp.10—17 (2007)
45. Blythe, J., Jain, S., Deelman, E., Gil, Y., Vahi, K., Mandal, A., Kennedy, K.: Task scheduling strategies for workflow-based applications in grids. In: 5th IEEE International Symposium on Cluster Computing and the Grid (CCGrid 2005), vol. 2, pp. 759–767 (2005)
46. Bjerregaard, T., Mahadevan, S.: A Survey of Research and Practices of Network-on-Chip. ACM Computing Surveys, New York (2006)

47. De Micheli, G., Benini, L.: Network on Chip: A New Paradigm for System-on-Chip Design., pp. 7–78 (2002)
48. Dally, W.J., Towles, B.: Route Packets, not Wires: on Chip Interconnection Networks. In: Proceedings of the Design Automation Conference (DAC), pp. 684–689 (2001)
49. Saastamoinen, I., Sigüenza-Tortosa, D., Nurmi, J.: Interconnect IP Node for Future System-on-Chip Designs. In: IEEE International workshop on Electronic design, Test, and Applications, New Zealand, pp. 116–120 (2002)
50. Sgroi, M., Sheets, M., Mihal, A., Keutzer, K., Malik, R.J., Sangiovanni-Vincentelli, A.: Addressing the System-on-Chip Interconnect Woes through Communication-based Design. In: The Design Automation Conference (DAC), pp. 667–672 (2001)
51. Lei, T., Kumar, S.: A Two Step Genetic Algorithm for Mapping Task Graphs to Network on Chip Architecture. In: Proceedings of the 3rd International Conference DSD 2003, Turkey, pp. 180–187 (2003)
52. Murali, S., Micheli, G.D.: Bandwidth-Constrained Mapping of Cores on to NoC Architectures. In: 4th International Conference on DATE 2004, pp. 896–901 (2004)
53. Shen, W.T., Chao, C.H., Lien, Y.K., Wu, A.Y.: A new Binomial Mapping and Optimization Algorithm for Reduced-Complexity Mesh-Based On-Chip Network. In: 1st IEEE International Symposium on Networks-on-Chip (NOCS 2007), New Jersey, pp. 317–322 (2007)
54. Moein-darbari, F., Khademzadeh, A., Gharooni-fard, G.: Evaluating the Performance of Chaos Genetic Algorithm for Solving the Network-on-Chip Mapping Problem. In: IEEE International Conference on Computational Science and Engineering, Vancouver, Canada, vol. 2, pp. 366–373 (2009)
55. Hu, J., Marculescu, R.: Energy-Aware Mapping for Tile-based NoC Architectures under Performance Constraints. ASP-DAC, 233–239 (2003)
56. Gharooni-fard, G., Khademzade, A., Moein-darbari, F.: Evaluating the Performance of Chaotic Maps in Network-on-Chip Mapping Problem. IEICE Electronic Express 6(12), 811–817 (2009)

Index